Optical Metrology

Third Edition

Optical Metrology
Third Edition

Kjell J. Gåsvik
Spectra Vision AS, Trondheim, Norway

JOHN WILEY & SONS, LTD

Other Wiley Editorial Offices

John Wiley & Sons Inc., 111 River Street, Hoboken, NJ 07030, USA

Jossey-Bass, 989 Market Street, San Francisco, CA 94103-1741, USA

Wiley-VCH Verlag GmbH, Boschstr. 12, D-69469 Weinheim, Germany

John Wiley & Sons Australia Ltd, 33 Park Road, Milton, Queensland 4064, Australia

John Wiley & Sons (Asia) Pte Ltd, 2 Clementi Loop #02-01, Jin Xing Distripark, Singapore 129809

John Wiley & Sons Canada Ltd, 22 Worcester Road, Etobicoke, Ontario, Canada M9W 1L1

British Library Cataloguing in Publication Data

A catalogue record for this book is available from the British Library

ISBN 0-470-84300-4

Typeset in 10/12pt Times by Laserwords Private Limited, Chennai, India
Printed and bound in Great Britain by Antony Rowe Ltd, Chippenham, Wiltshire
This book is printed on acid-free paper responsibly manufactured from sustainable forestry
in which at least two trees are planted for each one used for paper production.

Contents

Preface to the Third Edition

This edition of *Optical Metrology* contains a new chapter about computerized optical processes, including digital holography and digital speckle photography. Chapter 2, on Gaussian optics, and Chapter 5, on light sources and detectors, are greatly expanded to include descriptions of standard imaging systems, light-emitting diodes and solid-state detectors. Separate new sections on optical coherence tomography, speckle correlation, the Fast Fourier Transform, temporal phase unwrapping and fibre Bragg sensors are included. Finally, a new appendix about Fourier series is given. Solutions to the end-of-chapter problems can be found at http://www.wiley.co.uk/opticalmetrology.

Since the previous edition, the electronic camera has taken over more and more as the recording medium. The word 'digital' is becoming a prefix to an increasing number of techniques. I think this new edition reflects this trend.

It gives me great pleasure to acknowledge the many stimulating discussions with Professor H.M. Pedersen at The Norwegian University of Science and Technology. Thanks also to John Petter Gåsvik for designing many of the new figures.

1

Basics

1.1 INTRODUCTION

Before entering into the different techniques of optical metrology some basic terms and definitions have to be established. Optical metrology is about light and therefore we must develop a mathematical description of waves and wave propagation, introducing important terms like wavelength, phase, phase fronts, rays, etc. The treatment is kept as simple as possible, without going into complicated electromagnetic theory.

1.2 WAVE MOTION. THE ELECTROMAGNETIC SPECTRUM

Figure 1.1 shows a snapshot of a harmonic wave that propagates in the z-direction. The disturbance $\psi(z, t)$ is given as

$$\psi(z, t) = U \cos\left[2\pi\left(\frac{z}{\lambda} - vt\right) + \delta\right] \qquad (1.1)$$

The argument of the cosine function is termed the phase and δ the phase constant. Other parameters involved are

U = the amplitude
λ = the wavelength
v = the frequency (the number of waves per unit time)
$k = 2\pi/\lambda$ the wave number

The relation between the frequency and the wavelength is given by

$$\lambda v = v \qquad (1.2)$$

where

v = the wave velocity

$\psi(z, t)$ might represent the field in an electromagnetic wave for which we have

$$v = c = 3 \times 10^8 \text{ m/s}$$

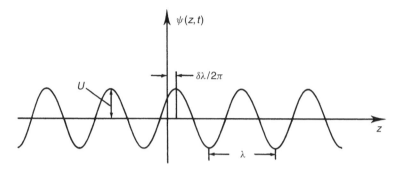

Figure 1.1   Harmonic wave

Table 1.1   The electromagnetic spectrum (From Young (1968))

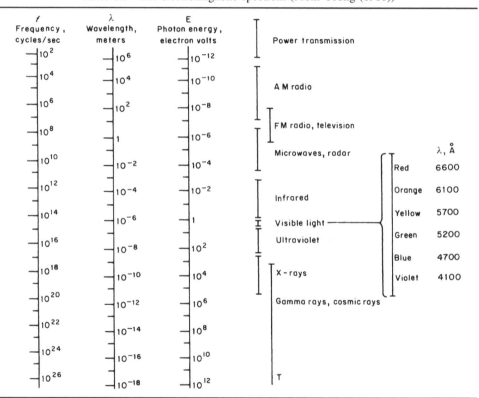

The ratio of the speed c of an electromagnetic wave in vacuum to the speed v in a medium is known as the absolute index of refraction n of that medium

$$n = \frac{c}{v} \tag{1.3}$$

The electromagnetic spectrum is given in Table 1.1.

Although it does not really affect our argument, we shall mainly be concerned with visible light where

$\lambda = 400-700$ nm (1 nm $= 10^{-9}$ m)
$\nu = (4.3-7.5) \times 10^{14}$ Hz

1.3 THE PLANE WAVE. LIGHT RAYS

Electromagnetic waves are not two dimensional as in Figure 1.1, but rather three-dimensional waves. The simplest example of such waves is given in Figure 1.2 where a plane wave that propagates in the direction of the **k**-vector is sketched. Points of equal phase lie on parallel planes that are perpendicular to the propagation direction. Such planes are called phase planes or phase fronts. In the figure, only some of the infinite number of phase planes are drawn. Ideally, they should also have infinite extent.

Equation (1.1) describes a plane wave that propagates in the z-direction. ($z = $ constant gives equal phase for all x, y, i.e. planes that are normal to the z-direction.) In the general case where a plane wave propagates in the direction of a unit vector **n**, the expression describing the field at an arbitrary point with radius vector $\mathbf{r} = (x, y, z)$ is given by

$$\psi(x, y, z, t) = U \cos[k\mathbf{n} \cdot \mathbf{r} - 2\pi \nu t + \delta] \qquad (1.4)$$

That the scalar product fulfilling the condition $\mathbf{n} \cdot \mathbf{r} = $ constant describes a plane which is perpendicular to **n** is shown in the two-dimensional case in Figure 1.3. That this is correct also in the three-dimensional case is easily proved.

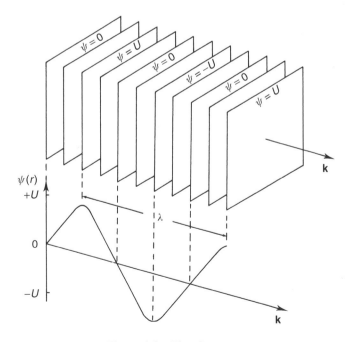

Figure 1.2   The plane wave

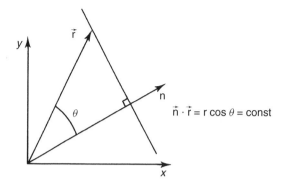

Figure 1.3

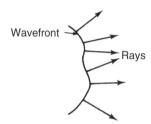

Figure 1.4

Next we give the definition of light rays. They are directed lines that are everywhere perpendicular to the phase planes. This is illustrated in Figure 1.4 where the cross-section of a rather complicated wavefront is sketched and where some of the light rays perpendicular to the wavefront are drawn.

1.4 PHASE DIFFERENCE

Let us for a moment turn back to the plane wave described by Equation (1.1). At two points z_1 and z_2 along the propagation direction, the phases are $\phi_1 = kz_1 - 2\pi vt + \delta$ and $\phi_2 = kz_2 - 2\pi vt + \delta$ respectively, and the phase difference

$$\Delta\phi = \phi_1 - \phi_2 = k(z_1 - z_2) \tag{1.5}$$

Hence, we see that the phase difference between two points along the propagation direction of a plane wave is equal to the geometrical path-length difference multiplied by the wave number. This is generally true for any light ray. When the light passes a medium different from air (vacuum), we have to multiply by the refractive index n of the medium, such that

optical path length = $n \times$ (geometrical path length)

phase difference = $k \times$ (optical path length)

1.5 COMPLEX NOTATION. COMPLEX AMPLITUDE

The expression in Equation (1.4) can be written in complex form as

$$\psi(x, y, z, t) = \text{Re}\{U e^{i(\phi - 2\pi v t)}\} \tag{1.6a}$$

where
$$\phi = k\mathbf{n} \cdot \mathbf{r} + \delta \tag{1.6b}$$

is the spatial dependent phase. In Appendix A, some simple arithmetic rules for complex numbers are given.

In the description of wave phenomena, the notation of Equation (1.6) is commonly adopted and 'Re' is omitted because it is silently understood that the field is described by the real part.

One advantage of such complex representation of the field is that the spatial and temporal parts factorize:

$$\psi(x, y, z, t) = U e^{i(\phi - 2\pi v t)} = U e^{i\phi} e^{-i2\pi v t} \tag{1.7}$$

In optical metrology (and in other branches of optics) one is most often interested in the spatial distribution of the field. Since the temporal-dependent part is known for each frequency component, we therefore can omit the factor $e^{-i2\pi v t}$ and only consider the spatial complex amplitude

$$u = U e^{i\phi} \tag{1.8}$$

This expression describes not only a plane wave, but a general three-dimensional wave where both the amplitude U and the phase ϕ may be functions of x, y and z.

Figure 1.5(a, b) shows examples of a cylindrical wave and a spherical wave, while in Figure 1.5(c) a more complicated wavefront resulting from reflection from a rough surface is sketched. Note that far away from the point source in Figure 1.5(b), the spherical wave is nearly a plane wave over a small area. A point source at infinity, represents a plane wave.

1.6 OBLIQUE INCIDENCE OF A PLANE WAVE

In optics, one is often interested in the amplitude and phase distribution of a wave over fixed planes in space. Let us consider the simple case sketched in Figure 1.6 where a plane wave falls obliquely on to a plane parallel to the xy-plane a distance z from it. The wave propagates along the unit vector $\mathbf{n}$ which is lying in the xz-plane (defined as the plane of incidence) and makes an angle θ to the z-axis. The components of the $\mathbf{n}$- and $\mathbf{r}$-vectors are therefore

$\mathbf{n} = (\sin\theta, 0, \cos\theta)$
$\mathbf{r} = (x, y, z)$

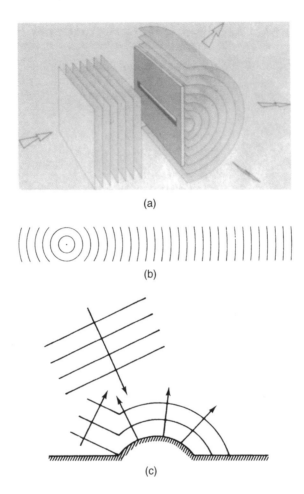

(a)

(b)

(c)

Figure 1.5 ((a) and (b) from Hecht & Zajac (1974), Figures 2.16 and 2.17. Reprinted with permission.)

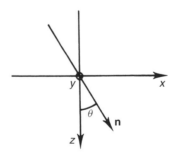

Figure 1.6

These expressions put into Equation (1.6) (Re and temporal part omitted) give

$$u = U e^{ik(x \sin \theta + z \cos \theta)} \tag{1.9a}$$

For $z = 0$ (the xy-plane) this reduces to

$$u = U e^{ikx \sin \theta} \tag{1.9b}$$

1.7 THE SPHERICAL WAVE

A spherical wave, illustrated in Figure 1.5(b), is a wave emitted by a point source. It should be easily realized that the complex amplitude representing a spherical wave must be of the form

$$u = \frac{U}{r} e^{ikr} \tag{1.10}$$

where r is the radial distance from the point source. We see that the phase of this wave is constant for $r = $ constant, i.e. the phase fronts are spheres centred at the point source. The r in the denominator of Equation (1.10) expresses the fact that the amplitude decreases as the inverse of the distance from the point source.

Consider Figure 1.7 where a point source is lying in the x_0, y_0-plane at a point of coordinates x_0, y_0. The field amplitude in a plane parallel to the $x_0 y_0$-plane at a distance z then will be given by Equation (1.10) with

$$r = \sqrt{z^2 + (x - x_0)^2 + (y - y_0)^2} \tag{1.11}$$

where x, y are the coordinates of the illuminated plane. This expression is, however, rather cumbersome to work with. One therefore usually makes some approximations, the first of which is to replace z for r in the denominator of Equation (1.10). This approximation cannot be put into the exponent since the resulting error is multiplied by the very large

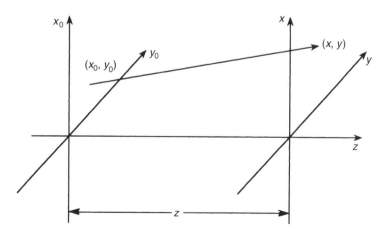

Figure 1.7

number k. A convenient means for approximation of the phase is offered by a binomial expansion of the square root, viz.

$$r = z\sqrt{1 + \left(\frac{x - x_0}{z}\right)^2 + \left(\frac{y - y_0}{z}\right)^2} \approx z\left[1 + \frac{1}{2}\left(\frac{x - x_0}{z}\right)^2 + \frac{1}{2}\left(\frac{y - y_0}{z}\right)^2\right] \quad (1.12)$$

where r is approximated by the two first terms of the expansion.

The complex field amplitude in the xy-plane resulting from a point source at x_0, y_0 in the $x_0 y_0$-plane is therefore given by

$$u(x, y, z) = \frac{U}{z}e^{ikz}e^{i(k/2z)[(x-x_0)^2+(y-y_0)^2]} \quad (1.13)$$

The approximations leading to this expression are called the Fresnel approximations. We shall here not discuss the detailed conditions for its validity, but it is clear that $(x - x_0)$ and $(y - y_0)$ must be much less than the distance z.

1.8 THE INTENSITY

With regard to the registration of light, we are faced with the fact that media for direct recording of the field amplitude do not exist. The most common detectors (like the eye, photodiodes, multiplication tubes, photographic film, etc.) register the irradiance (i.e. effect per unit area) which is proportional to the field amplitude absolutely squared:

$$I = |u|^2 = U^2 \quad (1.14)$$

This important quantity will hereafter be called the intensity.

We mention that the correct relation between U^2 and the irradiance is given by

$$I = \frac{\varepsilon v}{2}U^2 \quad (1.15)$$

where v is the wave velocity and ε is known as the electric permittivity of the medium. In this book, we will need this relation only when calculating the transmittance at an interface (see Section 9.5).

1.9 GEOMETRICAL OPTICS

For completeness, we refer to the three laws of geometrical optics:

(1) Rectilinear propagation in a uniform, homogeneous medium.

(2) Reflection. On reflection from a mirror, the angle of reflection is equal to the angle of incidence (see Figure 1.8). In this context we mention that on reflection (scattering) from a rough surface (roughness $>\lambda$) the light will be scattered in all directions (see Figure 1.9).

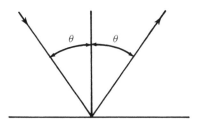

Figure 1.8   The law of reflection

Figure 1.9   Scattering from a rough surface

(3) Refraction. When light propagates from a medium of refractive index n_1 into a medium of refractive index n_2, the propagation direction changes according to

$$n_1 \sin \theta_1 = n_2 \sin \theta_2 \tag{1.16}$$

where θ_1 is the angle of incidence and θ_2 is the angle of emergence (see Figure 1.10). From Equation (1.16) we see that when $n_1 > n_2$, we can have $\theta_2 = \pi/2$. This occurs for an angle of incidence called the critical angle given by

$$\sin \theta_1 = \frac{n_2}{n_1} \tag{1.17}$$

This is called total internal reflection and will be treated in more detail in Section 9.5.

Finally, we also mention that for light reflected at the interface in Figure 1.10, when $n_1 < n_2$, the phase is changed by π.

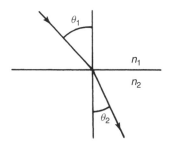

Figure 1.10   The law of refraction

1.10 THE SIMPLE CONVEX (POSITIVE) LENS

We shall here not go into the general theory of lenses, but just mention some of the more important properties of a simple, convex, ideal lens. For more details, see Chapter 2 and Section 4.6.

Figure 1.11 illustrates the imaging property of the lens. From an object point P_o, light rays are emitted in all directions. That this point is imaged means that all rays from P_o which pass the lens aperture D intersect at an image point P_i.

To find P_i, it is sufficient to trace just two of these rays. Figure 1.12 shows three of them. The distance b from the lens to the image plane is given by the lens formula

$$\frac{1}{a} + \frac{1}{b} = \frac{1}{f} \tag{1.18}$$

and the transversal magnification

$$m = \frac{h_i}{h_o} = \frac{b}{a} \tag{1.19}$$

In Figure 1.13(a), the case of a point source lying on the optical axis forming a spherical diverging wave that is converted to a converging wave and focuses onto a point on the optical axis is illustrated. In Figure 1.13(b) the point source is lying on-axis at a distance

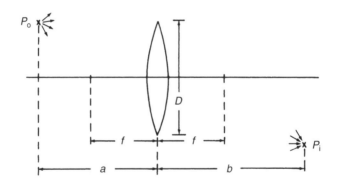

Figure 1.11

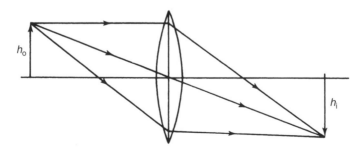

Figure 1.12

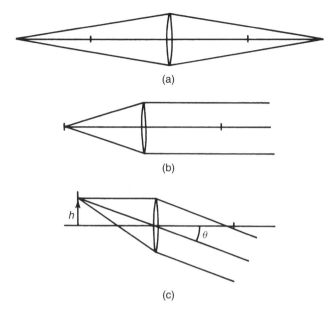

(a)

(b)

(c)

Figure 1.13

from the lens equal to the focal length f. We then get a plane wave that propagates along the optical axis. In Figure 1.13(c) the point source is displaced along the focal plane a distance h from the optical axis. We then get a plane wave propagating in a direction that makes an angle θ to the optical axis where

$$\tan \theta = h/f \tag{1.20}$$

1.11 A PLANE-WAVE SET-UP

Finally, we refer to Figure 1.14 which shows a commonly applied set-up to form a uniform, expanded plane wave from a laser beam. The laser beam is a plane wave with a small cross-section, typically 1 mm. To increase the cross-section, the beam is first directed through lens L_1, usually a microscope objective which is a lens of very short focal length f_1. A lens L_2 of greater diameter and longer focal length f_2 is placed as shown in the figure. In the focal point of L_1 a small opening (a pinhole) of diameter typically 10 μm is placed. In that way, light which does not fall at the focal point is blocked. Such stray light is due to dust and impurities crossed by the laser beam on its

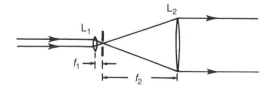

Figure 1.14   A plane wave set-up

way via other optical elements (like mirrors, beamsplitters, etc.) and it causes the beam not to be a perfect plane wave.

PROBLEMS

1.1 How many 'yellow' light waves ($\lambda = 550$ nm) will fit into a distance in space equal to the thickness of a piece of paper (0.1 mm)? How far will the same number of microwaves ($\nu = 10^{10}$ Hz, i.e 10 GHz, and $v = 3 \times 10^8$ m/s) extend?

1.2 Using the wave functions

$$\psi_1 = 4 \sin 2\pi (0.2z - 3t)$$
$$\psi_2 = \frac{\sin(7z + 3.5t)}{2.5}$$

determine in each case (a) the frequency, (b) wavelength, (c) period, (d) amplitude, (e) phase velocity and (f) direction of motion. Time is in seconds and z in metres.

1.3 Consider the plane electromagnetic wave (in SI units) given by the expressions $U_x = 0$, $U_y = \exp i[2\pi \times 10^{14}(t - x/c) + \pi/2]$, and $U_z = 0$.

What is the frequency, wavelength, direction of propagation, amplitude and phase constant of the wave?

1.4 A plane, harmonic light wave has an electric field given by

$$U_z = U_0 \exp i \left[\pi 10^{15} \left(t - \frac{x}{0.65c} \right) \right]$$

while travelling in a piece of glass. Find

(a) the frequency of the light,

(b) its wavelength,

(c) the index of refraction of the glass.

1.5 Imagine that we have a non-absorbing glass plate of index n and thickness Δz which stands between a source and an observer.

(a) If the unobstructed wave (without the plate present) is $U_u = U_0 \exp i\omega(t - z/c)$, ($\omega = 2\pi\nu$) show that with the plate in place the observer sees a wave

$$U_p = U_0 \exp i\omega \left[t - \frac{(n-1)\Delta z}{c} - \frac{z}{c} \right]$$

(b) Show that if either $n \approx 1$ or Δz is very small, then

$$U_p = U_u + \frac{\omega(n-1)\Delta z}{c} U_u e^{-i\pi/2}$$

The second term on the right may be interpreted as the field arising from the oscillating molecules in the glass plate.

1.6 Show that the optical path, defined as the sum of the products of the various indices times the thicknesses of media traversed by a beam, that is, $\sum_i n_i x_i$, is equivalent to the length of the path in vacuum which would take the same time for that beam to travel.

1.7 Write down an equation describing a sinusoidal plane wave in three dimensions with wavelength λ, velocity v, propagating in the following directions:

(a) $+z$-axis

(b) Along the line $x = y$, $z = 0$

(c) Perpendicular to the planes $x + y + z = $ const.

1.8 Show that the rays from a point source S that are reflected by a plane mirror appear to be coming from the image point S′. Locate S′.

1.9 Consider Figure P1.1. Calculate the deviation Δ produced by the plane parallel slab as a function of n_1, n_2, t, θ.

1.10 The deviation angle δ gives the total deviation of a ray incident onto a prism, see Figure P1.2. It is given by $\delta = \delta_1 + \delta_2$. Minimum deviation occurs when $\delta_1 = \delta_2$.

(a) Show that in this case δ_m, the value of δ, obeys the equation

$$\frac{n_2}{n_1} = \frac{\sin \frac{1}{2}(\alpha + \delta_m)}{\sin \frac{1}{2}\alpha}$$

(b) Find δ_m for $\alpha = 60°$ and $n_2/n_1 = 1.69$.

1.11 (a) Starting with Snell's law prove that the vector refraction equation has the form

$$n_2 \mathbf{k}_2 - n_1 \mathbf{k}_1 = (n_2 \cos \theta_2 - n_1 \cos \theta_1)\mathbf{u}_n$$

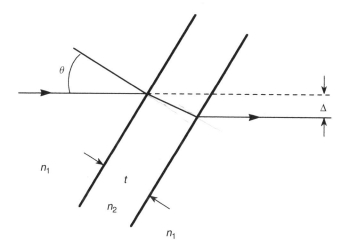

Figure P1.1

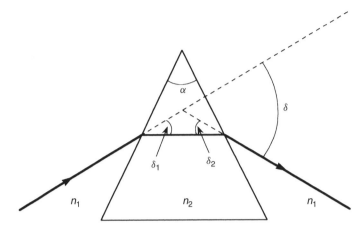

Figure P1.2

where $\mathbf{k}_1$, $\mathbf{k}_2$ are unit propagation vectors and $\mathbf{u}_n$ is the surface normal pointing from the incident to the transmitting medium.

(b) In the same way, derive a vector expression equivalent to the law of reflection.

2

Gaussian Optics

2.1 INTRODUCTION

Lenses are an important part of most optical systems. Good results in optical measurements often rely on the best selection of lenses. In this chapter we develop the relations governing the passage of light rays through imaging elements on the basis of the paraxial approximation using matrix algebra. We also mention the aberrations occurring when rays deviate from this ideal Gaussian behaviour. Finally we go through some of the standard imaging systems.

2.2 REFRACTION AT A SPHERICAL SURFACE

Consider Figure 2.1 where we have a sphere of radius R centred at C and with refractive index n'. The sphere is surrounded by a medium of refractive index n. A light ray making an angle α with the z-axis is incident on the sphere at a point A at height x above the z-axis. The ray is incident on a plane which is normal to the radius R and the angle of incidence θ is the angle between the ray and the radius from C. The angle of refraction is θ' and the refracted ray is making an angle α' with the z-axis. By introducing the auxiliary angle ϕ we have the following relations:

$$\phi = \theta' - \alpha' \tag{2.1a}$$

$$\phi = \theta - \alpha \tag{2.1b}$$

$$\sin \phi = \frac{x}{R} \tag{2.1c}$$

$$n \sin \theta = n' \sin \theta' \tag{2.1d}$$

The last equation follows from Snell's law of refraction. By assuming the angles to be small we have $\sin \phi \approx \phi$, $\sin \theta \approx \theta$, $\sin \theta' \approx \theta'$ and by combining Equations (2.1) we get the relation

$$\alpha' = \frac{n - n'}{n'R}x + \frac{n}{n'}\alpha = -\frac{P}{n'}x + \frac{n}{n'}\alpha \tag{2.2}$$

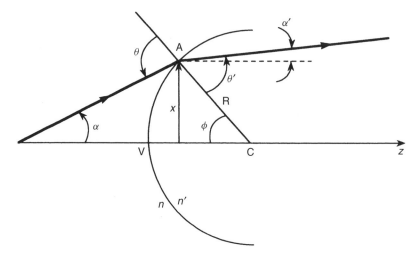

Figure 2.1   Refraction at a spherical interface

where

$$P = \frac{n' - n}{R} \tag{2.3}$$

is called the power of the surface.

The spherical surface in Figure 2.1 might be the front surface of a spherical lens. In tracing rays through optical systems it is important to maintain consistent sign conventions. It is common to define ray angles as positive counterclockwise from the z-axis and negative in the opposite direction. It is also common to define R as positive when the vertex V of the surface is to the left of the centre C and negative when it is to the right of C.

As can be realized, a ray is completely determined at any plane normal to the z-axis by specifying x, its height above the z-axis in that plane, and its angle α relative to the z-axis. A ray therefore can be specified by a column matrix

$$\begin{pmatrix} x \\ \alpha \end{pmatrix}$$

The two components of this matrix will be altered as the ray propagates through an optical system. At the point A in Figure 2.1 the height is unaltered, and this fact can be expressed as

$$x' = x \tag{2.4}$$

The transformation at this point can therefore be expressed in matrix form as

$$\begin{pmatrix} x' \\ \alpha' \end{pmatrix} = R \begin{pmatrix} x \\ \alpha \end{pmatrix} \tag{2.5}$$

where

$$R = \begin{pmatrix} 1 & 0 \\ -\dfrac{P}{n'} & \dfrac{n}{n'} \end{pmatrix} \qquad (2.6)$$

is the refraction matrix for the surface.

At this point it is appropriate to point out the approximations involved in reaching this formula. First, we have assumed the ray to lie in the xz-plane. To be general we should have considered the ray to lie in an arbitrary plane, taken its components in the xz- and yz-planes and introduced the component angles α and β relative to the z-axis. We then would have found that x and α at a given point depend only on x and α at other points, not on y and β. In other words, the pairs of variables (x, α) and (y, β) are decoupled from one another and may be treated independently. This is true only within the assumption of small angles. Because of this independence it is not necessary to perform calculations on both projections simultaneously. We do the calculations on the projection in the xz-plane and the answers will also apply for the yz-plane with the substitutions $x \to y$ and $\alpha \to \beta$. The xz projections behave as though y and β were zero. Such rays, which lie in a single plane containing the z-axis are called meridional rays.

In this theory we have assumed that an optical axis can be defined and that all light rays and all normals to refracting or reflecting surfaces make small angles with the axis. Such light rays are called paraxial rays. This first-order approximation was first formulated by C. F. Gauss and is therefore often termed Gaussian optics.

After these remarks we proceed by considering the system in Figure 2.2 consisting of two refracting surfaces with radii of curvature R_1 and R_2 separated by a distance D_{12}. The transformation at the first surface can be written as

$$\begin{pmatrix} x'_1 \\ \alpha'_1 \end{pmatrix} = R_1 \begin{pmatrix} x_1 \\ \alpha_1 \end{pmatrix} \quad \text{with} \quad R_1 = \begin{pmatrix} 1 & 0 \\ -\dfrac{P_1}{n'_1} & \dfrac{n_1}{n'_1} \end{pmatrix} \qquad (2.7)$$

where

$$P_1 = \frac{n'_1 - n_1}{R_1} \qquad (2.8)$$

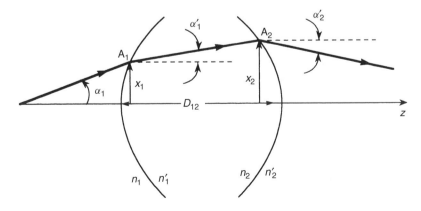

Figure 2.2   Ray tracing through a spherical lens

The translation from A_1 to A_2 is given by

$$x_2 = x_1' + D_{12}\alpha_1' \tag{2.9a}$$

$$\alpha_2 = \alpha_1' \tag{2.9b}$$

which can be written in matrix form as

$$\begin{pmatrix} x_2 \\ \alpha_2 \end{pmatrix} = T_{12} \begin{pmatrix} x_1' \\ \alpha_1' \end{pmatrix} \quad \text{with} \quad T_{12} = \begin{pmatrix} 1 & D_{12} \\ 0 & 1 \end{pmatrix} \tag{2.10}$$

The refraction at A_2 is described by

$$\begin{pmatrix} x_2' \\ \alpha_2' \end{pmatrix} = R_2 \begin{pmatrix} x_2 \\ \alpha_2 \end{pmatrix} \quad \text{with} \quad R_2 = \begin{pmatrix} 1 & 0 \\ -\dfrac{P_2}{n_2'} & \dfrac{n_2}{n_2'} \end{pmatrix} \tag{2.11}$$

where

$$P_2 = \frac{n_2' - n_2}{R_2} \tag{2.12}$$

These equations may be combined to give the overall transformation from a point just to the left of A_1 to a point just to the right of A_2:

$$\begin{pmatrix} x_2' \\ \alpha_2' \end{pmatrix} = M_{12} \begin{pmatrix} x_1 \\ \alpha_1 \end{pmatrix} \quad \text{with} \quad M_{12} = R_2 T_{12} R_1 \tag{2.13}$$

This process can be repeated as often as necessary. The linear transformation between the initial position and angle x, α and the final position and angle x', α' can then be written in the matrix form

$$\begin{pmatrix} x' \\ \alpha' \end{pmatrix} = M \begin{pmatrix} x \\ \alpha \end{pmatrix} \tag{2.14}$$

where M is the product of all the refraction and translation matrices written in order, from right to left, in the same sequence followed by the light ray.

The determinant of M is the product of all the determinants of the refraction and translation matrices. We see from Equation (2.10) that the determinant of a translation matrix is always unity and from Equation (2.6) that the determinant of a refraction matrix is given by the ratio of initial to final refractive indices. Thus the determinant of M is the product of the determinants of the separate refraction matrices and takes the form

$$\det M = \left(\frac{n_1}{n_1'} \right) \left(\frac{n_2}{n_2'} \right) \dots \tag{2.15}$$

But $n_1' = n_2$, $n_2' = n_3$ and so on, leaving us with

$$\det M = \frac{n}{n'} \tag{2.16}$$

where n is the index of the medium to the left of the first refracting surface, and n' is the index of the medium to the right of the last refracting surface.

2.2.1 Examples

(1) *Simple lens.* The matrix M is the same as M_{12} in Equation (2.13). By performing the matrix multiplication using $n'_1 = n_2$, $n_1 = n$, $n'_2 = n'$ and $D_{12} = d$, we get

$$M = \begin{pmatrix} 1 - \dfrac{P_1 d}{n_2} & \dfrac{nd}{n_2} \\ -\dfrac{P_2}{n'} + \dfrac{P_1 P_2 d}{n' n_2} - \dfrac{P_1}{n'} & \dfrac{n}{n'}\left(1 - \dfrac{P_2 d}{n_2}\right) \end{pmatrix} \tag{2.17}$$

(2) *Thin lens.* A thin lens is a simple lens with a negligible thickness d. If we let $d \to 0$ (i.e $d \ll R$) in Equation (2.17) we obtain

$$M = \begin{pmatrix} 1 & 0 \\ -\dfrac{P}{n'} & \dfrac{n}{n'} \end{pmatrix} \tag{2.18}$$

where the total power is given by (remember the sign convention for R)

$$P = P_1 + P_2 = \frac{n_2 - n}{R_1} + \frac{n' - n_2}{R_2} \tag{2.19}$$

Note that M has the same form for a thin lens as for a single refracting surface. Note also that the matrix elements $M_{11} = 1$ and $M_{12} = 0$. This means that we have $x' = x$, independently of the value of α.

2.3 THE GENERAL IMAGE-FORMING SYSTEM

In a general image-forming system (possibly consisting of several lens elements) an incoming ray at point B is outgoing from point B', shown schematically in Figure 2.3. The transformation matrix from B to B' is

$$M = \begin{pmatrix} M_{11} & M_{12} \\ M_{21} & M_{22} \end{pmatrix} \tag{2.20}$$

where the only requirement so far is

$$\det M = \frac{n}{n'} \tag{2.21}$$

We now ask if it is possible to find new reference planes instead of B and B' for which the general matrix M will take the form of that for a thin lens. These will turn out to

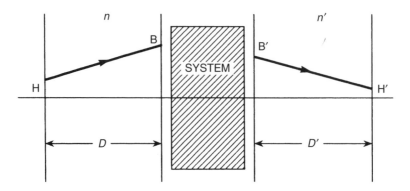

Figure 2.3

be the so-called principal planes and intersect the axis at H and H′ in Figure 2.3. The transformation matrix from the H-plane to the H′-plane can be written in terms of M by adding translation T and T′:

$$M_{\mathrm{HH'}} = T'MT = \begin{pmatrix} 1 & D' \\ 0 & 1 \end{pmatrix} \begin{pmatrix} M_{11} & M_{12} \\ M_{21} & M_{22} \end{pmatrix} \begin{pmatrix} 1 & D \\ 0 & 1 \end{pmatrix}$$

$$= \begin{pmatrix} M_{11} + D'M_{21} & M_{11}D + M_{12} + D'(M_{21}D + M_{22}) \\ M_{21} & M_{21}D + M_{22} \end{pmatrix} \tag{2.22}$$

The principal planes are defined as planes of unit magnification. Pairs of points in these planes are images of each other and planes with this property are called conjugate planes. Because of this requirement, the 1, 1 element of $M_{\mathrm{HH'}}$ must be unity and the 1, 2 element must be zero, giving

$$M_{\mathrm{HH'}} = \begin{pmatrix} 1 & 0 \\ M_{21} & \dfrac{n}{n'} \end{pmatrix} \tag{2.23}$$

We now equate the elements of the matrices in Equation (2.22) and (2.23)

$$11 : M_{11} + D'M_{21} = 1 \quad \text{i.e.} \quad D' = \frac{1 - M_{11}}{M_{21}} \tag{2.24a}$$

$$22 : M_{21}D + M_{22} = \frac{n}{n'} \quad \text{i.e.} \quad D = \frac{(n/n') - M_{22}}{M_{21}} \tag{2.24b}$$

These equations are meaningful only if the condition

$$M_{21} \neq 0 \tag{2.25}$$

is satisfied. This then becomes the requirement that our general Gaussian system be image-forming. (Identification of matrix element 12 gives the same condition.) To complete the final equivalence between our general image-forming system and a thin lens, it is only necessary to make the identification

$$-\frac{P}{n'} = M_{21} \tag{2.26}$$

Thus the image-formation condition, Equation (2.25) guarantees that our system has non-zero power. This means that all image forming systems have the same formal behaviour in Gaussian optics, as far as ray-tracing is concerned. It should be noted that for an afocal system like the plane wave set-up in Figure 1.14 where the two focal points coincide, $M_{21} = 0$. This is the same configuration as in a telescope where we only have angular magnification.

2.4 THE IMAGE-FORMATION PROCESS

We now want to move from the principal planes to other conjugate planes and determine the object-image relationships that result. This is done by translation transformations over the distances a and b in Figure 2.4. The overall transformation matrix from A to A' is given by

$$
M_{AA'} = \begin{pmatrix} 1 & b \\ 0 & 1 \end{pmatrix} \begin{pmatrix} 1 & 0 \\ -\dfrac{P}{n'} & \dfrac{n}{n'} \end{pmatrix} \begin{pmatrix} 1 & a \\ 0 & 1 \end{pmatrix}
$$

$$
= \begin{pmatrix} 1 - \dfrac{bP}{n'} & a - \dfrac{abP}{n'} + \dfrac{nb}{n'} \\ -\dfrac{P}{n'} & -\dfrac{aP}{n'} + \dfrac{n}{n'} \end{pmatrix} \tag{2.27}
$$

The image-formation condition is that the 1, 2 element of this matrix be zero:

$$a - \frac{abP}{n'} + \frac{nb}{n'} = 0 \tag{2.28}$$

that is

$$\frac{n}{a} + \frac{n'}{b} = P \tag{2.29}$$

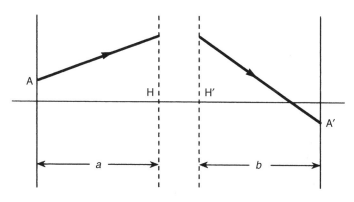

Figure 2.4

When the image is at $+\infty$, the object is in the first focal plane at a distance

$$a = \frac{n}{P} \equiv f \tag{2.30}$$

to the left of the first principal plane. When the object is at $+\infty$, the image is in the second focal plane at a distance

$$b = \frac{n'}{P} \equiv f' \tag{2.31}$$

to the right of the second principal plane. Thus Equation (2.29) may be written in the Gaussian form

$$\frac{n}{a} + \frac{n'}{b} = \frac{n}{f} = \frac{n'}{f'} \tag{2.32}$$

When the refractive indices in image and object space are the same ($n = n'$), this equation takes on the well known form

$$\frac{1}{a} + \frac{1}{b} = \frac{1}{f} \tag{2.33}$$

i.e. the lens formula.

When we have image formation, our matrix can be written

$$M_{AA'} = \begin{pmatrix} m_x & 0 \\ -\dfrac{P}{n'} & m_\alpha \end{pmatrix} \tag{2.34}$$

where the lateral magnification is

$$m_x = 1 - \frac{bP}{n'} = 1 - \frac{b}{f'} = -\frac{nb}{n'a} \tag{2.35}$$

and the ray angle magnification is

$$m_\alpha = -\frac{aP}{n'} + \frac{n}{n'} = -\frac{a}{b} \tag{2.36}$$

From the condition $\det M_{AA'} = n/n'$ we obtain the result

$$m_x m_\alpha = \frac{n}{n'} \tag{2.37}$$

In addition to the lateral (or transversal) magnification m_x, one might introduce a longitudinal (or axial) magnification defined as $\Delta b/\Delta a$. By differentiating the lens formula, we get $-\Delta a/a^2 - \Delta b/b^2 = 0$, which gives

$$\frac{\Delta b}{\Delta a} = -\left(\frac{b}{a}\right)^2 = -m_x^2 \tag{2.38}$$

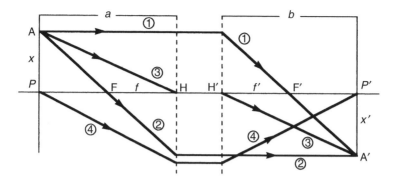

Figure 2.5   Principal planes with some key rays

It should be emphasized that the physical location of the principal planes could be inside one of the components of the image-forming system. Or they could be outside. The point to be made is that these are mathematical planes, and the rays behave as though they were deviated as shown in Figure 2.5. There is no *a priori* reason for the order of the principal planes. The plane H could be to the right of H′. The plane H will be to the right of F and H′ to the left of F′ if f and $f′$ are positive.

2.5 REFLECTION AT A SPHERICAL SURFACE

Spherical mirrors are used as elements in some optical systems. In this section we therefore develop transformations at a reflecting spherical surface.

In Figure 2.6 a light ray making an angle α with the z-axis is incident on the sphere at a point A at height x and is reflected at an angle $\alpha′$ to the z-axis. The sphere centre is

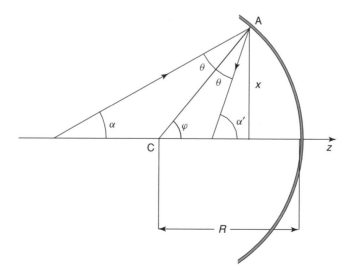

Figure 2.6   Reflection at a spherical surface

at C and therefore the reflection angle θ, equal to the angle of incidence, is as shown in the figure. From the geometry we see that

$$\alpha' = \phi + \theta$$

$$\phi = \alpha + \theta$$

which gives

$$\alpha' = 2\phi - \alpha \tag{2.39}$$

In the paraxial approximation we can put

$$\phi = x/R \tag{2.40}$$

When maintaining the same sign convention as in Section 2.2, R will be negative, and so also the angle α' (α' is positive clockwise from the negative z-axis). Put into Equation (2.39), this gives

$$\alpha' = \alpha + 2\frac{x}{R} \tag{2.41}$$

The transformation at point A therefore can be written as

$$\begin{pmatrix} x' \\ \alpha' \end{pmatrix} = \begin{pmatrix} 1 & 0 \\ 2/R & 1 \end{pmatrix} \begin{pmatrix} x \\ \alpha \end{pmatrix} \tag{2.42}$$

Comparing this with the object–image transformation matrix, Equation (2.34), we get for the focal length of the spherical mirror

$$f = -\frac{R}{2} \tag{2.43}$$

Figure 2.7 shows four rays from an object point that can be used to find the location of the image point. Note that one of the rays goes through C and the image point. When approaching the mirror from beyond a distance $2f = R$, the image will gradually increase

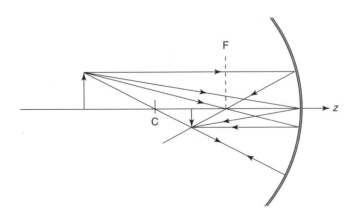

Figure 2.7 Imaging by a reflecting spherical surface

until at $2f$ it appears inverted and life-size. Moving still closer will cause the image to increase until it fills the entire mirror with an unrecognisable blur. Decreasing the distance further, the now erect, magnified image will decrease until the object rests on the mirror where the image is again life-size. The mirror in Figure 2.7 is concave. A mirror with opposite curvature is called convex. It is easily verified that a convex mirror forms a virtual image.

2.6 ASPHERIC LENSES

From school mathematics we learn that rays incident on a reflecting paraboloid parallel to its axis will be focused to a point on the axis. This comes from the mere definition of a parabola which is the locus of points at equal distance from a line and a point. The paraboloid and other non-spherical surfaces are called aspheric surfaces. The equation for the circular cross-section of a sphere is

$$x^2 + (z - R)^2 = R^2 \tag{2.44}$$

where the centre C is shifted from the origin by one radius R: see Figure 2.8. From this we can solve for z:

$$z = R \pm \sqrt{R^2 - x^2} \tag{2.45}$$

By choosing the minus sign, we concentrate on the left hemisphere, and by expanding z in a binomial series, we get

$$z = \frac{x^2}{2R} + \frac{1 \cdot x^4}{2^2 2! R^3} + \frac{1 \cdot 3 \cdot x^6}{2^3 \cdot 3! R^5} + \cdots \tag{2.46}$$

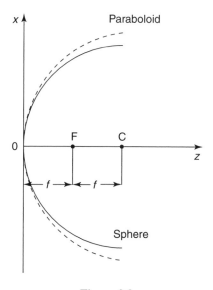

Figure 2.8

The equation for a parabola with its vertex at the origin and its focus a distance f to the right (see Figure 2.8) is

$$z = \frac{x^2}{4f} \tag{2.47}$$

By comparing these two formulas, we see that if $f = R/2$, the first contribution in the series can be thought of as being parabolic, while the remaining terms (in x^4 and higher) represent the deviation therefrom. Evidently this difference will only be appreciable when x is relatively large compared to R. In the paraxial region, i.e. in the immediate vicinity of the optical axis, these two configurations will be essentially indistinguishable. In practice, however, x will not be so limited and aberrations will appear. Moreover, aspherical surfaces produce perfect images only for pairs of axial points – they too will suffer from aberrations.

The best known aspherical element must be the antenna reflector for satellite TV reception. But the paraboloidal configuration ranges its present-day applications from flashlight and auto headlight reflectors to giant telescope antennas. There are several other aspherical mirrors of some interest, namely the ellipsoid and hyperboloid. So why are not aspheric lenses more commonly used? The first and most immediate answer is that, as we have seen, in the paraxial region there is no difference between a spherical and a paraboloidal surface. Secondly, paraboloidal glass surfaces are difficult to fabricate. We also might quote from Laikin (1991): 'The author's best advice concerning aspherics is that unless you have to, don't be tempted to use an aspheric surface'. An important exception is the video disk lens. Such lenses are small with high numerical aperture operating at a single laser wavelength; they cover a very small field and are diffraction limited. A recent trend in the manufacture of these lenses is to injection-mould them in plastic. This has the advantage of light weight and low cost (because of the large production volume) and an aspheric surface may be used.

2.7 STOPS AND APERTURES

Stops and apertures play an important role in lens systems.

The aperture stop is defined to be the aperture which physically limits the solid angle of rays passing through the system from an on-axis object point. A simple example is shown in Figure 2.9(a) where the hole in the screen limits the solid angle of rays from the object at P_0. The rays are cut off at A and B. The images of A and B are A′ and B′. To an observer looking back through the lens from a position near P_0' it will appear as if A′ and B′ are cutting off the rays. If we move the screen to the left of F, we have the situation shown in Figure 2.9(b). The screen is still the aperture stop, but the images A′, B′ of A and B are now to the right of P_0'. To an observer who moves sufficiently far to the right it still appears as if the rays are being cut off by A′ and B′.

A 'space' may be defined that contains all physical objects to the right of the lens plus all points conjugate to physical objects that are to the left of the lens. It is called the image space. In Figures 2.9(a, b) all primed points are in image space. The image of the aperture stop in image space is called the exit pupil. To an observer in image space it appears either as if the rays converging to an on-axis image P_0' are limited in solid angle

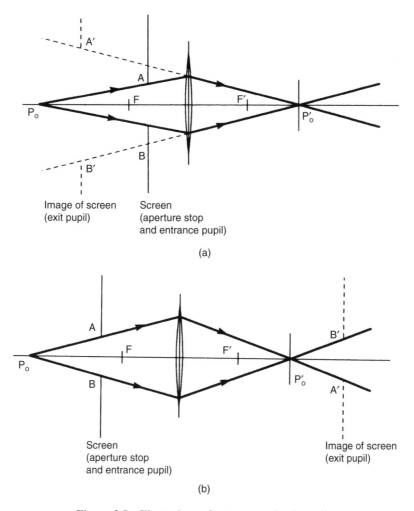

Figure 2.9 Illustrations of entrance and exit pupils

by the exit pupil A'B' as in Figure 2.9(a) or as if the rays diverging from P'_o are limited in solid angle by A'B' as in Figure 2.9(b).

By analogy to the image space, a space called the object space may be defined that contains all physical objects to the left of the lens plus all points conjugate to any physical object that may be to the right of the lens. In Figure 2.9(a, b) all unprimed objects are in the object space. The image of the aperture stop in the object space is defined as the entrance pupil. The aperture stop in Figure 2.9(a, b) is already in the object space, hence it is itself the entrance pupil.

In a multilens system some physical objects will be neither in the object nor in the image space but in between the elements. If a given point is imaged by all lens elements to its right, it will give an image in the image space; if imaged by all elements to its left, it will give an image in the object space. A systematic method of finding the entrance pupil is to image all stops and lens rims to the left through all intervening refracting elements of the system into the object space and find the solid angle subtended by each at P_o. The one with the smallest solid angle is the entrance pupil, and the physical object corresponding to it is the aperture stop. Alternatively we may image all stops and lens rims to the right through all intervening refractive elements into the image space and determine the solid angle subtended by each image at P_o'. The one with the smallest solid angle is the exit pupil, and the corresponding real physical object is the aperture stop.

2.8 LENS ABERRATIONS. COMPUTER LENS DESIGN

The ray-tracing equations used in the theory of Gaussian optics are correct to first order in the inclination angles of the rays and the normals to refracting or reflecting surfaces. When higher-order approximations are used for the trigonometric functions of the angles, departures from the predictions of Gaussian optics will be found. No longer will it be generally true that all the rays leaving a point object will exactly meet to form a point image or that the magnification in a given transverse plane is constant. Such deviations from ideal Gaussian behaviour are known as lens aberrations. In addition, the properties of a lens system may be wavelength- dependent, known as chromatic aberrations.

Monochromatic aberrations may be treated mathematically in lowest order by carrying out the ray-tracing calculations to third order in the angles. The resulting 'third-order theory' is itself valid only for small angles and for many real systems calculations must be carried out to still higher order, say fifth or seventh. (For a centred system with rotational symmetry, only odd powers of the angles will appear in the ray-tracing formulas.)

Most compound lens systems contain enough degrees of freedom in their design to compensate for aberrations predicted by the third-order theory. For real systems the residual higher-order aberrations would still be present, and there are not enough design parameters to eliminate all of them as well. The performance of a lens system must be judged according to the intended use. The criteria for a telescope objective and for a camera lens for close-ups are quite different.

Third-order monochromatic aberrations can be divided into two subgroups. Those belonging to the first are called spherical aberrations, coma and astigmatism and will deteriorate the image, making it unclear. The second type cover field curvature and distortion, which deform the image. Here we will not treat lens aberrations in any detail. Figure 2.10 illustrates spherical aberration, and in Section 10.4.1 distortion is treated in

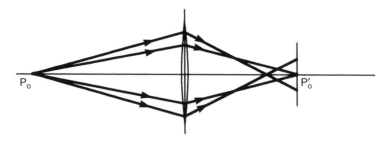

Figure 2.10 Spherical aberration. The focus of the paraxial rays is at P_o'. The marginal rays focus at a point closer to the lens

some detail. Because of the complexity of the higher-order aberrations they are usually treated numerically. Now lens design computer programs are available commercially. Such programs trace a lot of different rays through the system and the points where they intersect the image plane is called a spot diagram. By changing the design parameters, the change in the spot diagram can be observed. Some computer programs do such analyses automatically. The computer is given a quality factor (or merit function) of some sort, which means how much of each aberration is tolerated. Then a roughly designed system which, in the first approximation, meets the particular requirements is given as input. The computer will then trace several rays through the system and evaluate the image errors. After perhaps twenty or more iterations, it will have changed the initial configuration so that it now meets the specified limits on aberrations. However, a quality factor is somewhat like a crater-pocked surface in a multidimensional space. The computer will carry the design from one hole to the next until it finds one deep enough to meet the specifications. There is no way to tell if that solution corresponds to the deepest hole without sending the computer out again and again meandering along totally different routes.

2.9 IMAGING AND THE LENS FORMULA

Before studying specific lens systems, let us have a closer look at the imaging process and the lens formula. We have found that a general imaging system is characterized by the focal length f and the positions of the two principal planes H and H′ which determine the four cardinal points F, F′, H and H′: see Figure 2.11. Imaging takes place between conjugate planes in object and image space, and the object and image planes are related by the lens formula

$$\frac{1}{a} + \frac{1}{b} = \frac{1}{f} \tag{2.48}$$

where a and b are measured from the principal planes. Note that both a and b can assume values between $-\infty$ and ∞. If the object plane lies to the right of the vertex of the first refracting surface, we have no real object point, but rays that converge to a virtual object point behind the first refracting surface: see Figure 2.12(c). In the same way we have a virtual image plane if the image lies to the left of the last vertex of the lens system: Figure 2.12(b). The rays diverge as if coming from this virtual image point, but they do not intersect there. Only if rays really intersect at the image point do we have a real image point and that happens only if the image plane lies to the right of the last vertex

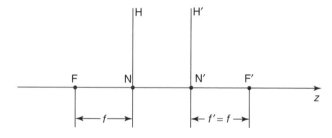

Figure 2.11   Principal points

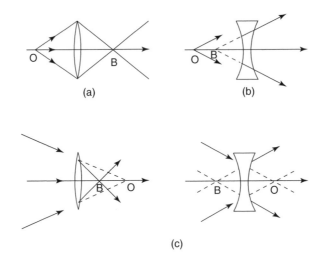

Figure 2.12   Real and virtual object (O) and image (I) points: (a) real object, real image; (b) real object, virtual image; (c) virtual object, real image; and (d) virtual object, virtual image

of the system. The focal length can also assume values in the range $[-\infty, \infty]$. When $f > 0$, we have a positive (or collecting) lens, and when $f < 0$ we have a negative lens: see Figure 2.12(b). For a negative lens, F is to the right of H, while F' is to the left of H'.

In addition to the above-mentioned cardinal points, we also have the so-called nodal points N and N' on the axis: see Figure 2.11. A ray incident on N in the object space leaves N' in the image space in the same direction. Rays through nodal points therefore are parallel, which means that the angular magnification between N and N' is unity. With the same refractive index in front and behind the lens ($n = n'$), we get $m_x m_\alpha = 1$, which means that the nodal points must lie in the principal planes. With unequal indices, the nodal points move away from the principal planes.

2.10 STANDARD OPTICAL SYSTEMS

It should be remembered that the systems described below are visual instruments of which the eye of the observer is an integral part.

2.10.1 Afocal Systems. The Telescope

An afocal system has zero power P. This can be realized by two lenses separated by a distance t equal to the sum of the individual focal lengths, $t = f_1 + f_2$: see Figure 2.13. The system matrix becomes

$$M = \begin{bmatrix} 1 & 0 \\ -1/f_2 & 1 \end{bmatrix} \begin{bmatrix} 1 & (f_1 + f_2) \\ 0 & 1 \end{bmatrix} \begin{bmatrix} 1 & 0 \\ -1/f_1 & 1 \end{bmatrix} = \begin{bmatrix} -f_2/f_1 & (f_1 + f_2) \\ 0 & -f_1/f_2 \end{bmatrix} \qquad (2.49)$$

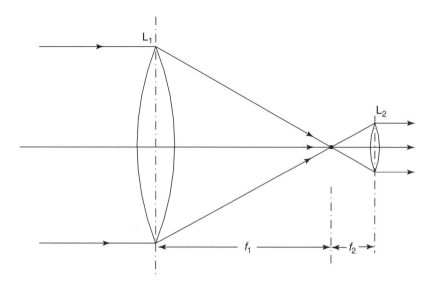

Figure 2.13 The telescope

We see that the M_{21}-element is zero, which means $P = 0$. Computing the transformation from a plane a distance d in front of the first lens to a plane a distance d' behind the second lens gives

$$
M_{dd'} = \begin{bmatrix} 1 & d' \\ 0 & 1 \end{bmatrix} \begin{bmatrix} -f_2/f_1 & (f_1 + f_2) \\ 0 & -f_1/f_2 \end{bmatrix} \begin{bmatrix} 1 & d \\ 0 & 1 \end{bmatrix}
$$

$$
= \begin{bmatrix} -f_2/f_1 & [(f_1 + f_2) - f_2 d/f_1 - f_1 d'/f_2] \\ 0 & -f_1/f_2 \end{bmatrix} \tag{2.50}
$$

Assuming d and d' to be the object and image planes, the $(1,2)$-element must be zero, and we get

$$
M_{dd'} = \begin{bmatrix} -f_2/f_1 & 0 \\ 0 & -f_1/f_2 \end{bmatrix} = \begin{bmatrix} m_x & 0 \\ 0 & m_\alpha \end{bmatrix} \tag{2.51}
$$

Contrary to other lens systems, the lateral magnification

$$
m_x = -f_2/f_1 \tag{2.52}
$$

is constant and independent of the object and image distances. This implies that an afocal system does not have principal planes with mutual unit magnification. The object–image relation is also very different from the usual lens formula:

$$
(f_2/f_1)d + (f_1/f_2)d' = f_1 + f_2 \tag{2.53a}
$$

or

$$d' = (f_1 + f_2)(f_1/f_1) - (f_2/f_1)^2 d \qquad (2.53b)$$

A telescope is an afocal system with $f_1 > |f_2|$ giving $|m_x| < 1$. The reason for this seeming paradox is that when $d \to \infty$ it is the angular magnification $m_\alpha = -f_1/f_2$ that determines how large the image looks. The virtual image is demagnified by a factor $m_x = -f_2/f_1$, but this is contrasted by being focused at a distance $d' \approx dm_\alpha^2$ and is moved closer by a factor $m_x^2 = (f_2/f_1)^2$. The angular magnification then becomes $m_x/m_x^2 = 1/m_x = -(f_1/f_2) > 1$.

Since negative lenses have virtual focal points and the focal points in an afocal system must coincide, the lens with the longest focal length must always be positive. The lens with the shortest focal length can be either negative, giving an erect image with $m_x > 0$ (Galileo's telescope, the theatre telescope), or positive, giving an inverted image. In binoculars, the image is erected by inverting the image in two total reflecting prisms. It should be noted that when observing faint stellar objects, large angular magnification is not sufficient if the irradiance is too low. The light-collecting capacity is determined by the front lens. Therefore, when judging the quality of a stellar telescope, the diameter of the front lens is a more important parameter than the magnification. However, large-aperture lenses inevitably give more aberrations. Since large-aperture corrected mirrors are easier to fabricate than lenses, stellar telescopes are often equipped with mirrors as front objectives. Figure 2.14 shows some of the most common designs.

2.10.2 The Simple Magnifier

The unaided eye focuses on an object when the object distance is larger than about $d_o = 25$ cm. The angular resolution (determined by the rods and cones) is about $0.5' = 0.5/60 = 1/120$ deg $= 1/7000$ radian. At a distance of 25 cm we therefore cannot distinguish object details less than 0.07 mm. To observe smaller objects we can use a magnifier.

In Figure 2.15 the object of height h is placed at a distance $a < f$, where f is the focal length of the magnifier. The resulting virtual image is located a distance b in front of the lens, given by the lens formula $1/a + 1/b = 1/f$. Since d_o is the closest distance the eye can focus, we put $b = -d_o$ (b is negative), giving

$$a = \frac{d_o f}{d_o + f} \qquad (2.54)$$

and the magnification

$$m = \frac{d_o}{a} = \frac{d_o + f}{f} = \frac{d_o}{f} + 1 \qquad (2.55)$$

For a magnifier with $f = 5$ cm, the effective magnification is about 5–6 depending on how the observer focuses.

A simple uncorrected magnifier has rather poor imaging qualities. Similar but well-corrected systems are applied as oculars in visual instruments. An ocular is a well-corrected magnifier for visual observation of intermediate images in optical systems. Since an intermediate image can be virtual, negative lenses can also be used as oculars.

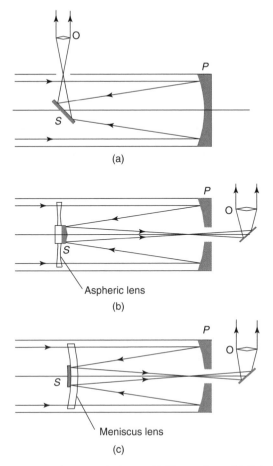

Figure 2.14   Some common telescope designs: (a) Newtonian; (b) Schmidt–Cassegrain; and (c) Maksutov–Cassegrain. P = primary mirror, S = secondary mirror, O = ocular

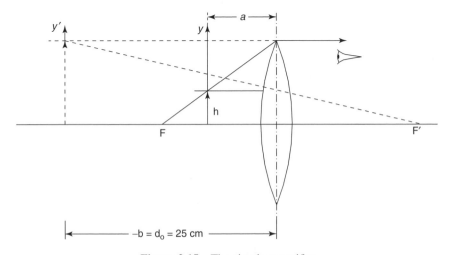

Figure 2.15   The simple magnifier

2.10.3 The Microscope

A microscope is used for observation of very small objects where the magnification of the viewing angle is so large that the assumptions of paraxial optics are no longer valid. The magnification can be several hundreds, the focal length lies in the millimetre range and the objective lens is composed of several elements (compound lens). Microscopes are specialized and standardized instruments consisting of exchangeable objectives with various focal lengths f_{ob}, but which focus an intermediate image at a fixed distance $b = T = 16$ cm (the tubus length). The magnification of the objective is therefore given by $m_{ob} \approx T/f_{ob}$. For a 40× objective we therefore get $f_{ob} \approx 16$ cm$/40 = 4$ mm. The magnified intermediate image is observed by the ocular, which focuses at infinity, giving a magnification of the viewing angle equal to $d_o/f_{oc} \approx 25$ cm$/f_{oc}$. A 10× ocular therefore has a focal length f_{oc} equal to 2.5 cm. The overall magnification becomes $m_{ob} \cdot m_{oc} \approx T d_o/(f_{ob} f_{oc})$, which in our example gives $40 \times 10 = 400$.

PROBLEMS

2.1 Verify directly by matrix methods that use of the matrix in Equation (2.34) will yield values of (x', α') for rays 1, 2, 3, 4 in Figure 2.5 so that they behave as shown.

2.2 Consider the system shown in Figure P2.1 where the focal lengths of the first system are f_1, f_1' and those of the second f_2, f_2'. The respective powers are

$$P_1 = \frac{n_1}{f_1} = \frac{n_1'}{f_1'}$$

$$P_2 = \frac{n_2}{f_2} = \frac{n_2'}{f_2'}$$

(a) Find the transformation matrix $M_{H_1 H_2'}$ between the first principal plane of the first system and the second principal plane of the second system.
We denote the principal planes of the whole system by H and H' and the distances $HH_1 = D$ and $H_2' H' = D'$.

(b) Express the transformation matrix $M_{HH'}$ between H and H' in terms of the total power P and n and m'.

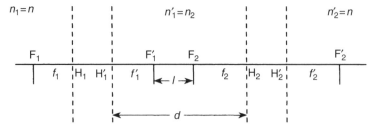

Figure P2.1

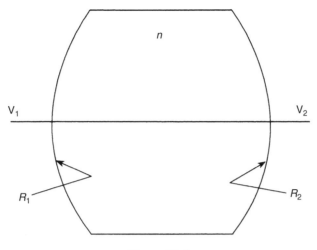

Figure P2.2

(c) Find the total power of the system.

(d) Find D and D'.

2.3 A doublet consists of two lenses with principal plane separation $d = f_1' + f_2 + l$, see Figure P2.1. We set $n_2 = n_2' = 1$.

(a) Find the power P of the doublet in terms of P_1, P_2 and l.

(b) Find the first and second focal lengths.

2.4 Find the power and the locations of the principal planes for a combination of two thin lenses each with the same focal length $f > 0$ separated by a distance d: (a) where $d = f$, (b) where $d = 3f/4$.

2.5 Show that the combination of two lenses having equal and opposite powers a finite, positive, distance d apart has a net positive power P, and find P as a function of d.

2.6 A thick lens as shown in Figure P2.2, is used in air. The first and second radii of curvature are $R_1 > 0$ and $R_2 < 0$, the index is $n > 1$, and the thickness $|V_1 V_2|$ is d. What will be the aperture stop for this lens for an axial object at a general distance S_1 to the left of V_1? Is the aperture stop always the same? (No calculation is necessary to solve this problem.)

2.7 A thin lens L_1 with a 5.0 cm diameter aperture and focal length +4.0 cm is placed 4.0 cm to the left of another lens L_2 4.0 cm in diameter with a focal length of +10.0 cm. A 2.0 cm high object is located with its centre on the axis 5 cm in front of L_1. There is a 3.0 cm diameter stop centered halfway between L_1 and L_2. Find the position and size of (a) the entrance pupil, (b) the exit pupil, (c) the image. Make a brief sketch to scale.

3

Interference

3.1 INTRODUCTION

The superposition principle for electromagnetic waves implies that, for example, two overlapping fields u_1 and u_2 add to give $u_1 + u_2$. This is the basis for interference. Because of the slow response of practical detectors, interference phenomena are also a matter of averaging over time and space. Therefore the concept of coherence is intimately related to interference. In this chapter we will investigate both topics. A high degree of coherence is obtained from lasers, which therefore have been widely used as light sources in interferometry. In recent years, lack of coherence has been taken to advantage in a technique called low-coherence or white-light interferometry, which we will investigate at the end of the chapter.

3.2 GENERAL DESCRIPTION

Interference can occur when two or more waves overlap each other in space.

Assume that two waves described by

$$u_1 = U_1 e^{i\phi_1} \tag{3.1a}$$

and

$$u_2 = U_2 e^{i\phi_2} \tag{3.1b}$$

overlap.

The electromagnetic wave theory tells us that the resulting field simply becomes the sum, viz.

$$u = u_1 + u_2 \tag{3.2}$$

The observable quantity is, however, the intensity, which becomes

$$I = |u|^2 = |u_1 + u_2|^2 = U_1^2 + U_2^2 + 2U_1 U_2 \cos(\phi_1 - \phi_2)$$
$$= I_1 + I_2 + 2\sqrt{I_1 I_2} \cos \Delta\phi \tag{3.3}$$

where

$$\Delta\phi = \phi_1 - \phi_2 \tag{3.4}$$

As can be seen, the resulting intensity does not become merely the sum of the intensities $(= I_1 + I_2)$ of the two partial waves. One says that the two waves interfere and $2\sqrt{I_1 I_2} \cos \Delta\phi$ is called the interference term. We also see that when

$$\Delta\phi = (2n + 1)\pi, \quad \text{for} \quad n = 0, 1, 2, \ldots$$

$\cos \Delta\phi = -1$ and I reaches its minima. The two waves are in antiphase which means that they interfere destructively.

When

$$\Delta\phi = 2n\pi, \quad \text{for} \quad n = 0, 1, 2, \ldots$$

$\cos \Delta\phi = 1$ and the intensity reaches its maxima. The two waves are in phase which means that they interfere constructively.

For two waves of equal intensity, i.e. $I_1 = I_2 = I_0$, Equation (3.3) becomes

$$I = 2I_0[1 + \cos \Delta\phi] = 4I_0 \cos^2 \left(\frac{\Delta\phi}{2}\right) \tag{3.5}$$

where the intensity varies between 0 and $4I_0$.

3.3 COHERENCE

Detection of light (i.e. intensity measurement) is an averaging process in space and time. In developing Equation (3.3) we did no averaging because we tacitly assumed the phase difference $\Delta\phi$ to be constant in time. That means that we assumed u_1 and u_2 to have the same single frequency. Ideally, a light wave with a single frequency must have an infinite length. Mathematically, even a pure sinusoidal wave of finite length will have a frequency spread according to the Fourier theorem (see Appendix B). Therefore, sources emitting light of a single frequency do not exist.

One way of illustrating the light emitted by real sources is to picture it as sinusoidal wave trains of finite length with randomly distributed phase differences between the individual trains.

Assume that we apply such a source in an interference experiment, e.g. the Michelson interferometer described in Section 3.6.2. Here the light is divided into two partial waves of equal amplitudes by a beamsplitter whereafter the two waves are recombined to interfere after having travelled different paths.

In Figure 3.1 we have sketched two successive wave trains of the partial waves. The two wave trains have equal amplitude and length L_c, with an abrupt, arbitrary phase difference. Figure 3.1(a) shows the situation when the two partial waves have travelled equal path lengths. We see that although the phase of the original wave fluctuates randomly, the phase difference between the partial waves 1 and 2 remains constant in time. The resulting intensity is therefore given by Equation (3.3). Figure 3.1(c) shows the situation when partial wave 2 has travelled a path length L_c longer than partial wave 1. The head of the wave trains in partial wave 2 then coincide with the tail of the corresponding wave trains in partial wave 1. The resulting instantaneous intensity is still given by Equation (3.3), but now the phase difference fluctuates randomly as the successive wave trains pass by. As

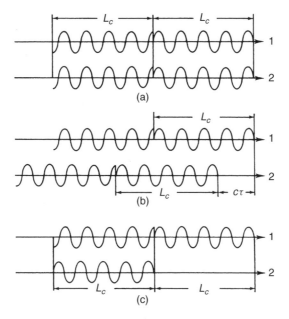

Figure 3.1

a result, $\cos \Delta\phi$ varies randomly between $+1$ and -1. When averaged over many wave trains, $\cos \Delta\phi$ therefore becomes zero and the resulting, observable intensity will be

$$I = I_1 + I_2 \tag{3.6}$$

Figure 3.1(b) shows an intermediate case where partial wave 2 has travelled a path length l longer than partial wave 1, where $0 < l < L_c$. Averaged over many wave trains, the phase difference now varies randomly in a time period proportional to $\tau = l/c$ and remains constant in a time period proportional to $\tau_c - \tau$ where $\tau_c = L_c/c$. The result is that we still can observe an interference pattern according to Equation (3.3), but with a reduced contrast. To account for this loss of contrast, Equation (3.3) can be written as

$$I = I_1 + I_2 + 2\sqrt{I_1 I_2}|\gamma(\tau)| \cos \Delta\phi \tag{3.7}$$

where $|\gamma(\tau)|$ means the absolute value of $\gamma(\tau)$.

To see clearly that this quantity is related to the contrast of the pattern, we introduce the definition of contrast or visibility

$$V = \frac{I_{max} - I_{min}}{I_{max} + I_{min}} \tag{3.8}$$

where I_{max} and I_{min} are two neighbouring maxima and minima of the interference pattern described by Equation (3.7). Since $\cos \Delta\phi$ varies between $+1$ and -1 we have

$$I_{max} = I_1 + I_2 + 2\sqrt{I_1 I_2}|\gamma(\tau)| \tag{3.9a}$$

$$I_{min} = I_1 + I_2 - 2\sqrt{I_1 I_2}|\gamma(\tau)| \tag{3.9b}$$

which, put into Equation (3.8), gives

$$V = \frac{2\sqrt{I_1 I_2}|\gamma(\tau)|}{I_1 + I_2} \tag{3.10}$$

For two waves of equal intensity, $I_1 = I_2$, and Equation (3.10) becomes

$$V = |\gamma(\tau)| \tag{3.11}$$

which shows that in this case $|\gamma(\tau)|$ is exactly equal to the visibility. $\gamma(\tau)$ is termed the complex degree of coherence and is a measure of the ability of the two wave fields to interfere. From the previous discussions we must have

$$|\gamma(0)| = 1 \tag{3.12a}$$

$$|\gamma(\tau_c)| = 0 \tag{3.12b}$$

$$0 \le |\gamma(\tau)| \le 1 \tag{3.12c}$$

where Equations (3.12a) and (3.12b) represent the two limiting cases of complete coherence and incoherence respectively, while inequality (3.12c) represents partial coherence.

Of more interest is to know the value of τ_c, i.e. at which path length difference $|\gamma(\tau)| = 0$. In Section 5.4.9 we find that in the case of a two-frequency laser this happens when

$$\tau = \tau_c = \frac{L_c}{c} = \frac{1}{\Delta \nu} \tag{3.13}$$

where $\Delta \nu$ is the difference between the two frequencies. It can be shown that this relation applies to any light source with a frequency distribution of width $\Delta \nu$. L_c is termed the coherence length and τ_c the coherence time.

We see that Equation (3.13) is in accordance with our previous discussion where we argued that sources of finite spectral width will emit wave trains of finite length. This is verified by the relation

$$\Delta \nu = \frac{\Delta \lambda_c}{\lambda^2} \tag{3.14}$$

which can be derived from Equation (1.2).

As given in Section 1.2, the visible spectrum ranges from 4.3 to 7.5×10^{14} Hz which gives a spectral width roughly equal to $\Delta \nu = 3 \times 10^{14}$ Hz. From Equation (3.13), the coherence time of white light is therefore about 3×10^{-15} s, which corresponds to a coherence length of about 1 μm. In white-light interferometry it is therefore difficult to observe more than two or three interference fringes. This condition can be improved by applying colour filters at the cost of decreasing the intensity.

Ordinary discharge lamps have spectral widths corresponding to coherence lengths of the order of 1 μm while the spectral lines emitted by low-pressure isotope lamps have coherence lengths of several millimetres.

By far the most coherent light source is the laser. A single-frequency laser can have coherence lengths of several hundred metres. This will be analysed in more detail in Section 5.4.9.

So far we have been discussing the coherence between two wave fields at one point in space. This phenomenon is termed temporal or longitudinal coherence. It is also possible to measure the coherence of a wave field at two points in space. This phenomenon is called spatial or transverse coherence and can be analysed by the classical Young's double slit (or pinhole) experiment (see Section 3.6.1). Here the wave field at two points P_1 and P_2 is analysed by passing the light through two small holes in a screen S_1 at P_1 and P_2 and observing the resulting interference pattern on a screen S_2 (see Figure 3.13(a)). In the same way as the temporal degree of coherence $\gamma(\tau)$ is a measure of the fringe contrast as a function of time difference τ, the spatial degree of coherence γ_{12} is a measure of the fringe contrast of the pattern on screen S_2 as a function of the spatial difference D between P_1 and P_2. Note that since γ_{12} is the spatial degree of coherence for $\tau = 0$, it is the contrast of the central fringe on S_2 that has to be measured.

To measure the spatial coherence of the source itself, screen S_1 has to be placed in contact with the source. It is immediately clear that for an extended thermal light source, $|\gamma_{12}| = 0$ unless $P_1 = P_2$, which gives $|\gamma_{11}| = 1$. On the other hand, if we move S_1 away from this source, we observe that $|\gamma_{12}|$ might be different from zero, which shows that a wave field increases its spatial coherence by mere propagation. We also observe that $|\gamma_{12}|$ increases by stopping down the source by, for example, an aperture until $|\gamma_{12}| = 1$ for a pinhole aperture. The distance D_c between P_1 and P_2 for which $|\gamma_{12}| = 0$ is called the spatial coherence length. It can be shown that D_c is inversely proportional to the diameter of the aperture in analogy with the temporal coherence length, which is inversely proportional to the spectral width. Moreover, it can be shown that $|\gamma_{12}|$ is the Fourier transform of the intensity distribution of the source and that $|\gamma(\tau)|$ is the Fourier transform of the spectral distribution of the source (see Section 3.7).

An experimentalist using techniques like holography, moiré, speckle and photoelasticity need not worry very much about the details of coherence theory. Both in theory and experiments one usually assumes that the degree of coherence is either one or zero. However, one should be familiar with fundamental facts such as:

(1) Light from two separate sources does not interfere.

(2) The spatial and temporal coherence of light from an extended thermal source is increased by stopping it down and by using a colour filter respectively.

(3) The visibility function of a multimode laser exhibits maxima at an integral multiple of twice the cavity length (see Section 5.4.9).

3.4 INTERFERENCE BETWEEN TWO PLANE WAVES

Figure 3.2(a) shows two plane waves u_1, u_2 with propagation directions $\mathbf{n}_1$, $\mathbf{n}_2$ that lie in the xz-plane making the angles θ_1 and θ_2 to the z-axis. We introduce the following quantities (see Figure 3.2(b)): $\alpha =$ the angle between $\mathbf{n}_1$ and $\mathbf{n}_2$, $\theta =$ the angle between the line bisecting α and the z-axis. The complex amplitude of the two plane waves then becomes (see Equation (1.9a))

$$u_1 = U_1 e^{i\phi_1} \tag{3.15}$$

$$u_2 = U_2 e^{i\phi_2} \tag{3.16}$$

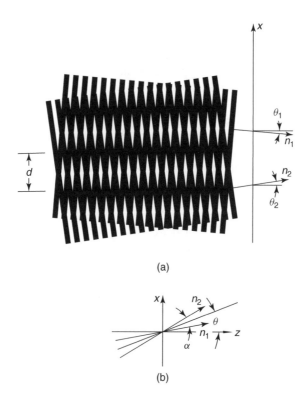

Figure 3.2 Interference between two plane waves

where

$$\phi_1 = k\left[x\sin\left(\theta - \frac{\alpha}{2}\right) + z\cos\left(\theta - \frac{\alpha}{2}\right)\right] \tag{3.17}$$

$$\phi_2 = k\left[x\sin\left(\theta + \frac{\alpha}{2}\right) + z\cos\left(\theta + \frac{\alpha}{2}\right)\right] \tag{3.18}$$

The intensity is given by the general expression in Equation (3.3) by inserting

$$\Delta\phi = \phi_1 - \phi_2 = k\left\{x\left[\sin\left(\theta - \frac{\alpha}{2}\right) - \sin\left(\theta + \frac{\alpha}{2}\right)\right] + z\left[\cos\left(\theta - \frac{\alpha}{2}\right) - \cos\left(\theta + \frac{\alpha}{2}\right)\right]\right\}$$

$$= 2k\sin\frac{\alpha}{2}\{-x\cos\theta + z\sin\theta\} \tag{3.19}$$

The interference term is therefore of the form

$$\cos\frac{2\pi}{d}(z\sin\theta - x\cos\theta) \tag{3.20}$$

By comparing this expression with the real part of Equation (1.9a), we see that Equation (3.20) can be regarded as representing a plane wave with its propagation direction lying in the xz-plane making an angle θ with the x-axis as depicted in Figure 3.3,

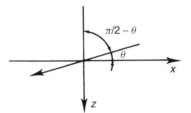

Figure 3.3

Figure 3.4

and with a wavelength equal to

$$d = \frac{\lambda}{2 \sin(\alpha/2)} \tag{3.21}$$

This is also clearly seen from Figure 3.2. From Equation (3.21) we see that the distance between the interference fringes (the wavelength d) is dependent only on the angle between $\mathbf{n}_1$ and $\mathbf{n}_2$. By comparing Figures 3.2 and 3.4 we see how d decreases as α increases. The diagram in Figure 3.5 shows the relation between d and α and $f = 1/d$ and α according to Equation (3.21). Here we have put $\lambda = 0.6328 \ \mu$m, the wavelength of the He–Ne laser.

The intensity distribution across the xy-plane is found by inserting $z = 0$ into Equation (3.19):

$$I = I_1 + I_2 + 2\sqrt{I_1 I_2} \cos\left[2kx \sin\frac{\alpha}{2}\cos\theta\right] \tag{3.22}$$

From the maxima (or minima) of this equation, we find the inter-fringe distance measured along the x-axis to be

$$d_x = \frac{1}{\sin\theta_2 - \sin\theta_1} = \frac{\lambda}{2\sin\dfrac{\alpha}{2}\cos\theta} = d/\cos\theta \tag{3.23}$$

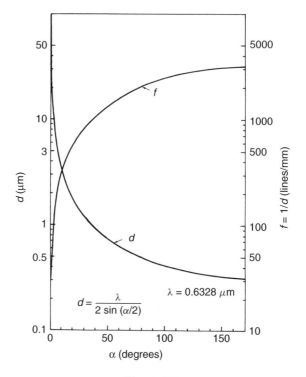

Figure 3.5

The second equality of this expression is found by trigonometric manipulation of the angles (see Figure 3.2(b)). Accordingly, the spatial frequency becomes

$$f_x = 1/d_x = \frac{2 \sin \dfrac{\alpha}{2} \cos \theta}{\lambda} = \cos \theta / d \qquad (3.24)$$

For completeness, we also quote the definition of the instantaneous frequency of a sinusoidal grating with phase $\phi(x)$ at a point x_0.

$$f_x(x = x_0) = \left. \frac{d\phi(x)}{dx} \right|_{x=x_0} \qquad (3.25)$$

The intensity distribution given in Equation (3.22) is sketched in Figure 3.6. We see that it varies between

$$I_{max} = I_1 + I_2 + 2\sqrt{I_1 I_2} \qquad (3.26)$$

and

$$I_{min} = I_1 + I_2 - 2\sqrt{I_1 I_2} \qquad (3.27)$$

with a mean value equal to

$$I_0 = I_1 + I_2 \qquad (3.28)$$

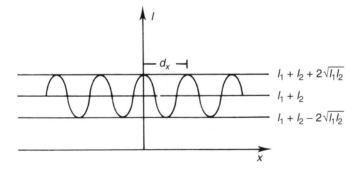

Figure 3.6 Intensity distribution in the xy-plane from interference between two plane waves

When Equations (3.26) and (3.27) are put into the expression for the visibility or contrast defined in Section 3.3, Equation (3.8), they give

$$V = \frac{I_{\max} - I_{\min}}{I_{\max} + I_{\min}} = \frac{2\sqrt{I_1 I_2}}{I_1 + I_2} \tag{3.29}$$

V is equal to the amplitude of the distribution divided by the mean value and varies between 0 and 1. We see that

$$V = 1 \quad \text{for} \quad I_1 = I_2$$
$$V = 0 \quad \text{for} \quad \text{either } I_1 \text{ or } I_2 = 0$$

3.4.1 Laser Doppler Velocimetry (LDV)

As the name (also termed laser Doppler anemometry (Durst *et al.* 1991), LDA) indicates, this is a method for measuring the velocity of, for example, moving objects or particles. It is based on the Doppler effect, which explains the fact that light changes its frequency (wavelength) when detected by a stationary observer after being scattered from a moving object.

This is in analogy with the classical example for acoustical waves when the whistle from a train changes from a high to a low tone as the train passes by.

Here we give an alternative description of the method. Consider Figure 3.7 where a particle is moving in a test volume where two plane waves are interfering at an angle α. In Section 3.4 it was found that these two waves will form interference planes which are parallel to the bisector of α and separated by a distance equal to (cf. Equation (3.21))

$$d = \frac{\lambda}{2 \sin \alpha/2} \tag{3.30}$$

As the particle moves through the test volume, it will scatter light when it is passing a bright interference fringe and scatter no light when it is passing a dark interference fringe. The resulting light pulses can be recorded by a detector placed as in Figure 3.7.

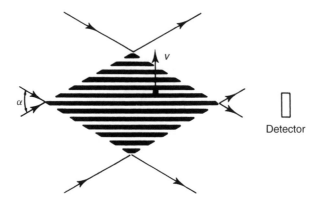

Figure 3.7   Laser Doppler velocimetry

For a particle moving in the direction normal to the interference planes with a velocity v, the time lapse between successive light pulses becomes

$$t_D = \frac{d}{v} \tag{3.31}$$

and thus the frequency

$$f_D = 1/t_D = \frac{2v \sin \alpha/2}{\lambda} \tag{3.32}$$

If there are many particles of different velocities, one will get many different frequencies. They can be recorded on a frequency analyser and the resulting frequency spectrum will tell how the particles are distributed among the different velocities.

This method does not distinguish between particles moving in opposite directions. If the direction of movement is unknown, one can modulate the phase of one of the plane waves (by means of, for example, an acousto-optic modulator) thereby making the interference planes move parallel to themselves with a known velocity. This velocity will then be subtracted when the particles are moving in the same direction and added when moving in the opposite direction.

In Figure 3.7, the particles pass between the light source and the detector. If the particles scatter enough light, the detector can also be placed on the same side of the test volume as the light source (the laser). Many other configurations of the light source and the detector are described in the literature. For example, one of the two waves can be directly incident on the detector, or it is possible to have one single wave and many detectors.

Laser Doppler velocimetry can be applied for measurement of the velocity of moving surfaces, turbulence in liquids and gases, etc. In the latter cases, the liquid or gas must be seeded with particles. Examples are measurements of stream velocities around ship propellers, velocity distributions of oil drops in combustion and diesel engines, etc.

3.5 INTERFERENCE BETWEEN OTHER WAVES

Figure 3.8 shows the geometric configuration of the fringe pattern in the xz-plane when two spherical waves from two point sources P_1 and P_2 on the z-axis interfere. From

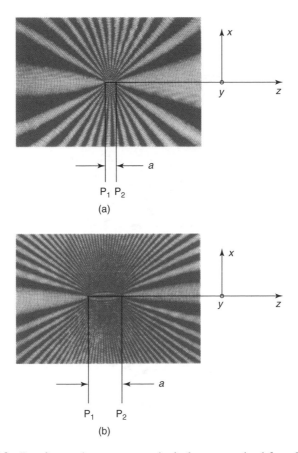

Figure 3.8   Interference between two spherical waves emitted from P_1 and P_2

Figures 3.8(a) and 3.8(b) we see how the density of the fringes increases as the distance between P_1 and P_2 increases. Note that the figure shows the situation for the actual wavelength. The distance between the point sources in Figure 3.8(a) is about seven wavelengths, which for light with $\lambda = 0.5$ μm would give 3.5 μm. Two real point sources separated by the same distance as in the figure therefore would have resulted in a pattern of much higher density; but the form of pattern would be the same.

Figure 3.9 shows the interference pattern in the xz-plane when a spherical wave from a point P on the z-axis interferes with a plane wave propagating in the z-direction.

In the same way as in the case of two plane waves, we can observe the intensity distribution over a plane of arbitrary orientation in space. A special distribution can be observed over the xy-plane in Figure 3.9. This is further illustrated in Figure 3.10 which shows the case of a spherical wave and a plane wave. The intensity distribution in the xy-plane is given as

$$I = I_1 + I_2 + 2\sqrt{I_1 I_2} \cos(\beta r^2) \tag{3.33}$$

with $r^2 = x^2 + y^2$, $\beta = $ constant. This is a sinusoidal pattern of linearly increasing frequency and is called a circular zone plate pattern.

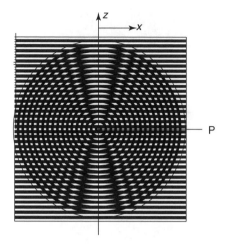

Figure 3.9 Interference between a plane wave and a spherical wave

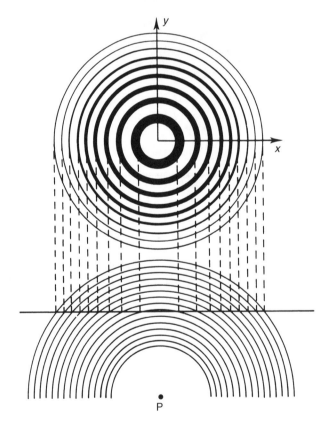

Figure 3.10 The circular zone plate pattern

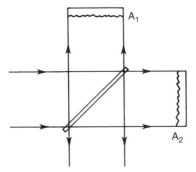

Figure 3.11

So far we have been dealing with interference between combinations of spherical waves and plane waves. An important point to note is that, by measuring the distance between interference fringes over selected planes in space, quantities such as the angle between the propagation directions of plane waves, the distance between point sources and the distance from a point source to the plane of observation can be determined. One further step would be to apply the same type of experiment to more complicated waveforms, such as a wave reflected from a rough surface as indicated in Figure 1.5(c). By observing the interference between the reflected wave and a plane wave, one should in principle be able to determine the topography of the surface. However, for surfaces of roughness greater than the wavelength, phenomena such as interference between light scattered from different points on the surface, multiple scattering and diffraction effects will occur. In this case, it therefore becomes impossible to derive the surface topography from a given interference pattern. For smoother surfaces, however, such as optical components (lenses, mirrors, etc.) where tolerances of the order of fractions of a wavelength are to be measured, that kind of interferometry is quite common.

Figure 3.11 shows a Michelson interferometer (see Section 3.6.2, Figure 3.15) where the mirrors are exchanged for two non-optical surfaces A_1 and A_2. If these surfaces are identical, it should be possible to observe interference between the waves scattered from A_1 and A_2 regardless of the complexity of the scattered wavefronts. In the case of a plastic deformation of, for example, surface A_2, it should be possible to do interferometric measurement of the resulting surface height difference between A_1 and A_2. The problem is, however, that the phrasing 'identical surfaces' in this context must be taken literally, i.e. the microstructure of the two surfaces must be identical. This has to do with the mutual spatial coherence of the two scattered waves. This requirement on the two surfaces makes this interference experiment impracticable. However, when we learn later on about holography, we shall see that this type of measurement becomes more than an imaginary experiment.

3.6 INTERFEROMETRY

Interference phenomena can be observed in interferometers. As stated in Section 3.3, light waves can interfere only if they are emitted by the same source. Most interferometers therefore consist of the following elements, shown schematically in Figure 3.12.

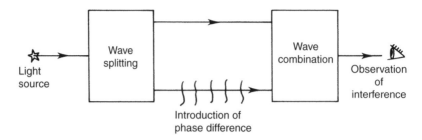

Figure 3.12

- light source;

- element for splitting the light into two (or more) partial waves;

- different propagation paths where the partial waves undergo different phase contributions;

- element for superposing the partial waves;

- detector for observation of the interference.

Depending on how the light is split, interferometers are commonly classified as wavefront-dividing or amplitude-dividing interferometers; but there are configurations which fall outside this classification.

3.6.1 Wavefront Division

As an example of a wavefront-dividing interferometer, consider the oldest of all interference experiments due to Thomas Young (1801) (Figure 3.13). The incident wavefront is divided by passing through two small holes at P_1 and P_2 in a screen S_1. The emerging spherical wavefronts from P_1 and P_2 will interfere, and the resulting interference pattern is observed on the screen S_2. This is in analogy with the case of two point sources in Figure 3.8 with the plane of observation oriented parallel to the yz-plane.

The geometric path length difference s of the light reaching an arbitrary point x on S_2 from P_1 and P_2 is found from Figure 3.13(b). When the distance z between S_1 and S_2 is much greater than the distance D between P_1 and P_2, we have, to a good approximation,

$$\frac{s}{D} = \frac{x}{z}, \quad \text{that is} \quad s = \frac{D}{z}x$$

The phase difference therefore becomes

$$\Delta\phi = \frac{2\pi}{\lambda}s = \frac{2\pi D}{\lambda z}x \tag{3.34}$$

which, inserted into the general expression for the resulting intensity distribution, Equation (3.3), gives

$$l(x) = 2I\left(1 + \cos\left(2\pi\frac{D}{\lambda z}x\right)\right) \tag{3.35}$$

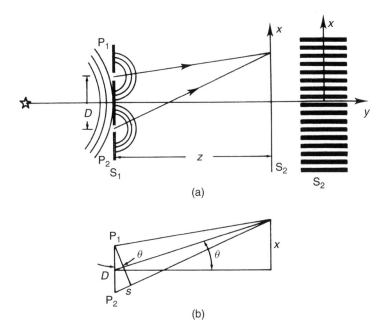

Figure 3.13 Young's interferometer

We get interference fringes parallel to the y-axis with a spatial period $\lambda z/D$ which decreases as the distance between P_1 and P_2 increases.

Here we have assumed that the waves from P_1 and P_2 are fully coherent. As stated in Section 3.3, this is an ideal case and becomes more and more difficult to fulfil as the distance D between P_1 and P_2 is increased. The contrast of the interference fringes on S_2 is a measure of the degree of coherence. As will be shown in Section 3.7 there is a Fourier transform relationship between the degree of coherence and the intensity distribution of the light source. By assuming the source to be an incoherent circular disc of uniform intensity, one can find the diameter of the source by increasing the separation between P_1 and P_2 until the contrast of the central interference fringe on S_2 vanishes. This is utilized in Michelson's stellar interferometer (Figure 3.14(c)) to measure the diameter of distant stars. This is an extension of Young's interferometer, where a mirror arrangement is used to make the effective distance D sufficiently long.

Other types of wavefront-dividing interferometers are shown in Figure 3.14.

3.6.2 Amplitude Division

The most well-known amplitude-dividing interferometer is the Michelson interferometer sketched in Figure 3.15. Here the amplitude of the incident light field is divided by the beamsplitter BS which is partly reflecting. The reflected and the transmitted partial waves propagate to the mirrors M_1 and M_2 respectively, from where they are reflected back and recombine to form the interference distribution on the detector D.

The path-length difference between the two partial waves can be varied by moving one of the mirrors, e.g. M_2, which might be mounted on a movable object. A displacement x

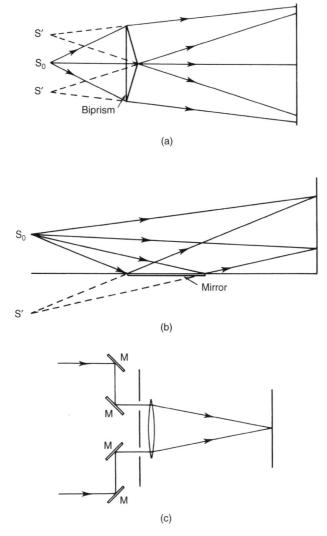

Figure 3.14   Examples of wavefront-dividing interferometers: (a) Fresnel biprism; (b) Lloyd's mirror; and (c) Michelson's stellar interferometer

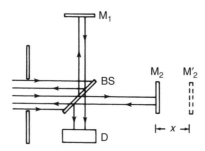

Figure 3.15   Michelson's interferometer

of M_2 gives a path length difference $2x$ and a phase difference equal to $\Delta\phi = (2\pi/\lambda)2x$. This results in an intensity distribution given by

$$I(x) = 2I\left(1 + \cos\frac{4\pi x}{\lambda}\right) \tag{3.36}$$

As M_2 moves, its displacement is measured by counting the number of light maxima registered by the detector. By counting the numbers of maxima per unit time, one can find the speed of the object.

Other types of amplitude-dividing interferometers are shown in Figure 3.16.

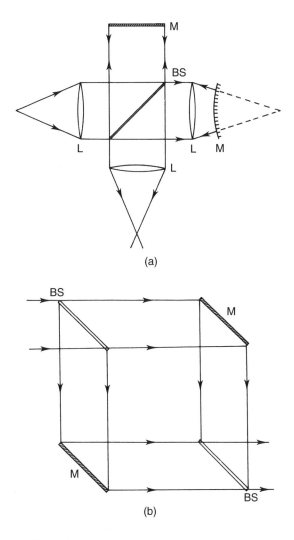

Figure 3.16 Examples of amplitude-dividing interferometers. (a) Twyman-Green interferometer and (b) Mach-Zehnder interferometer

3.6.3 The Dual-Frequency Michelson Interferometer

In Section 3.3 we stated that two waves of different frequencies do not produce observable interference. By combining two plane waves

$$\psi_1 = e^{i2\pi[(z/\lambda_1)-\nu_1 t]} \tag{3.37a}$$

and

$$\psi_2 = e^{i2\pi[(z/\lambda_2)-\nu_2 t]} \tag{3.37b}$$

of different frequencies, the resulting intensity becomes

$$I = 2\left[1 + \cos 2\pi\left(\left(\frac{1}{\lambda_1} - \frac{1}{\lambda_2}\right)z - (\nu_1 - \nu_2)t\right)\right] \tag{3.38}$$

If the frequency difference $\nu_1 - \nu_2$ is very small and constant, this variation in I with time can be detected. This is utilized in the dual-frequency Michelson interferometer for length measurement.

Such an interferometer designed by Hewlett-Packard is given in Figure 3.17. The light source is a single-mode laser where the frequency is Zeeman-split into two components f_1 and f_2. (In the figure, frequencies are denoted by f instead of ν) with a frequency difference of about 2 MHz. This is achieved by means of an applied magnetic field which splits the light into two orthogonally polarized waves. A polarization-sensitive beamsplitter transmits the component of frequency f_1 to the movable mirror and the

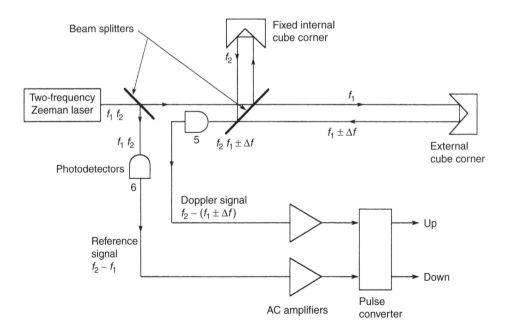

Figure 3.17   The Hewlett-Packard interferometer

component of frequency f_2 to the fixed mirror. These mirrors are cube-corner reflectors in which the reflected beam is parallel to the incoming beam independent of the angle of incidence. If the movable mirror moves with a velocity v_1, the frequency of the reflected light will be Doppler-shifted by an amount $\Delta f = 2v/\lambda_1$. The two light waves interfere on photodetector 5, which therefore gives an electrical signal of frequency $f_2 - (f_1 \pm \Delta f)$ according to Equation (3.38). For two orthogonally polarized waves to interfere, a polarizer has to be placed in front of the detector. A fraction of the laser light of frequencies f_1 and f_2 is sent to the detector 6, producing a reference signal of frequency $f_2 - f_1$. After being amplified, the two electric signals are fed to two frequency counters. The recorded frequencies are then subtracted, giving Δf, which determines the displacement

$$s = vt = \Delta f \lambda t / 2 \tag{3.39}$$

This type of interferometer, also called an a.c. (alternating current) interferometer, gives the displacement in terms of variations in frequency instead of variations in intensity as in the standard Michelson interferometer. Therefore this interferometer is more immune to disturbances like turbulent air in the optical path.

In normal operation the Hewlett-Packard interferometer can measure distances up to 60 m with a resolution of $\lambda/4(\sim 0.16~\mu\text{m})$ which can be extended electronically by factors as high as 100. The accuracy is 5×10^{-7} and the velocity can be measured up to 0.3 m/s. By means of a double cube-corner and an additional beamsplitter, angles and surface flatness can be measured. A new acquisition technique was proposed by Gaal et al. (1993).

3.6.4 Heterodyne (Homodyne) Detection

The dual-frequency Michelson interferometer is an example of so-called heterodyne detection. This technique is a well-known principle in telecommunication and optical communication. Here the optical signal is mixed with a so-called local oscillator signal at the detector. If we neglect the spatial term (which is constant at the detector), the electric fields of the optical signal ψ_S and of the local oscillator ψ_{LO} are given as

$$\psi_S = U_S \exp[i2\pi(\nu_S t + \phi(t))] \tag{3.40a}$$

$$\psi_{LO} = U_{LO} \exp[i2\pi((\nu_S + \nu_{IF})t)] \tag{3.40b}$$

where ν_S is the carrier frequency and $\phi(t)$ contains the frequency-modulating information-carrying signal. The local oscillator frequency $\nu_{LO} = \nu_S + \nu_{IF}$ is offset from the carrier frequency by the intermediate frequency ν_{IF}. In optical communication ν_{IF} is normally in the radio-frequency range from a few tens to hundreds of megahertz. A homodyne detection system results if there is no offset, i.e. $\nu_{IF} = 0$.

The intensity at the detector becomes

$$I_d = |\psi_S + \psi_{LO}|^2 = U_S^2 + U_{LO}^2 + 2U_S U_{LO} \cos 2\pi(\nu_{IF} t - \phi(t)) \tag{3.41}$$

The signal I_d from the detector can be demodulated by so-called synchronous demodulation consisting of first splitting it into two and multiplying by $\cos 2\pi \nu_{IF} t$ and $\sin 2\pi \nu_{IF} t$,

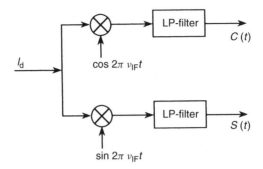

Figure 3.18   The principle of synchronous demodulation

see Figure 3.18:

$$I_d \cos 2\pi \nu_{IF} t = (U_S^2 + U_{LO}^2) \cos 2\pi \nu_{IF} t + U_S U_{LO} \cos 2\pi (2\nu_{IF} t - \phi(t))$$
$$+ U_S U_{LO} \cos 2\pi \phi(t)$$
$$I_d \sin 2\pi \nu_{IF} t = (U_S^2 + U_{LO}^2) \sin 2\pi \nu_{IF} t + U_S U_{LO} \sin 2\pi (2\nu_{IF} t - \phi(t))$$
$$+ U_S U_{LO} \sin 2\pi \phi(t) \tag{3.42}$$

We see that these two expressions both contain two high-frequency terms ν_{IF} and $2\nu_{IF}$. By low-pass filtering of the signals, one is left with

$$C(t) = U_S U_{LO} \cos 2\pi \phi(t) \tag{3.43a}$$
$$S(t) = U_S U_{LO} \sin 2\pi \phi(t) \tag{3.43b}$$

from which the modulating function is given by

$$\phi(t) = \frac{1}{2\pi} \tan^{-1} \left[\frac{S(t)}{C(t)} \right] \tag{3.44}$$

This principle is used in so-called superheterodyne radio receivers. In the dual-frequency Michelson interferometer, $\nu_{IF} = \nu_1 - \nu_2$ and $\phi(t) = 2\nu t / \lambda_1$. We shall see later that the heterodyne principle is used both in interferometry and moirè, and that the signal frequencies can be in the spatial domain rather than the time domain.

3.7 SPATIAL AND TEMPORAL COHERENCE

Having been introduced to interferometers and interferometry, we can consider the topic of coherence a bit further. The following material also demands some knowledge of Fourier transforms, which are treated in Chapter 4 and Appendix B.

In the treatment of the Young's interferometer in Section 3.6.1 we assumed the light to be incident from a point source. Consider the Young set-up in Figure 3.19 where the light is incident from an extended, incoherent, quasimonochromatic light source. In the

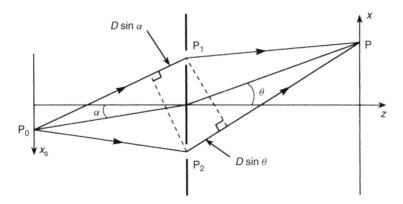

Figure 3.19   Young's interferometer with extended light source

figure, two light rays from a point P_0 on the source to the point of observation P via the holes P_1 and P_2 in the screen S_1 is drawn. The path-length difference between these two rays is seen to be

$$s = s_1 - s_2 = D \sin \theta - D \sin \alpha \approx D(\theta - \alpha) \qquad (3.45)$$

where in the last equality we have approximated the sines by the angles.

Assume that the intensity at P when only P_1 is open is $I(\alpha)/2$. Then because of the small path-length difference, the intensity at P when only P_2 is open also will be $I(\alpha)/2$. The intensity at P due to the light from P_0 therefore can be written as

$$\Delta I = \{1 + \cos[kD(\theta - \alpha)]\} \qquad (3.46)$$

To find the total intensity at P we have to integrate Equation (3.46) over the range of α from $-\alpha_m/2$ to $\alpha_m/2$. Since $I(\alpha)$ is zero outside the source, we can change the integration limits to $-\infty, \infty$. We then get

$$I(P) = \int_{-\infty}^{\infty} \Delta I \, d\alpha = \int_{-\infty}^{\infty} I(\alpha) \, d\alpha + \int_{-\infty}^{\infty} I(\alpha) \cos[kD(\theta - \alpha)] \, d\alpha$$

$$= I_t \left\{ 1 + \frac{1}{I_t} \int_{-\infty}^{\infty} I(\alpha) \cos[kD(\theta - \alpha)] \, d\alpha \right\} \qquad (3.47)$$

where

$$I_t = \int_{-\infty}^{\infty} I(\alpha) \, d\alpha \qquad (3.48)$$

is the total intensity with no interference.

The integrand in the last integral can be written to read

$$I(\alpha) \cos[kD(\theta - \alpha)] = I(\alpha) \, \mathrm{Re}\{e^{ikD(\theta - \alpha)}\} = \mathrm{Re}\{e^{ikD\theta} I(\alpha) e^{-ikD\alpha}\} \qquad (3.49)$$

This, put onto the last integral of Equation (3.47), gives

$$\frac{1}{I_t} \int_{-\infty}^{\infty} \mathrm{Re}\{e^{ikD\theta} I(\alpha) e^{-ikD\alpha}\} \, d\alpha = \frac{1}{I_t} \mathrm{Re}\left\{e^{ikD\theta} \int_{-\infty}^{\infty} I(\alpha) e^{-ikD\alpha} \, d\alpha\right\}$$

$$= \mathrm{Re}\{e^{ikD\theta}\gamma_{12}\} = |\gamma_{12}| \cos(kD\theta + \psi) \qquad (3.50)$$

where

$$\gamma_{12} = |\gamma_{12}| e^{i\psi} = \frac{\displaystyle\int_{-\infty}^{\infty} I(\alpha) e^{ikD\alpha} \, d\alpha}{\displaystyle\int_{-\infty}^{\infty} I(\alpha) \, d\alpha} \qquad (3.51)$$

This, inserted into Equation (3.47), gives

$$I(P) = I_t\{1 + |\gamma_{12}| \cos[kD\theta + \psi]\} \qquad (3.52)$$

From the discussion in Section 3.3 we recognize γ_{12} as the complex degree of spatial coherence. We see that γ_{12} is equal to the normalized Fourier transform of the intensity distribution of the source, with the frequency coordinate equal to D/λ.

For a square disc of uniform intensity

$$I(\alpha) = \frac{I_t}{\Delta\alpha} \, \mathrm{rect}\left(\frac{\alpha}{\Delta\alpha}\right) \qquad (3.53)$$

we get

$$|\gamma_{12}| = \left|\mathrm{sinc}\left(\frac{D\Delta\alpha}{\lambda}\right)\right| \qquad (3.54)$$

from which we find that the visibility has its first zero for

$$D = D_c = \frac{\lambda}{\Delta\alpha} \qquad (3.55)$$

For a circular disc of uniform intensity

$$I(\alpha) = \frac{I_t}{\Delta\alpha} \mathrm{circ}\left(\frac{\alpha}{\Delta\alpha}\right) \qquad (3.56)$$

we get

$$|\gamma_{12}| = \left|\frac{J_1\left(\frac{\pi D\Delta\alpha}{\lambda}\right)}{\frac{\pi D\Delta\alpha}{\lambda}}\right| \qquad (3.57)$$

with the first zero occurring for

$$D = D_c = 1.22\frac{\lambda}{\Delta\alpha} \qquad (3.58)$$

In an experiment conducted by A.A. Michelson at the Mount Wilson Laboratory, California, in December 1920, the fringes formed in the stellar interferometer (see

Figure 3.14(c)) by the light from the star Betelgeuse were made to vanish at $D = 121$ inches (3070 mm). With $\lambda = 570$ nm, $\Delta\alpha = 22.6 \times 10^{-8}$ rad, and from its known distance determined from parallax measurements, the star's diameter turned out to be about 380 million km or roughly 280 times that of the Sun.

In the description of the Michelson interferometer in Section 3.6.2 we assumed the light source to be monochromatic and found the output intensity (assuming a 50/50 beamsplitter) to be:

$$I = \Delta I \left(1 + \cos\left(\frac{2\pi \nu d}{c} \right) \right) = \Delta I (1 + \cos(2\pi \nu \tau)) \tag{3.59}$$

where ΔI is the total intensity of the incoming beam, d is the path-length difference and $\tau = d/c$ is the time difference between the two paths. When using a light source of N distinct frequencies, each of intensity I_n, we will, since each component can interfere with itself only, get an output intensity equal to

$$I = \sum_n I_n (1 + \cos(2\pi \nu_n \tau)) \tag{3.60}$$

Ordinary light sources have a continuous distribution of frequencies $I(\nu)$, in which case the sum in Equation (3.60) is converted into the integral

$$I = \int_0^\infty I(\nu)(1 + \cos(2\pi \nu \tau))\, d\nu = I_0 \left[1 + \int_0^\infty P(\nu) \cos(2\pi \nu \tau)\, d\nu \right] \tag{3.61}$$

where

$$I_0 = \int_0^\infty I(\nu)\, d\nu \tag{3.62}$$

is the total intensity and

$$P(\nu) = \frac{I(\nu)}{I_0} \tag{3.63}$$

is known as the normalized spectral distribution function of the source. Since light frequencies are positive, the lower integration limit of Equation (3.61) is zero. However, if the spectral distribution is peaked about $\nu = \nu_0$, we can write

$$P(\nu) = S(\nu - \nu_0) \tag{3.64}$$

which gives

$$\int_0^\infty P(\nu) \cos(2\pi \nu \tau)\, d\nu = \mathrm{Re} \left\{ \int_{-\infty}^\infty S(\nu - \nu_0) e^{-i2\pi \nu \tau}\, d\nu \right\}$$

$$= \mathrm{Re} \left\{ e^{-i2\pi \nu_0 \tau} \int_{-\infty}^\infty S(\nu) e^{-i2\pi \nu \tau}\, d\nu \right\}$$

$$= |\gamma(\tau)| \cos(2\pi \nu_0 \tau - \varphi) \tag{3.65}$$

This, put into Equation (3.61), gives

$$I = I_0[1 + |\gamma(\tau)| \cos(2\pi \nu_0 \tau - \varphi)] \tag{3.66}$$

where

$$\gamma(\tau) = \int_{-\infty}^{\infty} S(\nu) e^{-i2\pi \nu \tau} \, d\nu = |\gamma(\tau)| e^{i\varphi} \tag{3.67}$$

is recognized as the complex temporal degree of coherence and is equal to the Fourier transform of the spectral distribution function of the source with the origin of the frequency axis moved to ν_0. Therefore $\gamma(\tau)$ and $S(\nu)$ form a Fourier transform pair

$$S(\nu) = \int_{-\infty}^{\infty} \gamma(\tau) e^{i2\pi \nu \tau} \, d\tau \tag{3.68}$$

At this point it should be appropriate to introduce a precise definition of the coherence time τ_c (see Section 3.3). A multitude of definitions of τ_c in terms of $\gamma(\tau)$ exists, but the most frequently used one is

$$\tau_c = \int_{-\infty}^{\infty} |\gamma(\tau)|^2 \, d\tau \tag{3.69}$$

As an example, consider a source with a Gaussian spectral distribution function

$$P(\nu) = \frac{1}{\sqrt{2\pi}\sigma} \exp\left\{-\frac{1}{2}\left(\frac{\nu - \nu_0}{\sigma}\right)^2\right\} \tag{3.70}$$

where σ is the standard deviation from the mean frequency ν_0. $P(\nu)$ is commonly expressed in terms of its FWHM (Full Width Half Maximum) value $\Delta\nu$, in which case it becomes

$$P(\nu) = \frac{2\sqrt{\ln 2}}{\sqrt{\pi}\Delta\nu} \exp\left\{-\left(2\sqrt{\ln 2}\frac{\nu - \nu_0}{\Delta\nu}\right)^2\right\} \tag{3.71}$$

This, put into Equation (3.67), gives

$$\gamma(\tau) = \exp\left\{-\left(\frac{\pi\Delta\nu\tau}{2\sqrt{\ln 2}}\right)^2\right\} \exp\{-i2\pi\nu_0\tau\} \tag{3.72}$$

which gives for the intensity

$$I = I_0\left[1 + \exp\left\{-\left(\frac{\pi\Delta\nu\tau}{2\sqrt{\ln 2}}\right)^2\right\} \cos(2\pi\nu_0\tau)\right] \tag{3.73}$$

This intensity distribution is illustrated in Figure 3.20. From Equation (3.73) we see that the envelope of the sinusoidal fringes becomes narrower with increasing spectral width

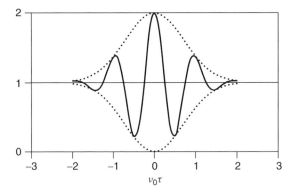

Figure 3.20 Interference from a Gaussian source. The ratio $\nu_0/\Delta\nu = \pi/2\sqrt{\ln 2} = 1.888$

$\Delta\nu$, i.e. the fringe visibility drops faster with increasing τ. From the definition of the coherence time (Equation (3.69)) we get

$$\tau_c = \sqrt{\frac{2\ln 2}{\pi}} \frac{1}{\Delta\nu} = \frac{0.664}{\Delta\nu} \tag{3.74}$$

The coherence length $L_c = c\tau_c$ is therefore inversely proportional to $\Delta\nu$. For a source covering the whole visible spectrum, $\Delta\nu \approx 3 \times 10^{14}$ Hz, we get a coherence length $L_c = 664$ nm, which means that only the zeroth-order fringe has full visibility, while the visibility of the first-order fringe has dropped to almost zero.

In the development of the spatial degree of coherence, we assumed a quasimonochromatic source, and in the development of the temporal degree of coherence we assumed a plane wave. Coherence theory can, however, be developed without these assumptions, and it can be shown that we can introduce a generalized coherence function

$$\gamma_{12}(\tau, D) \approx \gamma_{12}(\tau, 0)\gamma_{12}(0, D) = \gamma(\tau)\gamma_{12} \tag{3.75}$$

This equation is known as the reduction property of the complex degree of coherence. In the Young interferometer experiment, we can in principle measure $\gamma(\tau, D)$, but due to the limited region of overlap between the waves from P_1 and P_2 it is difficult to obtain a value of τ significantly different from zero.

3.8 OPTICAL COHERENCE TOMOGRAPHY

As seen in the previous section, when using a light source with a broad spectral distribution in interference experiments, acceptable visibility is obtained only for the zeroth-order fringe. This is of course very unfortunate in standard interferometry and therefore high-coherence lasers are often used in such experiments. However, this effect is taken to advantage in a technique called low-coherence interferometry (or low-coherence reflectometry, LCR).

Consider Figure 3.21 which is a Michelson interferometer with a low-coherence light source S. If the optical path lengths from the beamsplitter to the mirrors M_1 and M_2

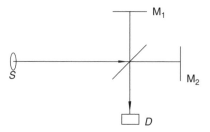

Figure 3.21

are equal, we get a bright zero-order fringe at the detector. If we move M_2, the fringe visibility and thereby the intensity at the detector drops rapidly. Assuming the intensities of the waves from each mirror to both be equal to I, the intensity at the detector drops from $4I$ (full coherence) to $2I$ (no coherence). By moving M_1 until the intensity again reaches maximum, the optical paths are again equal and the unknown movement of M_2 is equal to the known movement of M_1. This is the operating principle of LCR. Here the mirror M_2 is replaced by the object under investigation. This technique is especially suited for measurement on semi-transparent materials such as biological tissues.

When the beam of light is directed onto such a material, it is reflected from boundaries between different tissues of differing optical properties. By a scanned motion of the reference mirror M_1, intensity maxima at the detector are found and thereby the depth of such boundaries can be measured. This principle is analogous to ultrasound imaging or

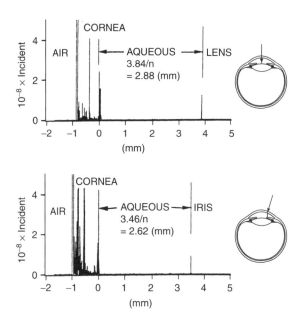

Figure 3.22 Measurement of the anterior chamber depth using optical low-coherence interferometry. The graph displays the magnitude of the reflected intensity as a function of distance. From Puliafito, C.A., *et al.* (1996) *Optical Coherence Tomography of Ocular Diseases*, SLACK, Inc., NJ. Reproduced by permission of SLACK, Incorporated

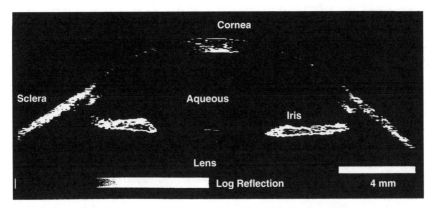

Figure 3.23 Grey scale Optical Coherence Tomography image of the anterior chamber of a human eye obtained *in vivo*. The image is displayed using a logarithmic mapping of the measured optical signal to brightness. From Puliafito, C.A., *et al.* (1996) *Optical Coherence Tomography of Ocular Diseases*, SLACK, Inc., NJ. Reproduced by permission of SLACK, Incorporated

radar which relies on measuring the time-of-flight of reflected echoes. Figure 3.22 shows an example of axial range measurements performed in the anterior chamber of the eye. The graph shows the intensity at the detector as a function of the position of the reference mirror. The intensity is a measure of the discontinuity of the optical properties of the tissue. To determine the actual depth of the various boundaries, the distance between the echoes has to be multiplied by the index of refraction of the tissue.

This method can be developed further by scanning the light beam in the transverse direction. This technique is called optical coherence tomography (OCT). Figure 3.23 shows an example of a tomographic image of the anterior chamber of the eye displayed in grey scale. The optical beam was scanned in the transverse direction and 200 axial measurements were performed. The image clearly shows the structure in the anterior eye. Fibre optic technology has made it possible to engineer compact, robust and low-cost OCT systems. Figure 3.24 shows a schematic representation of a fibre optic version of the interferometer with a superluminiscent LED (see Section 5.4.4) as the light source.

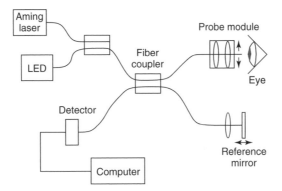

Figure 3.24 Fibre-optic interferometer OCT system. From Puliafito, C.A., *et al.* (1996) *Optical Coherence Tomography of Ocular Diseases*, SLACK, Inc., NJ. Reproduced by permission of SLACK, Incorporated

PROBLEMS

3.1 Find the resultant $U = U_1 + U_2$, where $U_1 = U_0 \exp i(kz + \omega t)$ and $U_2 = U_0 \exp i(kz - \omega t)$. Describe the resultant wave.

3.2 A carrier wave of frequency ω_c is amplitude modulated by a sine wave of frequency ω_m, that is

$$U = U_0(1 + \alpha \cos \omega_m t)e^{i\omega_c t}$$

Show that this is equivalent to the superposition of three waves of frequencies ω_c, $\omega_c + \omega_m$, and $\omega_c - \omega_m$. When a number of modulating frequencies are present, we write U as a Fourier series and sum over all values of ω_m. The terms $\omega_c + \omega_m$ constitute what is called the upper sideband while all of the $\omega_c - \omega_m$ terms form the lower sideband. What bandwidth would you need in order to transmit the complete audible range?

3.3 Calculate the coherence length of the following sources:

(a) filtered thermic radiation of bandwidth 1 nm at wavelength 600 nm;

(b) a multimode He−Ne laser with bandwidth 1 GHz;

(c) a monomode He−Ne laser with bandwidth 10 kHz.

3.4 (a) In a magnetic-field technique an He−Ne laser can be stabilized to 2 parts in 10^{10}. At 632.8 nm, what would be the coherence length of a laser with such a frequency stability? (b) Imagine that we chop a continuous laser beam (assumed to be monochromatic at $\lambda_0 = 632.8$ nm) into 0.1 ns pulses using some sort of shutter. Compute the resultant linewidth $\Delta\lambda$, bandwidth and coherence length. Find the bandwidth and linewidth which would result if we could chop at 10^{15} Hz.

3.5 In a Lloyd's mirror experiment with light of wavelength $= 500$ nm, it is found that the bright fringes on a screen 1 m away from the source are 1 mm apart. Calculate the (perpendicular) height of the source slit from the mirror.

3.6 Show that a for the Fresnel biprism of Figure P3.1 is given by $a = 2d(n - 1)\alpha$.

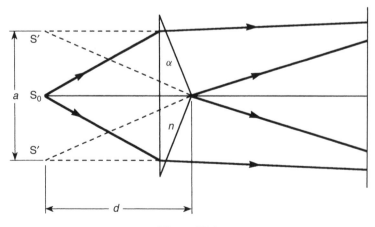

Figure P3.1

3.7 The Fresnel biprism is used to obtain fringes from a point source which is placed 2 m from the screen and the prism is midway between the source and the screen. Let the light have a wavelength $\lambda_0 = 500$ nm, and the index of refraction of the glass be $n = 1.5$. What is the prism angle if the separation of the fringes is 0.5 mm?

3.8 As mentioned in the text, the spatial coherence function γ_{12} is equal to the Fourier transform of the intensity distribution of the source.

Take the Sun as the light source. For simplicity, assume that the Sun is square and that its intensity distribution is constant. The angle subtended by the Sun is $\Delta\alpha = 9.3$ millirad ($\approx 0.53°$).

(a) Find the spatial degree of coherence $|\gamma_{12}|$ expressed by the spatial separation D, $\Delta\alpha$ and λ.

(b) Find the spatial coherence length D_c and the coherence area $\pi(D_c/2)^2$. Use $\lambda_0 = 500$ nm.

Next assume that we use the reflected sunlight from a window of area 1 m × 1 m a distance 1 km away as the light source.

(c) Find the spatial coherence length and area in this case.

3.9 Consider a light source consisting of two point sources with an angular separation equal to $\Delta\alpha$. Repeat (a) and (b) of Problem 3.8 and compare.

4
Diffraction

4.1 INTRODUCTION

When light passes an edge, it will deviate from rectilinear propagation. This phenomenon (which is a natural consequence of the wave nature of light) is known as diffraction and plays an important role in optics. The term diffraction has been conveniently defined by Sommerfeld as 'any deviation from rectilinear paths which cannot be interpreted as reflection or refraction'. A rigorous theory of diffraction is quite complicated. Here we develop expressions for the diffracted field based on Huygens' principle of secondary spherical wavelets. It is a very fortunate coincidence that the aperture and the diffracted far field are connected by a Fourier transform relationship. Because of that, optics and electrical engineering have for a long time shared a common source of mathematical theory.

As a consequence of diffraction, a point source cannot be imaged as a point. An imaging system without aberrations is therefore said to be diffraction limited.

4.2 DIFFRACTION FROM A SINGLE SLIT

Figure 4.1 shows a plane wave which is partly blocked by a screen S_1 before falling onto a screen S_2. According to geometrical optics, a sharp edge is formed by the shadow at point A. By closer inspection, however, one finds that this is not strictly correct. The light distribution is not sharply bounded, but forms a pattern in a small region around A. This must be due to a bending of the light around the edge of S_1. This bending is called diffraction and the light pattern seen on S_2 as a result of interference between the bent light waves is called a diffraction pattern.

Another example of this phenomenon can be observed by sending light through a small hole. If this hole is made small enough, the light will not propagate as a narrow beam but as a spherical wave from the centre of the hole (see Figure 4.2). This is evidence of Huygens' principle which says that every point on a wavefront can be regarded as a source of secondary spherical wavelets. By adding these wavelets and calculating the intensity distribution over a given plane, one finds the diffraction pattern in that plane. This simple principle has proved to be very fruitful and constitutes the foundation of the classical diffraction theory.

With this simple assumption, we shall try to calculate the diffraction pattern from a long, narrow slit (see Figure 4.3). The slit width a in the x_0-direction is much smaller than

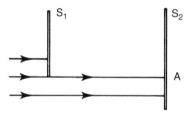

Figure 4.1

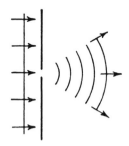

Figure 4.2

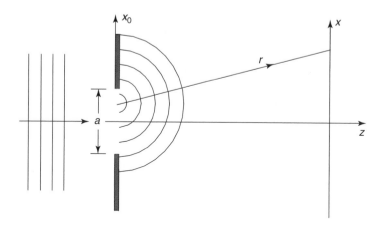

Figure 4.3

the slit length in the y_0-direction. We therefore consider the problem as one-dimensional. From the left, a plane wave with unit amplitude is falling normally onto the slit. According to Huygens' principle, the contribution $\Delta u(x)$ to the field at a point x from an arbitrary point x_0 inside the slit is equal to the field of a spherical wave with its centre at x_0:

$$\Delta u(x) = \frac{e^{ikr}}{r} \tag{4.1}$$

To calculate the total field at the point x, we have to sum the Huygens' wavelets from all points inside the slit. This sum turns into the integral

$$u(x) = \int_{-a/2}^{a/2} \frac{e^{ikr}}{r} \, dx_0 \tag{4.2}$$

Applying the Fresnel approximation (Equation (1.12)), this yields

$$u(x) = \frac{e^{ikz}}{z} \int_{-a/2}^{a/2} e^{i\frac{k}{2z}(x-x_0)^2} \, dx_0 = \frac{e^{ikz}}{z} e^{i\frac{k}{2z}x^2} \int_{-a/2}^{a/2} e^{i\frac{k}{2z}x_0^2} e^{-i\frac{k}{z}xx_0} \, dx_0 \tag{4.3}$$

By moving the observation plane away from the slit such that

$$z \gg \frac{kx_{0\,\text{max}}^2}{2}$$

the quadratic phase factor inside the integral can be set to unity. We then get the simple expression

$$u(x) = \frac{K}{z} \int_{-a/2}^{a/2} e^{-i\frac{k}{z}xx_0} \, dx_0 = \frac{K}{-ikx} [e^{-i\frac{k}{z}xx_0}]_{-a/2}^{a/2} = \frac{Ka}{z} \frac{\sin\left(\dfrac{\pi a}{z\lambda}x\right)}{\left(\dfrac{\pi ax}{z\lambda}\right)} \tag{4.4}$$

where we have collected the phase factors outside the integral into a constant K. The intensity becomes proportional to

$$I(x) = |u(x)|^2 = a^2 \frac{\sin^2\left(\dfrac{\pi a}{z\lambda}x\right)}{\left(\dfrac{\pi ax}{z\lambda}\right)^2} \tag{4.5}$$

In deriving Equation (4.4), we have made some approximations. These are called the Fraunhofer approximation in optics. To justify this approximation, the observation plane must be moved far away from the diffracting object. A simple way of fulfilling this condition is to observe the diffraction pattern in the focal plane of a lens (see Section 4.3.2).

In Figure 4.4, the Fraunhofer diffraction pattern from a single slit according to Equation (4.5) is shown. The distribution constitutes a pattern of light and dark fringes. From Equation (4.5) we find the distance between adjacent minima to be

$$\Delta x = \frac{\lambda z}{a} \tag{4.6}$$

We see that Δx is inversely proportional to the slit width. It is easily shown that the diffraction pattern from an opaque strip will be the same as from a slit of the same width.

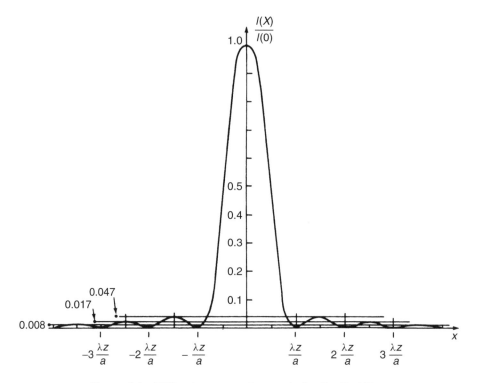

Figure 4.4   Diffraction pattern from a single slit of width a

It should be mentioned that according to a more rigorous diffraction theory, the field at a point P behind a diffracting screen is given by

$$u(P) = \frac{1}{i\lambda} \iint_{\Sigma} u(P_0) \frac{e^{ikr}}{r} \cos \Omega \, ds \tag{4.7}$$

where Σ denotes the open aperture of the screen, ds is the differential area, $u(P_0)$ is the field incident on the screen and Ω is the angle between the incident and the diffracted rays at point P_0. Equation (4.7) is known as the Rayleigh-Sommerfeld diffraction formula. When putting $u(P_0) = 1$ (normally incident plane wave of unit amplitude) and $\Omega = 0$, this formula becomes equal to Equation (4.2) except for the factor $1/i\lambda$ which becomes unimportant for our purposes, since we will be mostly concerned with relative field amplitudes.

4.3 DIFFRACTION FROM A GRATING

4.3.1 The Grating Equation. Amplitude Transmittance

Figure 4.5 shows a plane wave normally incident on a grating with a grating period equal to d. The grating lines are so narrow that we can regard the light from each opening as cylindrical waves. In Figure 4.5(b) we have drawn three of these openings, A, B and C,

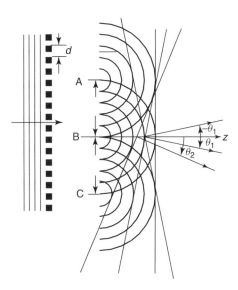

Figure 4.5 Diffraction from a square wave grating

each with five concentric circles separated by λ representing the cylindrical waves. The tangent to circle number 5 for all openings will represent a plane wave propagating in the z-direction.

The tangent to circle 5 from opening A, circle 4 from B and circle number 3 from C will represent a plane wave propagating in a direction making an angle θ_1 to the z-axis. From the figure we see that

$$\sin \theta_1 = \lambda / d$$

The tangent to circle 5 from opening A, circle 3 from B and circle 1 from C will represent a plane wave propagating in a direction making an angle θ_2 to the z-axis given by

$$\sin \theta_2 = 2\lambda / d$$

In the same manner we can proceed up to the plane wave number n making an angle θ_n to the z-axis given by

$$\sin \theta_n = n\lambda / d \tag{4.8}$$

Equation (4.8) is called the grating equation. Also in the same manner we can draw the tangent to circle 5 from opening C, circle 4 from B and circle 3 from A and so on. Therefore n in Equation (4.8) will be an integer between $-\infty$ and $+\infty$.

The grating in Figure 4.5 can be represented by the function $t(x)$ in Figure 4.6. This is a square-wave function discontinuously varying between 0 and 1. If the wave incident on the grating is represented by u_i, the wave just behind the grating is given by

$$u_u = t(x)u_i \tag{4.9}$$

Therefore, behind the grating plane $u_u = u_i$ where $t(x) = 1$, i.e. the light is transmitted and $u_u = 0$ wherever $t(x) = 0$, i.e. the light is blocked.

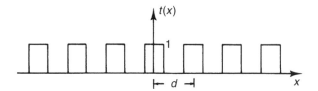

Figure 4.6   Amplitude transmittance t of a square wave grating

The function $t(x)$ is called the complex amplitude transmittance of the grating. We have seen that such a grating will diffract plane waves in directions given by Equation (4.8). If we turned the propagation direction $180°$ around for all these waves, it should not be difficult to imagine that they would interfere, forming an interference pattern with a light distribution given by $t(x)$ in Figure 4.6. In the same way we realize that a sinusoidal grating (which can be formed on a photographic film by interference between two plane waves) will diffract two plane waves propagating symmetrically around the z-axis when illuminated by a plane wave like the square wave grating in Figure 4.5. Diffraction from a sinusoidal (cosinusoidal) grating is therefore also described by Equation (4.8), but now n will assume the values -1, 0 and 1 only.

Further reasoning along the same lines tells us that a zone-plate pattern formed by registration (e.g. on a photographic film) of the interference between a plane wave and a spherical wave (see Figure 4.7(a)), will diffract two spherical waves. One of them will be a diverging spherical wave with its centre at P and the other will converge (focus) to a point P′ separated from the zone-plate by the same distance a as the point P (see Figure 4.7(b)). These arguments are perhaps not so easy to accept, but are nevertheless correct.

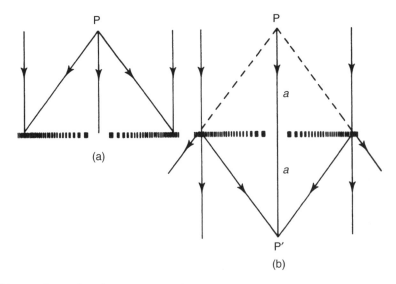

Figure 4.7   (a) Zone plate formed by interference between a spherical and a plane wave and (b) Diffraction of a plane wave from a zone plate

4.3.2 The Spatial Frequency Spectrum

Assume that we place a positive lens behind the grating in Figure 4.5 such as in Figure 4.8. In Section 1.10 (Equation (1.20)) we have shown that a plane wave in the xz-plane with propagation direction an angle θ to the optical axis (the z-axis) will focus to a point in the focal plane of the lens at a distance x_f from the z-axis given by

$$x_f = f \tan \theta \qquad (4.10)$$

where f is the focal length.

By substituting Equation (4.8) we get

$$x_f = n \frac{\lambda f}{d} = n \lambda f f_0 \qquad (4.11)$$

where we have used the approximation $\sin \theta = \tan \theta$ and inserted the grating frequency $f_0 = 1/d$. If we represent the intensity distribution in a focal point by an arrow, the intensity distribution in a focal plane in Figure 4.8 will be like that given in Figure 4.9(a). By exchanging the square-wave grating with a sinusoidal grating, the intensity distribution in the focal plane will be like that given in Figure 4.9(b).

Figure 4.10 is a reproduction of Figure 4.9 apart from a rescaling of the ordinate axis from x_f of dimension length to $f_x = x_f/\lambda f$ of dimension inverse length, i.e. spatial

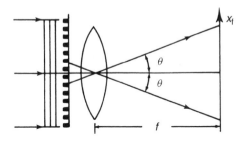

Figure 4.8

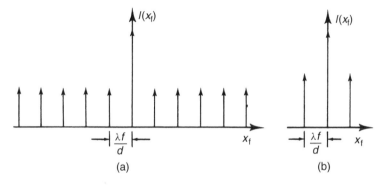

Figure 4.9   Diffraction patterns from (a) square wave grating and (b) sinusoidal grating

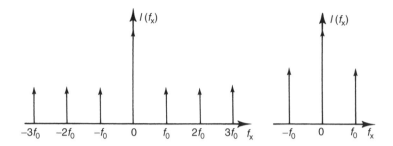

Figure 4.10 Spatial frequency spectra from (a) square wave grating and (b) sinusoidal grating

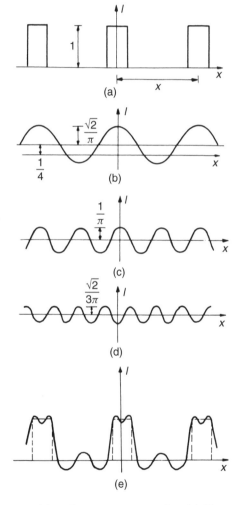

Figure 4.11 Fourier decomposition of a square wave grating. (a) The transmittance function of the grating; (b) The constant term and the first harmonic of the Fourier series; (c) The second harmonic; (d) The third harmonic; and (e) The sum of the four first terms of the series. The transmittance function of the grating is shown dashed

frequency. In that way we get a direct representation of the frequency content or plane-wave content of the gratings. We see that the sinusoidal grating contains the frequencies $\pm f_0$ and 0, while the square-wave grating contains all positive and negative integer multiples of f_0. The diagrams in Figure 4.10 are called spatial frequency spectra.

If we successively put into the set-up in Figure 4.8 sinusoidal gratings of frequencies $f_0, 2f_0, 3f_0, \ldots, nf_0$, and if we could add all the resulting spectra, we would get a spectrum like that given in Figure 4.10(a). This would be a proof of the fact that a square-wave grating can be represented by a sum of sinusoidal (cosinusoidal) gratings of frequencies which are integer multiples of the basic frequency f_0, in other words a Fourier series. This is further evidenced in Figure 4.11 where, in Figure 4.11(e) we see that the approximation to a square-wave grating is already quite good by adding the four first terms of the series. To improve the reproduction of the edges of the square wave grating, one has to include the higher-order terms of the series. Sharp edges in an object will therefore represent high spatial frequencies.

4.4 FOURIER OPTICS

Let us turn back to Section 4.2 where we found an expression for the field $u(x_f)$ in the x_f-plane diffracted from a single slit of width a in the x-plane at a distance z (see Equation (4.4)):

$$u(x_f) = \int_{-a/2}^{a/2} e^{-i2\pi f_x x} \, dx \tag{4.12}$$

where $f_x = x_f/\lambda z$ and where we have omitted a constant phase factor. The transmittance function for a single slit would be

$$t(x) = \begin{cases} 1 & \text{for } |x| < a/2 \\ 0 & \text{otherwise} \end{cases} \tag{4.13}$$

By putting $t(x)$ into the integral of Equation (4.12) we may let the limits of integration approach $\pm\infty$, and we get

$$u(x_f) = \int_{-\infty}^{\infty} t(x)e^{-i2\pi f_x x} \, dx \tag{4.14}$$

This is a Fourier integral and $u(x_f)$ is called the Fourier transform of $t(x)$. In the general case where the transmittance function varies both in the x- and y-direction, we get

$$u(x_f, y_f) = \int_{-\infty}^{\infty} \int t(x, y)e^{-i2\pi(f_x x + f_y y)} \, dx \, dy \tag{4.15}$$

which in shorthand notation can be written

$$u(x_f, y_f) = T(f_x, f_y) = \mathscr{F}\{t(x, y)\} \tag{4.16}$$

where $\mathscr{F}\{t(x, y)\}$ means 'the Fourier transform of $t(x, y)$'.

In deriving Equation (4.4) we assumed a plane wave of unit amplitude incident on the slit. If a light wave given by $u_i(x, y)$ falls onto an object given by the transmittance function $t(x, y)$, the field just behind the object is $u(x, y) = t(x, y)u_i(x, y)$ and the field in the x_f-plane becomes

$$u(x_f, y_f) = \frac{K}{i\lambda z}\mathscr{F}\{u(x, y)\} \tag{4.17}$$

Here K is a pure phase factor ($|K|^2 = 1$) which is unimportant when calculating the intensity. By the factor $(1/i\lambda z)$ we have brought Equation (4.17) into accordance with the Huygens – Fresnel diffraction theory.

As mentioned in Section 4.2, the approximations leading to Equation (4.4) and therefore Equation (4.17) are called the Fraunhofer approximation. To fulfil this, the plane of observation has to be far away from the object. A more practical way of fulfilling this requirement is to place the plane of observation in the focal plane of a lens as in Figure 4.8. z in Equation (4.17) then has to be replaced by the focal length f. Other practical methods are treated in Section 4.5.1. We also mention that by placing the object in the front focal plane (to the left of the lens) in Figure 4.8, we have $K = 1$, and we get a direct Fourier transform.

The way we have derived the general formula of Equation (4.17) is of course by no means a strict proof of its validity. Rigorous diffraction theory using the same approximations leads, however, to the same result. Equation (4.17) is a powerful tool in calculating diffraction patterns and analysis of optical systems. Some of its consequences are treated more extensively in Appendix B.

For example, the calculation of the frequency spectrum of a sinusoidal grating given by

$$t(x) = 1 + \cos 2\pi f_0 x \tag{4.18}$$

now becomes straightforward. By using Equation (4.15) we get

$$
\begin{aligned}
u(x_f) &= \int_{-\infty}^{\infty} (1 + \cos 2\pi f_0 x)e^{-i2\pi f_x x}\, dx \\
&= \int_{-\infty}^{\infty} (1 + \tfrac{1}{2}e^{i2\pi f_0 x} + \tfrac{1}{2}e^{-i2\pi f_0 x})e^{-i2\pi f_x x}\, dx \\
&= \int_{-\infty}^{\infty} (e^{-i2\pi f_x x} + \tfrac{1}{2}e^{-i2\pi (f_x - f_0)x} + \tfrac{1}{2}e^{-i2\pi (f_x + f_0)x})\, dx \\
&= \delta(f_x) + \tfrac{1}{2}\delta(f_x - f_0) + \tfrac{1}{2}\delta(f_x + f_0)
\end{aligned}
\tag{4.19}
$$

The last equality follows from the definition of the delta function given in Equation (B.11) in Appendix B.2. Equation (4.19) shows that the spectrum of a sinusoidal grating is given by the three delta functions, i.e. three focal points. These are the zero order at $f_x = 0$ and the two side orders at $f_x = \pm f_0$ (see Figure 4.10(b)).

4.5 OPTICAL FILTERING

Figure 4.12 shows a point source (1) placed in the focal plane of a lens (2) resulting in a plane wave falling onto a square wave grating (3) which lies in the object plane. A

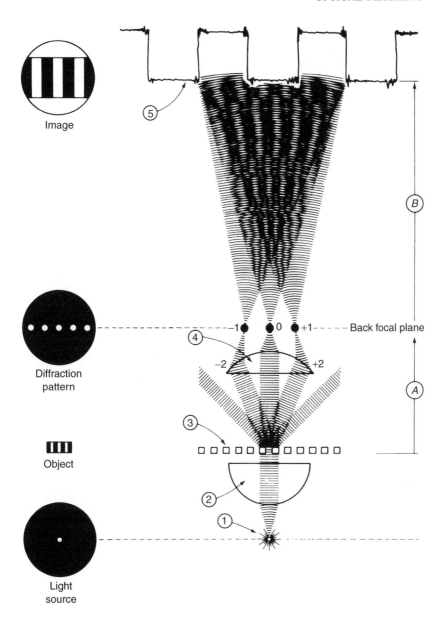

Figure 4.12 Optical filtering process. (From Jurgen R. Meyer-Arendt, Introduction to Classical and Modern Optics, © 1972, p. 393. Reprinted by permission of Prentice-Hall, Inc., Englewood Cliffs, New Jersey)

lens (4) placed a distance a from the object plane, images the square wave grating on to the image plane where the intensity distribution (5) of the image of the grating can be observed.

Although the figure shows that only the $\pm$ 1st side orders are accepted by the lens (4), we will in the following assume that all plane-wave components diffracted from the grating will go through the lens. We can therefore, in the same way as in Figure 4.8, observe the spectrum of the grating in the back focal plane of the lens.

Let us now consider two cases:

Case 1. We place a square-wave grating of basic frequency f_0 in the object plane. The distance between the focal points in the focal plane then becomes $\lambda f f_0$ (see Equation (4.11) and Figure 4.9). In the image plane we will see a square-wave grating of basic frequency f_i where $1/f_i = m(1/f_0)$ and $m = b/a$ is the magnification from the object plane to the image plane.

Case 2. We place a square wave grating of basic frequency $2 f_0$ in the object plane. The distance between the focal points now becomes $2\lambda f f_0$ and the basic frequency of the imaged grating will be $2 f_i$.

If in case 1 we placed in the focal plane a screen with holes separated by the distance $2\lambda f f_0$ and adjusted it until every second focal point in the spectrum was let through, then the situation in case 2 would be simulated. In other words, a grating of basic frequency f_0 in the object plane would have resulted in a grating of basic frequency $2 f_i$ in the image plane. Such a manipulation of the spectrum is called optical filtering and the back focal plane of lens (4) is called the filter plane. What we have here described is only one of many examples of optical filtering. The types of filters can have many variations and can be rather more complicated than a screen with holes. We shall in Section 4.7 consider a special type of filtering which we later will apply in practical problems.

4.5.1 Practical Filtering Set-Ups

In Section 4.3.1 we derived the grating equation

$$d \, \sin \theta_n = n\lambda \tag{4.20}$$

which applies to normal incidence on the grating. This equation implies that the phase difference between light from the different openings in the grating must be an integer number of wavelengths. When the light is obliquely incident upon the grating at an angle θ_i, this condition is fulfilled by the general grating equation:

$$d(\sin \theta_i + \sin \theta_n) = n\lambda \tag{4.21}$$

This is illustrated in Figure 4.13.

Figure 4.14 shows the same set-up as in Figure 4.8 apart from the grating being moved to the other side of the lens a distance s from the back focal plane. Assume that the first side order from an arbitrary point x on the grating is diffracted to the point x_f. The grating equation (4.21) applied to this light path gives

$$d(\sin \theta_i + \sin \theta_1) \simeq d(\tan \theta_i + \tan \theta_1) = d \left(\frac{x}{s} + \frac{x_f - x}{s} \right) = \lambda \tag{4.22}$$

Equation (4.22) implies that the first side order will be diffracted to the same point

$$x_f = \frac{\lambda s}{d} \tag{4.23}$$

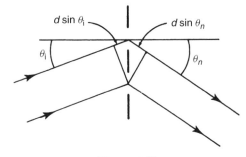

Figure 4.13

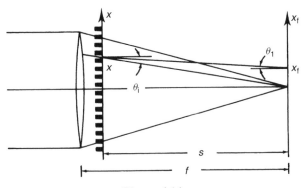

Figure 4.14

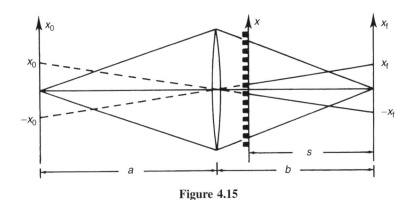

Figure 4.15

in the focal plane independent of x, i.e. from all points on the grating. The intensity distribution in the focal plane therefore becomes like that in Figure 4.9 except that the diffraction orders now are separated by a distance $\lambda s/d$.

The above arguments do not assume a plane wave incident on the lens. The result remains the same when a point source is imaged by the lens on to the optical axis as in Figure 4.15.

Our next question is: how will the light be distributed in the x_0-plane with a point source at $x_f = 0$ in the set-up of Figure 4.15? It should be easy to accept that the answer is given by removing the grating and placing two point sources at the positions $\pm x_f = \pm \lambda s/d$.

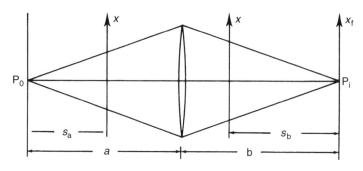

Figure 4.16

The first two diffraction orders are therefore found at the points $\pm x_0 = \pm(a/b)\lambda s/d$.

The conclusions of the above considerations are collected in Figure 4.16. Here, P_i is the image point of the point source P_0 on the optical axis. They are separated by distances given by the lens formula

$$\frac{1}{a} + \frac{1}{b} = \frac{1}{f} \tag{4.24}$$

where f = the lens focal length.

The important point to note is that independent of the positioning of the grating (or another transparent object) in the light path between P_0 and P_i, the spectrum will be found in the x_f-plane. The distance between the diffraction orders becomes: if we place the grating

(1) to the right of the lens at a distance s_b from P_i

$$x_f = \lambda s_b/d \tag{4.25}$$

(2) to the left of the lens at a distance s_a from P_0

$$x_f = (b/a)\lambda s_a/d = \frac{a - t_a}{a - f}\frac{\lambda f}{d} \tag{4.26}$$

where t_a is the distance from the grating to the lens.

These results make possible a lot of different filtering set-ups. Figure 4.17 shows the simplest of them all. Here the lens works as both the transforming and the imaging lens. The object in the x-plane is imaged to the x_i-plane while the filter plane is in the x_f-plane, the image plane of the point source P_0.

Figure 4.18 shows a practical filtering set-up often used for optical filtering of moiré and speckle photographs. The film is placed just to the right of the lens a distance s_b from the filter plane. By imaging the film through a hole in the filter plane a distance x_f from the optical axis, one is filtering out the first side order of a grating of frequency

$$f_x = \frac{x_f}{\lambda s_b} \tag{4.27}$$

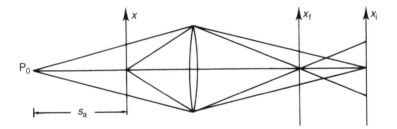

Figure 4.17 Optical filtering set-up with one single lens

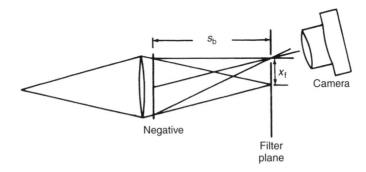

Figure 4.18 Practical filtering set-up

Owing, for example, to a deformation, this grating can be regarded as phase-modulated, and the side order therefore has been broadened, see Section 4.7. When the hole in the filter plane is made wide enough to let through this modulated side order, the image of the object on the film will be covered by moiré or speckle Fourier fringes. This will be treated in more detail in Chapters 7 and 8.

4.6 PHYSICAL OPTICS DESCRIPTION OF IMAGE FORMATION

We are now in a position to look more closely at a lens system from a physical optics point of view. In Section 4.5 we have shown that by placing a transparent diffracting object anywhere in the light path between P_0 and P_i of Figure 4.16, we obtain the spectrum (i.e the Fourier transform) of the object in the x_f-plane. But this argument applies also to the circular lens aperture itself. The Fourier transform of a circular opening given by (see definition, Equation (B.17) in Appendix B.2)

$$\text{circ}\left(\frac{r}{D/2}\right) \tag{4.28}$$

is given in Table B.1 (Appendix B.2) as

$$\mathscr{F}\left\{\text{circ}\left(\frac{r}{D/2}\right)\right\} = \left(\frac{D}{2}\right)^2 \frac{J_1(\pi D\rho)}{D\rho/2} \tag{4.29}$$

where D is the diameter of the lens, J_1 is the first-order Bessel function and where we also have made use of the similarity theorem, Equation (B.2b), Appendix B.1. The frequency coordinate in this case is equal to

$$\rho = \frac{r_f}{\lambda b} \tag{4.30}$$

where b is the image distance and

$$r_f = \sqrt{x_f^2 + y_f^2} \tag{4.31}$$

The intensity distribution in the x_f, y_f-plane in Figure 4.16 with a point source at P_0 therefore becomes proportional to

$$I(r_f) = \left[\frac{J_1(kDr_f/2b)}{(kDr_f/2b)} \right]^2 \tag{4.32}$$

This intensity distribution is generally referred to as the Airy pattern, after G. B. Airy who first derived it. Figure 4.19(a) shows a cross-section and Figure 4.19(b) shows a photograph of this Airy pattern. The first minimum of $J_1(\pi x)$ occurs for $x = 1.22$ which gives for the radius of the so-called Airy disc

$$\Delta r_f = 1.22 \frac{\lambda b}{D} \tag{4.33}$$

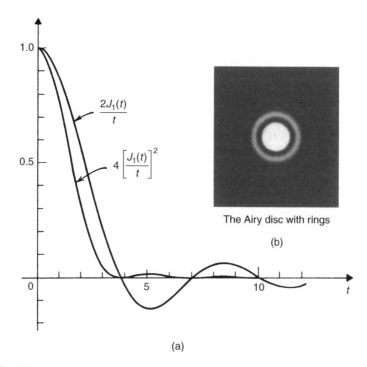

The Airy disc with rings

(b)

$\frac{2J_1(t)}{t}$

$4 \left[\frac{J_1(t)}{t} \right]^2$

(a)

Figure 4.19   The Airy pattern. (a) Intensity distribution and (b) Photograph of the pattern. (Reproduced from Klein and Furtak 1986 by permission of John Wiley & Sons Inc)

This result shows that even for an aberration-free system, an object point is not imaged as an image point predicted by geometrical optics, but as an intensity distribution given by Equation (4.32).

An imaging system is usually composed, not of a single lens, but of several lenses, perhaps some positive and some negative. To specify the properties of a lens system we consider it as a 'black box' with the entrance and exit pupils as 'terminals'. The entrance pupil represents a finite aperture (effective or real) through which light must pass to reach the imaging elements and the exit pupil represents a finite aperture through which the light must pass as it leaves the imaging elements on its way to the image plane. It is usually assumed that passage of light between the entrance and exit planes is adequately described by geometrical optics. To find the two pupils, one first has to identify the smallest (physical) aperture (the aperture stop) within the system. Then the entrance pupil is found by imaging (geometrically) this aperture through all the lens elements to the left (towards the object plane) and the exit pupil by imaging the aperture through the lens elements to the right, see Section 2.7. The lens system is said to be diffraction limited if a diverging spherical wave incident on the entrance pupil is mapped into a converging spherical wave at the exit pupil.

With this model of the imaging system, it is possible to associate all diffraction limitations with either the light propagation from object to entrance pupil or from the exit pupil to the image. In fact, the two points of view are entirely equivalent.

The view that regards diffraction effects as resulting from the finite entrance pupil was first examined by Ernst Abbé in 1873. According to his theory, only a limited number of diffracted orders from an object are intercepted by the finite entrance pupil (see e.g. Figure 4.12). The equivalent view of regarding diffraction effects as resulting from the finite exit pupil was presented by Lord Rayleigh in 1896. We will adopt this viewpoint here.

4.6.1 The Coherent Transfer Function

Generally, if the exit pupil is represented not merely by a circ-function but by some general aperture function $P(x, y)$, the field amplitude distribution in the image plane due to an on-axis point source in the object plane will be given by a function $h(x_i, y_i)$

$$h(x_i, y_i) = \mathscr{F}\{P(x, y)\} \tag{4.34}$$

Equation (4.29) is a special case of this formula. Here x, y are lens aperture coordinates, x_i, y_i are image plane coordinates and the Fourier frequency coordinates are given by $f_x = x_i/\lambda b$, $f_y = y_i/\lambda b$. With the object point moved off-axis to a point x_o, y_o in the object plane, the resulting field distribution in the image plane will be centred around a point of coordinates $x_i = -mx_o$, $y_i = -my_o$ where m is the transversal magnification. If the field amplitude of this object point is given by $\Delta u_o(x_o, y_o)$ (which ideally should be a delta-function $\delta(x_o - x_o', y_o - y_o')$), the field amplitude distribution in the image plane therefore is given by

$$\Delta u_i(x_i, y_i) = h(x_i - mx_o, y_i - my_o)\Delta u_o(x_o, y_o) \tag{4.35}$$

If, in the object plane there are many point sources which are mutually coherent, we have to add the contribution from each point according to Equation (4.35). Moreover, if in the object plane there is a continuous and coherent field amplitude distribution given by $u_o(x_o, y_o)$, this sum will be converted to an integral, viz.

$$u_i(x_i, y_i) = \iint_{-\infty}^{\infty} h(x_i - mx_o, y_i - mx_o)u_o(x_o, y_o)\,dx_o\,dy_o \qquad (4.36)$$

$h(x_i - mx_o, y_i - my_o)$ is termed the impulse response of the imaging system since it gives the response of an impulse in spatial coordinates, i.e. a delta function. For an ideal lens the impulse response is given by Equation (4.29). Such a lens is said to be diffraction limited since its performance is limited only by the size of the exit pupil.

Equation (4.36) is seen to be a convolution integral (see Equation (B.2e) in Appendix B.1). According to the convolution theorem (see Appendix B.3) this implies that

$$G_i(f_x, f_y) = H(f_x, f_y)G_o(f_x, f_y) \qquad (4.37)$$

where G_i, G_o and H are the Fourier transforms of u_i, u_o and h respectively. From this expression we see that the spatial frequency distribution $G_i(f_x, f_y)$ of the image is equal to the spatial frequency distribution $G_o(f_x, f_y)$ of the object times the function $H(f_x, f_y)$ which is referred to as the coherent transfer function. $H(f_x, f_y)$ acts as a filter, filtering out the spatial frequencies (the plane wave components) not accepted by the lens aperture. We have that, since

$$\mathscr{F}\mathscr{F}\{g(x, y)\} = g(-x, -y) \qquad (4.38)$$

$$H(f_x, f_y) = \mathscr{F}\{h(x_i, y_i)\} = \mathscr{F}\mathscr{F}\{P(x, y)\} = P(-\lambda b f_x, -\lambda b f_y) \qquad (4.39)$$

For a circularly symmetric aperture function we have $P(-\lambda b f_x, -\lambda b f_y) = P(\lambda b f_x, \lambda b f_y)$. Figure 4.20 shows a plot of the coherent transfer function for a diffraction limited lens with a circular aperture along the f_x-axis. We see that the spatial frequencies f_x are unaffected up to f_o, the maximum spatial frequency resolved by the lens given by

$$\lambda b f_o = \frac{D}{2}, \quad \text{that is} \quad f_o = \frac{D}{2b\lambda} \qquad (4.40)$$

which is called the cut-off frequency of the coherent transfer function.

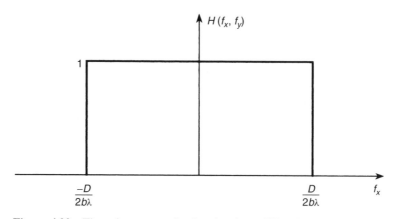

Figure 4.20 The coherent transfer function for a diffraction-limited system

4.6.2 The Incoherent Transfer Function

If the object plane contains a collection of point sources which are mutually incoherent, we get the light distribution in the image plane not by adding the field amplitudes, but by adding the intensities contributed by each point. In accordance with Equation (4.36) we then get for the intensity

$$I_i(x_i, y_i) = \iint_{-\infty}^{\infty} |h(x_i - mx_o, y_i - my_o)|^2 I_o(x_o, y_o)\, dx_o\, dy_o \qquad (4.41)$$

In analogy with Equation (4.39) we therefore can introduce an incoherent transfer function $\mathscr{H}(f_x, f_y)$ given by

$$\mathscr{H}(f_x, f_y) = \frac{\displaystyle\iint_{-\infty}^{\infty} |h(x_i, y_i)|^2 e^{-i2\pi(f_x x_i + f_y y_i)}\, dx_i\, dy_i}{\displaystyle\iint_{-\infty}^{\infty} |h(x_i, y_i)|^2\, dx_i\, dy_i} \qquad (4.42)$$

Where we by the denominator have normalized the transfer function, so that $\mathscr{H}(0, 0) = 1$.

The relationship between the coherent and the incoherent transfer function is therefore given by

$$H(f_x, f_y) = \mathscr{F}\{h\} \qquad (4.43a)$$

$$\mathscr{H}(f_x, f_y) = \frac{\mathscr{F}\{|h|^2\}}{\mathscr{F}\{|h|^2\}_{f_x=f_y=0}} \qquad (4.43b)$$

From the autocorrelation theorem (see Equation (B.2f), Appendix B.1) it follows that

$$\mathscr{H}(f_x, f_y) = \frac{\displaystyle\iint_{-\infty}^{\infty} H(\xi', \eta') H^*(\xi' + f_x, \eta' + f_y)\, d\xi'\, d\eta'}{\displaystyle\iint_{-\infty}^{\infty} |H(\xi', \eta')|^2\, d\xi'\, d\eta'} \qquad (4.44)$$

A simple change of variables

$$\xi = \xi' + \frac{f_x}{2} \qquad (4.45a)$$

$$\eta = \eta' + \frac{f_y}{2} \qquad (4.45b)$$

results in the symmetrical expression

$$\mathscr{H}(f_x, f_y) = \frac{\displaystyle\iint_{-\infty}^{\infty} H\left(\xi - \frac{f_x}{2}, \eta - \frac{f_y}{2}\right) H^*\left(\xi + \frac{f_x}{2}, \eta + \frac{f_y}{2}\right) d\xi\, d\eta}{\displaystyle\iint_{-\infty}^{\infty} |H(\xi, \eta)|^2\, d\xi\, d\eta} \qquad (4.46)$$

To this point, our discussion has been equally applicable to systems with and without abberations. We now consider the special case of a diffraction-limited incoherent system. Recall that for the coherent system we have, cf. Equation (4.39)

$$H(f_x, f_y) = P(\lambda b f_x, \lambda b f_y) \tag{4.47}$$

For the incoherent system it follows from Equation (4.46) (with a simple change of variables) that

$$\mathcal{H}(f_x, f_y) = \frac{\iint_{-\infty}^{\infty} P\left(\xi - \frac{\lambda b f_x}{2}, \eta - \frac{\lambda b f_y}{2}\right) P\left(\xi + \frac{\lambda b f_x}{2}, \eta + \frac{\lambda b f_y}{2}\right) d\xi \, d\eta}{\iint_{-\infty}^{\infty} |P(\xi, \eta)| \, d\xi \, d\eta} \tag{4.48}$$

where in the denominator, since $|P|$ equals either unity or zero, $|P|^2$ has been replaced by $|P|$.

The geometrical interpretation of Equation (4.48) is that the numerator represents the area of overlap of two displaced pupil functions, one centred at $(\lambda b f_x/2, \lambda b f_y/2)$ and the second centred at $(-\lambda b f_x/2, -\lambda b f_y/2)$. The denominator simply normalizes this area of overlap by the total area of the pupil. Note that this interpretation demonstrates that $\mathcal{H}(f_x, f_y)$ of a diffraction-limited system is always real and non-negative.

As an example, consider Figure 4.21 where the pupil is a square of width l. The area of overlap is

$$A(f_x, f_y) = \begin{cases} (l - \lambda b|f_x|)(l - \lambda b|f_y|) & |f_x| \le \dfrac{l}{\lambda b} \quad |f_y| \le \dfrac{l}{\lambda b} \\ 0 & \text{otherwise} \end{cases} \tag{4.49}$$

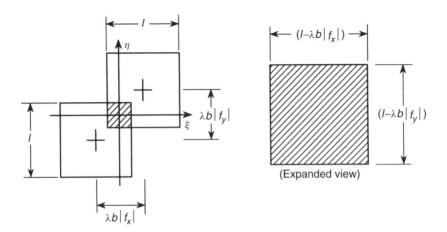

Figure 4.21 Geometry of the area of overlap when calculating the optical transfer function of a diffraction-limited system with a square pupil

When dividing by the total area l^2, the result becomes

$$\mathscr{H}(f_x, f_y) = \Lambda\left(\frac{\lambda b f_x}{l}\right)\Lambda\left(\frac{\lambda b f_y}{l}\right) = \Lambda\left(\frac{f_x}{2f_0}\right)\Lambda\left(\frac{f_y}{2f_0}\right) \tag{4.50}$$

where Λ is the triangle function (see Equation (B.15), Appendix B.2) and f_0 is the cut-off frequency of the same system used with coherent illumination

$$f_0 = \frac{l}{2\lambda b} \tag{4.51}$$

We therefore have that the cut-off frequency for the incoherent transfer function is twice that of the coherent transfer function.

$\mathscr{H}(f_x, f_y)$ is commonly known as the optical transfer function (OTF) of the system. Generally it is a complex quantity and its modulus $|\mathscr{H}|$ is known as the modulation transfer function (MTF). When aberrations are present, this can be accounted for by introducing a generalized pupil function

$$P(x, y) = |P(x, y)|e^{ikW(x,y)} \tag{4.52}$$

where $k = 2\pi/\lambda$. Here $P(x, y)$ represents a phase-shifting plate while $W(x, y)$ is an effective path-length error deforming the ideal, converging spherical wavefront representing the diffraction-limited case.

Since imaging systems are usually symmetric about the optical axis it is sufficient to display the MTF for positive frequencies along one axis (the f_x-axis). Figure 4.22 shows $|\mathscr{H}(f_x)|$ from Equation (4.50) representing the MTF for a diffraction-limited system (curve (a)). Curve (b)) shows the MTF for a 'soft' lens system. The term 'soft' comes from the fact that this lens will enhance low frequency components relative to high-frequency components. The image in such a system will appear soft. Curve (c) shows the MTF for a 'hard' lens system.

The MTF-concept is adopted from electrical engineering. It means that if we have an 'input' object with an intensity distribution

$$I_o = a(1 + b\cos 2\pi f_g x) \tag{4.53}$$

Its spectrum becomes

$$J_o = \mathscr{F}\{I_o\} = a\delta(f_x) + \frac{ab}{2}\delta(f_x - f_g) + \frac{ab}{2}\delta(f_x + f_g) \tag{4.54}$$

The spectrum of the 'output' image becomes

$$\mathscr{F}\{I_i\} = J_i = a\mathscr{H}\delta(f_x) + \frac{ab}{2}\mathscr{H}\delta(f_x - f_g) + \mathscr{H}\frac{ab}{2}\delta(f_x + f_g) \tag{4.55}$$

which by taking the inverse transform gives the 'output' image

$$I_i = a[1 + |\mathscr{H}(f_g)|b\cos(2\pi f_g x)] \tag{4.56}$$

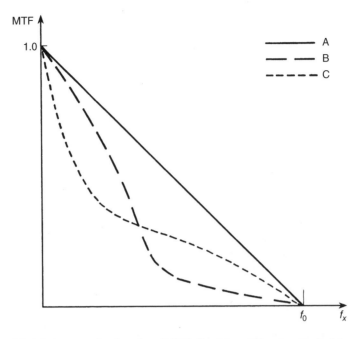

Figure 4.22 Modulation transfer function (MTF) for (a) a diffraction limited lens; (b) a 'soft' lens; and (c) a 'hard' lens

This means that the modulation (or the visibility, see Section 3.3) of the input sinusoidal grating has changed by a factor $|\mathscr{H}(f_g)|$, the value of the MTF at frequency f_g.

When comparing coherent and incoherent imaging, we see that the image intensity in the two cases are given by

$$\text{Incoherent}: \quad I_i = |h|^2 \otimes I_o = |h|^2 \otimes |u_o|^2 \qquad (4.57a)$$

$$\text{Coherent}: \quad I_i = |h \otimes u_o|^2 \qquad (4.57b)$$

From the definition of the autocorrelation integral (see Equation (B.2f), Appendix B.1) we therefore can write the frequency spectra of the image intensity in the two cases as

$$\text{Incoherent}: \quad \mathscr{F}\{I_i\} = [H \odot H][G_o \odot G_o] \qquad (4.58a)$$

$$\text{Coherent}: \quad \mathscr{F}\{I_i\} = HG_o \odot HG_o \qquad (4.58b)$$

4.6.3 The Depth of Focus

Another aspect of the imaging properties of a lens is the depth of focus. By this is meant the maximum distance by which the image plane can be moved away from the exact focus and still have an acceptable focused image. Consider Figure 4.23 where the rays passing through the edge of the lens aperture from an on-axis point source are drawn. The image plane is moved a distance Δb away from the exact image point. Let us take the depth of focus to be the distance Δb when the radius of the light spot on the image

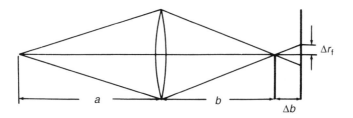

Figure 4.23

plane is equal to the radius of the Airy disc Δr_f. We then have

$$\frac{\Delta r_f}{\Delta b} = \frac{D/2}{b} \tag{4.59}$$

which by inserting the value of the radius of the Airy disc, Equation (4.33), gives

$$\Delta b = 2.44\lambda \left(\frac{b}{D}\right)^2 = 2.44\lambda \left(\frac{bF}{f}\right)^2 = 2.44\lambda(1+m)^2 F^2 \tag{4.60}$$

Here $F = f/D$ is the aperture number, m is the magnification and f is the focal length. From this expression we see that the depth of focus is inversely proportional to the square of the lens aperture or proportional to the square of the aperture number.

4.7 THE PHASE-MODULATED SINUSOIDAL GRATING

Consider a sinusoidal grating given by

$$t_1 = a(1 + \cos\phi_1) = a\left(1 + \cos\frac{2\pi x}{d}\right) \tag{4.61}$$

Assume that this grating undergoes a change whereafter it is given by

$$t_2 = a(1 + \cos\phi_2) = a\left[1 + \cos\left(\frac{2\pi x}{d} + \psi(x)\right)\right] \tag{4.62}$$

This change might result from different causes, which will be taken up later. We assume that $\psi(x)$ varies much more slowly than $2\pi x/d$. The spectrum of such a grating is indicated in Figure 4.24(b) which should be compared with the spectrum of the grating given in Equation (4.61) in Figure 4.24(a). We see that the side orders have become broader because of the slowly varying function $\psi(x)$. The grating is said to be phase-modulated and $\psi(x)$ is called the modulation function.

We then image these gratings onto a photographic film resulting in an intensity distribution (blackening) on the film given by

$$I(x) = t_1 + t_2 = a(1 + \cos\phi_1) + a(1 + \cos\phi_2)$$

$$= 2a\left(1 + \cos\frac{\phi_1 - \phi_2}{2}\cos\frac{\phi_1 + \phi_2}{2}\right) \tag{4.63}$$

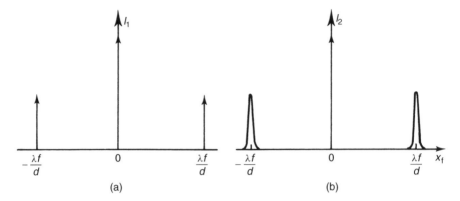

Figure 4.24 The spectrum of (a) a sinusoidal grating of period d and (b) the same grating phase-modulated

After development, $I(x)$ becomes equal to the transmittance function of the film. We then place this film in the object plane in the set-up of Figure 4.12. The intensity distribution in the image plane then becomes equal to Equation (4.63) with a scaling factor determined by the magnification m. Equation (4.63) can be written as

$$I(x) = 2a + \frac{a}{2}(e^{i\phi_1} + e^{i\phi_2}) + \frac{a}{2}(e^{-i\phi_1} + e^{-i\phi_2}) \tag{4.64}$$

where the first term represents the zeroth order and the second and third terms represent the two side orders. In the filter plane we place a screen with a hole a distance $\lambda f/d$ from the optical axis and with an opening wide enough to transmit the full width of one of the modulated side orders given in Figure 4.24(b). The wave field in the image plane then becomes (assuming a magnification $m = 1$)

$$u = \frac{a}{2}(e^{i\phi_1} + e^{i\phi_2}) \tag{4.65}$$

and the intensity

$$I = |u|^2 = \frac{a^2}{2}[1 + \cos(\phi_1 - \phi_2)] = \frac{a^2}{2}[1 + \cos \psi(x)] \tag{4.66}$$

From this equation we see that we have obtained an expression dependent on the modulation function alone. The grating is said to be demodulated.

Another procedure is to image the two gratings given in Equations (4.61) and (4.62) onto two separate negatives and then put them together in the image plane. The resulting transmittance function then becomes not the sum but the product

$$t(x) = t_1 t_2 = a(1 + \cos \phi_1)a(1 + \cos \phi_2)$$
$$= a^2(1 + \cos \phi_1 + \cos \phi_2 + \cos \phi_1 \cos \phi_2)$$
$$= a^2[1 + \cos \phi_1 + \cos \phi_2 + \tfrac{1}{2} \cos(\phi_1 - \phi_2) + \tfrac{1}{2} \cos(\phi_1 + \phi_2)]$$

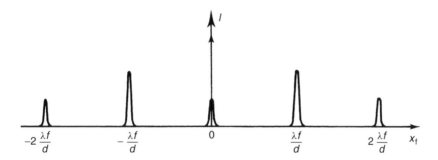

Figure 4.25 The spectrum of the gratings in Figure 4.24(a, b) multiplied

$$= a^2 \left[1 + \tfrac{1}{2} \cos \psi(x) + \cos \frac{2\pi x}{d} + \cos \left(\frac{2\pi x}{d} + \psi(x) \right) \right.$$

$$\left. + \tfrac{1}{2} \cos \left(\frac{4\pi x}{d} + \psi(x) \right) \right] \tag{4.67}$$

The spectrum of this function is sketched in Figure 4.25. The second term in Equation (4.67) gives the modulation function directly, i.e. it modulates the zeroth order. In practice, the zeroth order is often much stronger than the modulation function, resulting in a pattern of very low contrast. We then have the possibility of filtering out the first side order and obtaining the same result as in the first procedure.

PROBLEMS

4.1 Suppose that we have a laser emitting a diffraction-limited beam ($\lambda_o = 632.8$ nm) having a 2 mm diameter. How big a light spot would be produced on the surface of the moon a distance of 376×10^3 km away? Neglect any effects of the Earth's atmosphere.

4.2 Imagine that you are looking through a piece of square woven cloth at a point source ($\lambda_o = 600$ nm) 20 m away. If you see a square arrangement of bright spots located about the point source, each separated by an apparent nearest-neighbour distance of 12 cm, how close together are the strands of cloth?

4.3 Imagine an opaque screen containing thirty randomly located circular holes of the same diameter. The light source is such that every aperture is coherently illuminated by its own plane wave. Each wave in turn is completely incoherent with respect to all the others. Describe the resulting far-field diffraction pattern.

4.4 Determine the Fourier transform of the function

$$U(x) = \begin{cases} U_0 \sin 2\pi f_0 x & |x| \le L \\ 0 & |x| > L \end{cases}$$

Make a sketch of it.

4.5 Determine the Fourier transform of

$$f(x) = \begin{cases} \sin^2 2\pi f_0 x & |x| \le L \\ 0 & |x| > L \end{cases}$$

Make a sketch of it.

4.6 Suppose we have two functions, $f(x, y)$ and $h(x, y)$, where both have a value of 1 over a square region in the xy-plane and are zero everywhere else (Figure P4.1). Given that $g(x, y)$ is their convolution, make a plot of $g(x, 0)$.

4.7 Calculate and sketch the convolution between the two functions $f(x)$ and $h(x)$ depicted in Figure P4.2.

4.8 Make a sketch of the resulting function arising from the convolution of the two functions depicted in Figure P4.3.

4.9 Show (for normally incident plane waves) that if an aperture has a centre of symmetry, i.e. if the aperture function is even, then the diffracted field in the Fraunhofer case also possesses a centre of symmetry.

4.10 Another way to do the integral for Fraunhofer diffraction by a circular aperture is to stay in Cartesian coordinates. If the Fourier transform of $f(x, y)$ is $F(f_x, f_y)$ it is sufficient to calculate $F(f_x, 0)$ or $F(0, f_y)$ since we know that $F(f_x, f_y)$ also

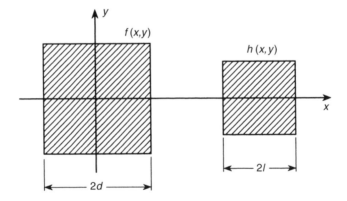

Figure P4.1

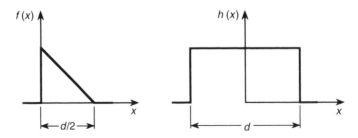

Figure P4.2

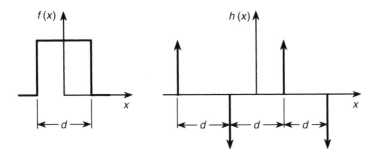

Figure P4.3

is circularly symmetric. In this way, calculate the Fourier transform of a circular aperture and show that

$$F(\rho) = 4r_0^2 \int_0^1 \sqrt{(1 - u^2)} \cos(tu) \, du$$

where $t = 2\pi\rho r_0$ and r_0 is the radius of the aperture. By comparison with Equation (4.29) we obtain another representation of the Bessel function $J_1(t)$:

$$J_1(t) = \frac{2t}{\pi} \int_0^1 \sqrt{(1 - u^2)} \cos(tu) \, du$$

4.11 A real function $f(x)$ can be decomposed into a sum of an even and an odd function.

(a) Show that $f_{even}(x) = \frac{1}{2}[f(x) + f(-x)]$ and $f_{odd}(x) = \frac{1}{2}[f(x) - f(-x)]$.

(b) Show that

$$\mathscr{F}\{f_{even}(x)\} = \text{Re}\{\mathscr{F}\{f(x)\}\}$$
$$\mathscr{F}\{f_{odd}(x)\} = i\,\text{Im}\{\mathscr{F}\{f(x)\}\}$$

4.12 The sign-function is defined as

$$\text{sgn}(x) = \begin{cases} 1 & x > 0 \\ 0 & x = 0 \\ -1 & x < 0 \end{cases}$$

Prove that

$$\mathscr{F}\{\text{sgn}(x)\text{sgn}(y)\} = \left(\frac{1}{i\pi f_x}\right)\left(\frac{1}{i\pi f_y}\right)$$

4.13 Prove that

$$\mathscr{F}\{\nabla^2 g(x, y)\} = -4\pi^2(f_x^2 + f_y^2)\mathscr{F}\{g(x, y)\}$$

where ∇^2 is the Laplacian operator

$$\nabla^2 = \frac{\partial^2}{\partial x^2} + \frac{\partial^2}{\partial y^2}$$

4.14 Prove that

$$\mathscr{F}\mathscr{F}\{g(x, y)\} = \mathscr{F}^{-1}\mathscr{F}^{-1}\{g(x, y)\} = g(-x, -y)$$

4.15 The Gaussian distribution function is given as

$$g(x) = Ce^{-ax^2}$$

where

$$C = \sqrt{\left(\frac{a}{x}\right)}, \quad \sigma_x = \frac{1}{\sqrt{(2a)}}$$

where σ_x is the standard deviation. Given that

$$\int_{-\infty}^{\infty} e^{-u^2} \, du = \sqrt{\pi}$$

prove that the Fourier transform of $g(x)$ is

$$G(f_x) = \exp\left[-\frac{(2\pi f_x)^2}{4a}\right]$$

4.16 Assuming a unit-amplitude, normally incident plane-wave illumination:

(a) Find the intensity distribution in the Fraunhofer diffraction pattern of the double-slit aperture shown in Figure P4.4.

(b) Sketch the intensity distribution along the x_f-axis of the observation plane. Let $X/\lambda z = 1$ m^{-1}, and $d/\lambda z = 3/2$ m^{-1}, where z is the distance to the observation plane and λ the wavelength.

4.17 (a) Sketch the aperture described by the transmittance function

$$t(x, y) = \left\{\left[\text{rect}\left(\frac{x}{X}\right)\text{rect}\left(\frac{y}{Y}\right)\right] \otimes \left[\frac{1}{d} \text{comb}\left(\frac{x}{d}\right)\delta(y)\right]\right\} \text{rect}\left(\frac{x}{Nd}\right)$$

where N is an odd integer and $d > X$.

(b) Find an expression for the intensity distribution in the Fraunhofer diffraction pattern of that aperture, assuming illumination by a normally incident plane wave and $N \gg 1$.

4.18 Find an expression for the intensity distribution in the Fraunhofer diffraction pattern of the aperture shown in Figure P4.5. Assume unit-amplitude, normally incident plane-wave illumination.

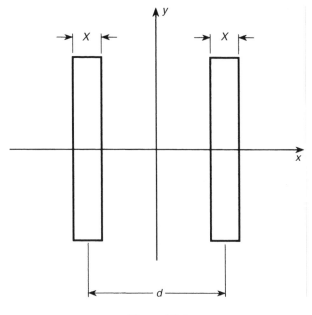

Figure P4.4

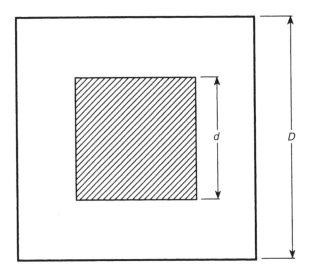

Figure P4.5

4.19 A normally incident, unit-amplitude, monochromatic ($\lambda = 1$ μm) plane wave illuminates a positive lens of 40 mm diameter and 2 m of focal length. An object is placed 1 m behind the lens and centred on the lens axis with amplitude transmittance

$$t(x, y) = \tfrac{1}{2}(1 + \cos 2\pi f_0 x) \, \text{rect}\left(\frac{x}{L}\right) \text{rect}\left(\frac{y}{L}\right)$$

Assuming $L = 10$ mm and $f_0 = 10$ lines/mm, sketch the intensity distribution across the x_f-axis of the focal plane. Indicate the numerical values of the distance between the diffracted components and the width (between first zeros) of the individual components.

4.20 An incoherent imaging system has a square pupil function of width D. A square stop, of width $D/2$, is placed at the centre of the exit pupil, as shown in Figure P4.5.

(a) Sketch a cross-section of the optical transfer function $\mathscr{H}(f_x, 0)$ along the f_x-axis with and without the stop present.

(b) Sketch the limiting form of the optical transfer function as the size of the stop approaches the size of the exit pupil.

4.21 A sinusoidal amplitude grating with amplitude transmittance

$$t(x_o, y_o) = \tfrac{1}{2} + \tfrac{1}{2}\cos 2\pi f_1 x_o$$

is placed in the object plane x_o, y_o and imaged to the image plane x_i, y_i by a lens (circular of diameter D, focal length f) and obliquely illuminated by a unit-amplitude monochromatic plane wave incident at an angle θ in the $x_o z$-plane. The object- and image distances are a and b respectively.

(a) Find the Fourier spectrum of the field-amplitude distribution transmitted by the object screen.

(b) Assuming $a = b = 2f$, what is the maximum angle θ_m for which any variations of intensity will appear in the image plane?

(c) By applying this maximum angle θ_m, what is the intensity distribution in the image plane, and how does it compare with the corresponding intensity distribution for $\theta = 0$?

(d) Assuming that the maximum angle θ_m is used. Find the maximum grating frequency f_1 that will give any variations of intensity in the image plane. Compare this frequency with the cutoff frequency when $\theta = 0$.

4.22 An object has an intensity transmittance given by

$$\tau(x, y) = \tfrac{1}{2}(1 + \cos 2\pi f_1 x)$$

and introduces a constant, uniform phase delay across the object plane. This object is imaged by a positive lens of diameter D and focal length f, with an object and image distance $a = b = 2f$. Compare the maximum frequencies f_1 transmitted by the system for the case of coherent and incoherent illumination.

4.23 An object has a periodic amplitude transmittance described by

$$t(x, y) = t(x)$$

where $t(x)$ is shown in Figure P4.6. This object is placed in the object plane of a lens with object- and image distance $a = b = 2f$. A small opaque stop is introduced

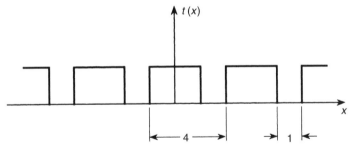

Figure P4.6

on the lens axis in the focal plane. Sketch the resulting intensity distribution in the image plane.

4.24 The so-called central dark ground method for observing phase objects is achieved by placing a small opaque stop in the back focal plane of the imaging lens to block the undiffracted light. Assuming that the phase delay $\phi(x, y)$ through the object is always much less than 1 radian, find the observed image intensity in terms of $\phi(x, y)$.

4.25 According to the so-called Rayleigh criterion of resolution, two incoherent point sources are just resolved by a diffraction-limited system when the centre of the Airy pattern generated by one source falls on the first zero of the Airy pattern generated by the second. Assume that the image points of two points just resolved are centred at $(s/2, 0)$ and $(-s/2, 0)$ in the image plane, where s is the minimum resolvable separation.

(a) Find s in terms of the exit pupil diameter D, the image distance b and the wavelength. The intensity distribution along the x-axis in the image plane will consist of two maxima on each side of a central dip.

(b) Calculate the ratio between the intensity of the central dip and the maximum. $J_1(0.61\pi) = 0.58$.

5

Light Sources and Detectors

5.1 INTRODUCTION

The most important 'hardware' in optical metrology is light sources and detectors. To appreciate the various concepts of these devices, we first introduce the different units and terms for the measurement of electromagnetic radiation. Then the laser is given a relatively comprehensive treatment. The description of detectors involves some understanding of semiconductor technology. Therefore a brief introduction to semiconductors is given in Appendix E. Because of the increasing use of the CCD camera in optical metrology, this device is described separately in Section 5.8.

5.2 RADIOMETRY. PHOTOMETRY

To compare light sources we have to make a brief introduction to units and terms for the measurement of electromagnetic radiation (Slater 1980; Klein and Furtak 1986; Longhurst 1967). Below we present the most common radiometric units.

Radiant energy, Q, is energy travelling in the form of electromagnetic waves, measured in joules.

Radiant flux, $\Phi = \partial Q/\partial t$ is the time rate of change, or rate of transfer, of radiant energy, measured in watts. Power is equivalent to, and often used instead of, flux. Radiant flux density at a surface, $M = E = \partial \Phi/\partial A$, is the radiant flux at a surface divided by the area of the surface. When referring to the radiant flux emitted from a surface it is called radiant exitance M. When referring to the radiant flux incident on a surface it is called irradiance E. Both are measured in watts per square metre. Note that in the rest of this book, we use the term intensity, which is proportional to irradiance.

Radiant intensity, $I = \partial \Phi/\partial \Omega$, of a source is the radiant flux proceeding from the source per unit solid angle in the direction considered, measured in watts per steradian.

Radiance, $L = \partial^2 \Phi/\partial \Omega \partial A \cos \theta$, in a given direction, is the radiant flux leaving an element of a surface and propagated in directions defined by an elementary cone containing the given direction, divided by the product of the solid angle of the cone and the area of the projection of the surface element on a plane perpendicular to the given direction. Figure 5.1 illustrates the concept of radiance. It is measured in watts per square metre and steradian.

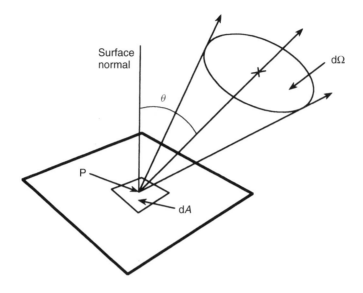

Figure 5.1   The concept of radiance

Table 5.1   Symbols, standard units and defining equations for fundamental radiometric and photometric quantities

Symbol	Radiometric quantity	Radiometric units	Defining equation	Photometric quantity	Photometric units
Q	Radiant energy	J		Luminous energy	lm s
Φ	Radiant flux	W	$\Phi = \partial Q/\partial t$	Luminous flux	lm
M	Radiant exitance	$W\,m^{-2}$	$M = \partial\Phi/\partial A$	Luminous exitance	$lm\,m^{-2}$
E	Irradiance	$W\,m^{-2}$	$E = \partial\Phi/\partial A$	Illuminance	$lm\,m^{-2}$
I	Radiant intensity	$W\,sr^{-1}$	$I = \partial\Phi/\partial\Omega$	Luminous intensity	$lm\,sr^{-1}$
L	Radiance	$W\,sr^{-1}\,m^{-2}$	$L = \partial^2\Phi/\partial\Omega\partial A \cos\theta$	Luminance	$lm\,sr^{-1}\,m^{-2}$

All of the radiometric terms have their photometric counterparts. They are related to how the (standard) human eye respond to optical radiation and is limited to the visible part of the spectrum. In Table 5.1 we list the radiometric and the corresponding photometric quantities.

To distinguish radiometric and photometric symbols they are given subscripts e and v respectively (e.g. L_e = radiance, L_v = luminance).

The radiometric quantities refer to total radiation of all wavelengths. A spectral version for each may be defined by adding the subscript λ (e.g. $M_{e\lambda}$ or simply M_λ) where for example a spectral flux $\Phi_\lambda\,d\lambda$ represents the flux in a wavelength interval between λ and $\lambda + d\lambda$, with units watts per nanometre $(W\,nm^{-1})$ or watt per micrometre $(W\,\mu m^{-1})$.

To represent the response of the human eye, a standard luminosity curve $V(\lambda)$ has been established, see Figure 5.2. It has a peak value of unity at $\lambda = 555$ nm. The conversion

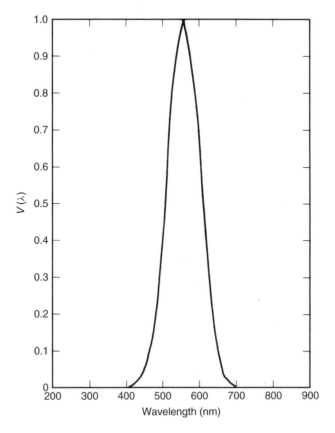

Figure 5.2   The standard luminosity curve

at $\lambda = 555$ nm is standardized to be

$$K_\mathrm{m} = 680 \text{ lm/W} \qquad (5.1)$$

This means that 1 W of flux at 555 nm gives the same physical sensation as 680 lm. For other wavelengths the conversion factor is

$$K = K_\mathrm{m} V(\lambda) = 680 \ V(\lambda) \text{lm/W} \qquad (5.2)$$

We may use Equation (5.2) to convert any radiometric quantity to the corresponding photometric quantity. For instance, if we have a spectral radiant flux $\Phi_{\mathrm{e}\lambda}$, the luminous flux is given by

$$\Phi_\mathrm{v} = 680 \int_0^\infty V(\lambda) \Phi_{\mathrm{e}\lambda} \, \mathrm{d}\lambda \qquad (5.3)$$

An average conversion factor is defined by

$$K_\mathrm{av} = \Phi_\mathrm{v} / \Phi_\mathrm{e} \qquad (5.4)$$

5.2.1 Lambertian Surface

A Lambertian surface is a perfectly diffuse reflecting surface defined as one which the radiance L is constant for any angle of reflection θ to the surface normal. Lambert's cosine law states that the intensity (flux per unit solid angle) in any direction varies as the cosine of the reflection angle:

$$I = I_0 \cos \theta \tag{5.5}$$

Since the projected area of the source also varies as $\cos \theta$, the radiance becomes independent of the viewing angle:

$$L = \frac{I}{\mathrm{d}A \cos \theta} = \frac{I_0}{\mathrm{d}A} \tag{5.6}$$

Assume that an elemental Lambertian surface $\mathrm{d}A$ is irradiated by E in W m^{-2} and that the radiant flux reflected in any direction θ to the surface normal is given by the basic equation

$$\mathrm{d}^2\Phi = L \, \mathrm{d}\Omega \, \mathrm{d}A \cos \theta \tag{5.7}$$

The solid angle $\mathrm{d}\Omega$ in spherical coordinates (see Figure 5.3) is given by

$$\mathrm{d}\Omega = (r \sin \theta \, \mathrm{d}\theta r \, \mathrm{d}\phi)/r^2 = \sin \theta \, \mathrm{d}\theta \, \mathrm{d}\phi \tag{5.8}$$

The total radiant flux reflected into the hemisphere therefore is given by

$$\mathrm{d}\Phi_\mathrm{h} = \int_0^{2\pi} \mathrm{d}\varphi \int_0^{\pi/2} L \, \mathrm{d}A \cos \theta \sin \theta \, \mathrm{d}\theta = \pi L \, \mathrm{d}A \tag{5.9}$$

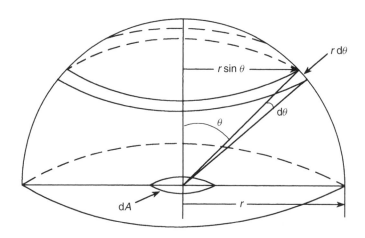

Figure 5.3

The ratio of the total reflected radiant flux to the incident radiant flux $d\Phi_i = E\,dA$ defines the diffuse reflectance of the surface

$$\frac{d\Phi_h}{d\Phi_i} = \rho = \frac{\pi L}{E} \tag{5.10}$$

The quantity ρE is the radiant flux density reflected from the surface which is equivalent to the radiant exitance M of a self-emitting source, giving

$$M = \pi L \tag{5.11}$$

for a Lambertian surface.

For non-Lambertian surfaces, L is a function of both θ and the azimuthal angle ϕ and therefore can not be taken outside the integral in Equation (5.9). Many natural surfaces show Lambertian characteristics up to $\theta = 40°$. In satellite observations, one has found snow and desert to be Lambertian up to about $50°$ or $60°$. Most naturally occurring surfaces depart significantly from the Lambertian case for θ greater than about $60°$, an exception is White Sands, the desert in New Mexico, which is nearly Lambertian for all angles.

5.2.2 Blackbody Radiator

A blackbody at a given temperature provides the maximum radiant exitance at any wavelength that any body in thermodynamic equilibrium at that temperature can provide. It follows that a blackbody is a Lambertian source and that it is a perfect absorber as well as a perfect radiator. The spectral radiant exitance M_λ from a blackbody is given by Planck's formula

$$M_\lambda = \frac{2\pi hc^2}{\lambda^5[\exp(hc/\lambda kT) - 1]} \tag{5.12}$$

where

h = Planck's constant = 6.6256×10^{-34} J s;
c = velocity of light = 2.997925×10^8 ms^{-1};
k = Boltzmann's constant = 1.38054×10^{-23} J K^{-1};
T = absolute temperature in kelvin;
λ = wavelength in metres.

which gives M_λ in $W\,m^{-2}\,\mu m^{-1}$. Figure 5.4 shows M_λ as a function of wavelength for different temperatures.

By integrating over all wavelengths we get the Stefan–Boltzmann law

$$M = \int_0^\infty M_\lambda\,d\lambda = \sigma T^4 \tag{5.13}$$

where $\sigma = (2\pi^5 k^4)/(15c^2 h^3) = 5.672 \times 10^{-8}$ W m^{-2} K^{-4} is called the Stefan–Boltzmann constant.

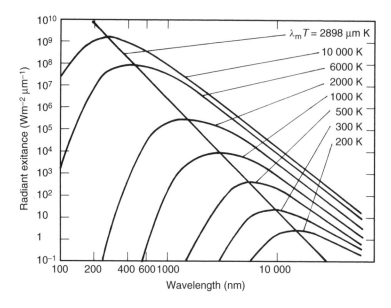

Figure 5.4 Spectral radiant exitance from a blackbody at various temperatures according to Planck's law

By differentiating Equation (5.12) we get the wavelength λ_m for which M_λ is peaked

$$\lambda_m T = 2897.8 \ \mu m \, K \tag{5.14}$$

This relation is called Wien's displacement law.

The blackbody is an idealization. In nature most radiators are selective radiators, i.e. the spectral distribution of the emitted flux is not the same as for a blackbody. Emissitivity is a measure of how a real source compares with a blackbody and is defined as

$$\varepsilon = M'/M \tag{5.15}$$

where M' is the radiant exitance of the source of interest and M is the radiant exitance of a blackbody at the same temperature. ε is a number between 0 and 1 and is in general both wavelength and temperature dependent. When ε is independent of wavelength the source is called a greybody. A more general form of Equation (5.15) can be written to take into account the spectrally varying quantities, thus ε, the emissivity for a selective radiator, as an average over all wavelengths is

$$\varepsilon = \frac{M'}{M} = \frac{\displaystyle\int_0^\infty \varepsilon(\lambda) M_\lambda \, d\lambda}{\displaystyle\int_0^\infty M_\lambda \, d\lambda} = \frac{1}{\sigma T^4} \int_0^\infty \varepsilon(\lambda) M_\lambda \, d\lambda \tag{5.16}$$

Consider two slabs of different materials A and B, and that each is of semi-infinite thickness and infinite area, forming a cavity as shown in Figure 5.5. Assume that A is a

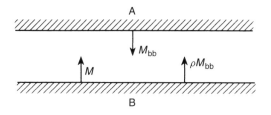

Figure 5.5   Radiant exitances between a blackbody A and another material B

blackbody and that B is a material with emissivity ε, reflectance ρ and absorbtance α, and that the materials and cavity are in thermal equilibrium. Because of the last assumption, the flux onto B must equal the flux leaving B toward A. Thus

$$M_{bb} = \rho M_{bb} + M \tag{5.17}$$

where M_{bb} and M are the radiant exitance of A (the blackbody) and B respectively. From the definition of emissivity we have

$$M/M_{bb} = \varepsilon = 1 - \rho \tag{5.18}$$

which is referred to as Kirchhoff's law.

Because of conservation of energy, the reflectance, transmittance and absorbance at a surface add up to unity. Since we have assumed semi-infinitely thick materials, the transmittance is zero and we have

$$M/M_{bb} = \varepsilon = 1 - \rho = \alpha \tag{5.19}$$

where α is the absorptance of material B. Equation (5.19) states that good emitters and absorbers are poor reflectors and vice versa. We can anticipate that Equation (5.19) holds for any given spectral interval which gives the more general form

$$M_\lambda/M_{\lambda bb} = \varepsilon(\lambda) = 1 - \rho(\lambda) = \alpha(\lambda) \tag{5.20}$$

5.2.3 Examples

Let us compare the light from a typical He−Ne laser and a blackbody with the same area as the output aperture of the laser. Assume this area to be 1 mm² and the blackbody temperature to be 3000 K, close to the temperature of the filament of an incandescent lamp. From Equation (5.13) we find the blackbody exitance to be 4.6×10^6 W m⁻² which gives a radiant flux of 4.6 W. An ordinary He−Ne laser has an output of about 1/1000th of this, not very impressive even if we take into account that most of the radiation from the blackbody is outside the visible region.

From Equation (5.11) we find the radiance from the blackbody to be

$$L = M/\pi = 1.46 \times 10^6 \text{ W m}^{-2} \text{ sr}^{-1}$$

The light beam from the laser has a diverging angle of about λ/d where λ is the wavelength and d is the output aperture diameter. This gives a solid angle of about λ^2/A where A is the aperture area. The radiance at the centre of the beam is therefore $(\cos\theta = 1)$

$$L = \frac{\Phi}{\Omega A} = \frac{\Phi}{\lambda^2} \tag{5.21}$$

With a radiant power (flux) $\Phi = 5$ mW and a wavelength $\lambda = 0.6328$ µm, this gives $L = 1.2 \times 10^{10}$ W m^{-2} sr^{-1}, a number clearly in favour of the laser. Note that the radiance of the blackbody is independent of its area. By decreasing the power of the laser by reducing its output aperture, the radiance decreases accordingly.

Figure 5.6 illustrates the imaging of an object of elemental area dA_o by a lens system with the entrance and exit pupils as sketched. We assume that the object is a Lambertian surface of radiance L_o. The flux incident over an annular element of the entrance pupil is given by

$$d^2\Phi = L_o\,dA_o\cos\theta\,d\Omega \tag{5.22}$$

where

$$d\Omega = 2\pi\sin\theta\,d\theta \tag{5.23}$$

If θ_m is the angle of the marginal ray passing through the entrance pupil, the flux incident over the entrance pupil is

$$d\Phi_o = 2\pi L_o\,dA_o\int_0^{\theta_m}\sin\theta\cos\theta\,d\theta = \pi L_o\,dA_o\sin^2\theta_m \tag{5.24}$$

Equation (5.24) is not the product of radiance, area and solid angle, or $2\pi L_o\,dA_o(1 - \cos\theta_m)$, as we might at first expect, because the cosine factor, which accounts for the projected area in any direction in the solid angle, has to be included in the integration.

We can write a similar expression for the flux $d\Phi_i$ incident over the exit pupil from a fictitious Lambertian source L_i, in the plane of the image. Then, evoking the principle of

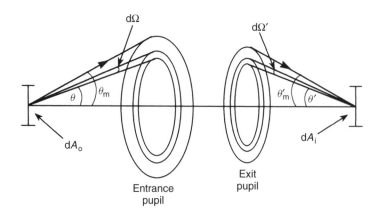

Figure 5.6   Geometry for determining the radiometry of an optical system

the reversibility of light, we can say that this flux, leaving the exit pupil in the direction of the image, gives rise to an image plane radiance L_i according to

$$d\Phi_i = \pi L_i \, dA_i \sin^2 \theta'_m \tag{5.25}$$

where dA_i is the image area and θ'_m is the inclination of the marginal ray in image space. For a perfect lossless system (that is, one without reflection, absorption and scattering losses), $d\Phi_o = d\Phi_i$, so

$$L_i \sin^2 \theta'_m \, dA_i = L_o \sin^2 \theta_m \, dA_o \tag{5.26}$$

Now

$$\frac{dA_i}{dA_o} = m^2 \tag{5.27}$$

the square of the lateral magnification. We assume that the lens is aplanatic, obeying the Abbe sine condition, that is, it exhibits zero spherical abberation and coma for objects near the axis, no matter how low the F-number. (Spherical abberation is illustrated in Figure 2.10). Thus,

$$n \sin \theta_m = mn' \sin \theta'_m \tag{5.28}$$

where n and n' are the refractive indices in object and image space which we set equal to unity. Then

$$\frac{\sin \theta_m}{\sin \theta'_m} = m \tag{5.29}$$

and we get $L_i = L_o$ which shows that the radiance is conserved in a lossless imaging system.

Equation (5.25) then gives for the image irradiance

$$E_i = d\Phi_i/dA_i = \pi L_o \sin^2 \theta'_m \tag{5.30}$$

As usual in paraxial optics, we approximate $\sin \theta'_m$ by $\tan \theta'_m$, giving

$$\sin \theta'_m \approx \tan \theta'_m = \frac{D_i}{2b} \tag{5.31}$$

where D_i is the diameter of the exit pupil and b is the image distance. By introducing the aperture number $F = f/D_i$ where f is the focal length, Equation (5.30) becomes

$$E_i = \pi L_o \left(\frac{D_i}{2b} \right)^2 = \frac{\pi L_o}{4F^2(1+m)^2} \tag{5.32}$$

With the object at infinity, $b = f$ and we get

$$E_i = \pi L_o \left(\frac{D_i}{2f} \right)^2 = \frac{\pi L_o}{4F^2} \tag{5.33}$$

Equation (5.33) or a similar form of it, is generally referred to as the 'camera equation'. It indicates that image irradiance is inversely proportional to the square of the F-(aperture) number. Therefore the diaphragm or stop openings for a lens are marked in a geometrical ratio of $2^{1/2}$.

Recall that we have assumed the object to be a Lambertian surface. For example, for a point source of radiant intensity I as the object, the flux intercepted by the entrance pupil is

$$d\Phi = I\, d\Omega = \frac{I\,S}{a^2} = I\pi \left(\frac{D}{2a} \right)^2 \tag{5.34}$$

where S is the area and D is the diameter of the entrance pupil, a is the object distance and where we for simplicity assume the entrance and exit pupils to have equal area S.

If we take the image area dA_i to be equal to the area of the Airy disc (see Section 4.6, Equation (4.33))

$$dA_i = \pi(\Delta r_i)^2 = 1.5\pi \left(\frac{\lambda b}{D} \right)^2 \tag{5.35}$$

we get for the image irradiance

$$E_i = \frac{d\Phi}{dA_i} = \frac{8}{3}\frac{I}{\lambda^2} \left(\frac{D}{2a} \right)^2 \left(\frac{D}{2b} \right)^2 = \frac{I}{6\lambda^2} \left(\frac{m}{F^2(1+m)^2} \right)^2 \tag{5.36}$$

From this expression we see that the image irradiance is dependent on both the object and image distances. In conclusion we might say that for a Lambertian surface we can not increase the image irradiance by placing the lens closer to the object, but for a point source we can, the maximum occurring at unit magnification, i.e. when $a = b = 2f$.

5.3 INCOHERENT LIGHT SOURCES

Most light sources are incoherent, from the candle light to the Sun. They all radiate light due to spontaneous emission (see Section 5.4.1). Here we will consider some sources often used in scientific applications. These are incandescent sources, low-pressure gas discharge lamps and high-pressure gas discharge-arc lamps. They are commonly rated, not according to their radiant flux, but according to their electric power consumption.

Tungsten halogen lamps

Quartz tungsten halogen lamps (QTH) produce a bright, stable, visible and infrared output and is the most commonly applied incandescent source in radiometric and photometric studies. It emits radiation due to the thermal excitation of source atoms or molecules. The spectrum of the emitted radiation is continuous and approximates a blackbody. Spectral distribution and total radiated flux depend on the temperature, area and emissivity. For a QTH lamp, the temperature lies above 3000 K and the emissivity varies around 0.4 in the visible region.

In all tungsten filament lamps, the tungsten evaporates from the filament and is deposited on the inside of the envelope. This blackens the bulb wall and thins the tungsten filament, gradually reducing the light output. With tungsten halogen lamps, the halogen gas effectively removes the deposited tungsten, and returns it to the hot filament, leaving the inside of the envelope clean, and greatly increases lamp life. This process is called the halogen cycle. A commercial 1000 W QTH lamp have a luminous flux of up to 30 000 lm with a filament size of 5 mm × 18 mm (Oriel Corporation 1994).

Low-pressure gas-discharge lamps

In these sources an electric current passes through a gas. Gas atoms or molecules become ionized to conduct the current. At low current density and pressures, electrons bound to the gas atoms become excited to well-defined higher-energy levels. Radiation is emitted as the electron falls to a lower energy level characteristic of the particular type of gas. The spectral distribution is then a number of narrow fixed spectral lines with little background radiation. The known wavelengths determined by the energy levels are useful for calibration of spectral instruments.

High-pressure gas-discharge arc lamps

High-current-density arc discharges through high-pressure gas are the brightest conventional sources of optical radiation. Thermal conditions in the arc are such that gas atoms (or molecules) are highly excited. The result is a volume of plasma. The hot plasma emits like an incandescent source, while ionized atoms emit substantially broadened lines. The spectral distribution of the radiation is a combination of both the continuum and the line spectra. The most common sources of this type are the Xenon (Xe) and mercury (Hg) short arc lamps. Xenon lamps have colour temperatures of about 6000 K, close to that of the Sun. A commercial 1000 W Hg lamp produces a luminous flux of 45 000 lm with an effective arc size of 3 mm × 2.6 mm. A commercial 1000 W Xe lamp is even brighter with luminous flux of 30 000 lumens with an effective arc size of 1 mm × 3 mm (Oriel Corporation 1994).

5.4 COHERENT LIGHT SOURCES

5.4.1 Stimulated Emission

Figure 5.7 shows an energy-level diagram for a fictive atom or molecule (hereafter called an atom). Here only four levels are shown. Assume that the atom by some process is raised to an excited state with energy E_3. Then the atom drops to energy levels E_2, E_1 and E_0 in successive steps. E_0 may or may not be the ground state of the atom. We do not specify the type of transition from E_3 to E_2 and from E_1 to E_0, but assume that the energy difference between E_2 and E_1 is released as electromagnetic radiation of frequency ν given by

$$E_2 - E_1 = h\nu \tag{5.37}$$

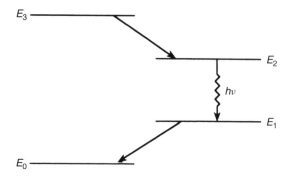

Figure 5.7   Energy-level diagram of a fictitious atom

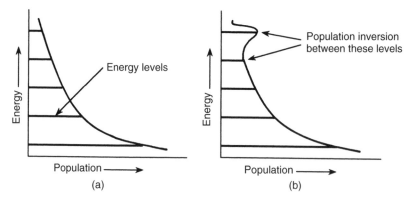

Figure 5.8   Distribution of populations among energy levels at (a) equilibrium and (b) during a population inversion

where $h = 6.6256 \times 10^{-34}$ J s is the Planck constant. This might be the situation in an ordinary light source (e.g. a discharge lamp) where the transition occurs spontaneously and the photon is therefore said to be created by spontaneous emission.

As postulated by Einstein, also another type of transition is possible: if a photon of frequency given by Equation (5.37) passes the atom it might trigger the transition from E_2 to E_1 thereby releasing a new photon of the same frequency by so-called stimulated emission. Under normal conditions known as thermodynamic equilibrium, the number of atoms in a state tends to decrease as its energy increases as shown in Figure 5.8(a). This means that there will be a larger population in the lower state of a transition than in the higher state. Therefore photons passing the atom are far more likely to be absorbed than to stimulate emission. Under these conditions, spontaneous emission dominates.

However, if the excitation of the atoms is sufficiently strong, the population of the upper level might become higher than that of the lower level. This is called population inversion and is illustrated in Figure 5.8(b). Then by passing of a photon of frequency given by Equation (5.37), it will be more likely to stimulate emission from the excited state than to be absorbed by the lower state. This is the condition that must be obtained in a laser and the result is laser gain or amplification, a net increase in the number of photons with the transition energy. Light amplification by the stimulated emission of radiation therefore has given rise to the acronym 'laser'. Laser gain is proportional to the difference between the chance of stimulated emission and the chance of absorption. Therefore the population of both the upper and lower levels of the laser transition are important. Thus if laser action is to be sustained, the lower level must be depopulated as the upper level is populated or the population inversion will end. That is indeed what happens in some pulsed lasers. Stimulated emission has the same wavelength as the original photon and it is in phase (or coherent) with the original light.

In the description given above, four energy levels are involved. This is the best condition for laser action and is called a four-level laser. But also three-level lasers exists which is the case when e.g. the lower transition level is the ground state. To maintain population inversion, it is easily realized that the lifetime of the E_2-level should be as long as possible and the lifetime of the E_1-level should be as short as possible. The process of raising the atom to the E_3 excited level is called pumping. The pumping mechanism is different for different laser types.

In the description given above, we have assumed laser transition between energy levels E_2 and E_1 only. Usually, stimulated emission can be obtained between many different energy levels in the same laser medium. Dependent on the construction of the laser, one can obtain lasing from a single transition or from a multitude of transitions.

Laser amplification can occur over a range of wavelengths because no transition is infinitely narrow. The range of wavelengths at which absorption and emission can occur is broadened by molecular motion (Doppler broadening) vibrational and rotational energy levels, and other factors.

To be more specific, let us consider the most familiar of all lasers, the helium–neon (He–Ne) laser.

Figure 5.9 shows the construction of a typical He–Ne laser. Inside a discharge tube is a gas mixture of helium and neon. Typically the mixture contains 5 to 12 times more helium than neon. By applying voltage to the electrodes, the resulting electric field will accelerate free electrons and ions inside the tube. These collide with helium atoms raising them to a higher energy level. By collision between helium and neon atoms, the latter are raised to a higher energy level. This constitute the pumping process. The neon atoms, which constitute the active medium, return to a lower energy level and the energy difference is released as electromagnetic radiation.

Figure 5.10 shows an energy-level diagram for an He–Ne laser emitting red light. Excited helium atoms in the 1s2s state transfer energy to neon atoms in the ground state by collisions, exciting the neon atoms to the 5s excited state. By returning to the 3p state the energy difference is released as light of wavelength 632.8 nm.

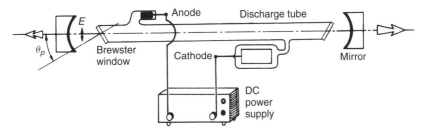

Figure 5.9   He–Ne laser. (Hecht & Zajac, Optics, © 1974, Addison-Wesley Publishing Company, Inc., Reading, Massachusetts. Figure 14.31. Reprinted with permission)

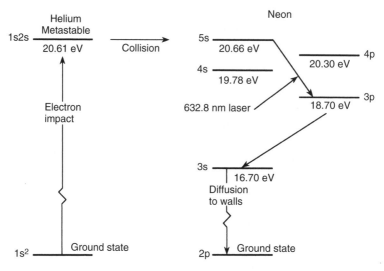

Figure 5.10   He–Ne laser energy levels

Population inversion alone is sufficient to produce 'light amplification by the stimulated emission of radiation', but the result is only a coherent monochromatic light bulb. In fact, population inversion is observed in the atmosphere of Mars. To get light oscillation however, the discharge tube is enclosed in an optical cavity or resonator which is two mirrors facing each other as in Figure 5.9. The result is that the light is reflected back and forth through the tube, stimulating emission again and again from neon atoms. Emission in other directions is lost out of the laser medium and the light is concentrated in a beam oscillating back and forth between the mirrors. The optical cavity therefore acts as an oscillator or feedback amplifier and is an essential part of a laser.

Below we give a short description of other lasers. Numerous lasers and laser media have been demonstrated in the laboratory. Here we concentrate on lasers which are available commercially. For further details, the excellent book by Hecht (1992) is highly recommended.

There are many potential criteria for classifying lasers, but the two most useful ones are the type of active medium and the way in which it is excited (pumped). Usual practice is to group most devices as gas, liquid, solid-state or semiconductor lasers. A few important lasers are exceptions. Liquid- and solid-state lasers are pumped optically, i.e. by means of a flashlamp or another laser. Semiconductor lasers are excited when charge carriers in a semiconductor recombine at the junction of regions doped with n- and p-type donor materials. Gas lasers can be pumped in various ways, including discharge excitation (cf. the He–Ne laser), radio frequency (RF) excitation, chemical and optical excitation and also by gas expansion (gas dynamics).

5.4.2 Gas Lasers

Helium–neon

Among the first lasers demonstrated and the first gas laser (Javan *et al.* 1961). The 632.8 nm line is the most important because it can give up to about 50 mW continuous wave (c.w.). Green, yellow, orange and multiline versions are being offered commercially. Advantages: Output beam of low divergence and high coherence.

Noble gas ion lasers

Emit on ionized rare gas lines. Pumping: Electrical discharge. The most important is argon with strong lines in the blue-green and weaker lines in the ultraviolet and near-infrared. The 514.5 nm line is the strongest in larger water-cooled lasers while the 488.0 nm line is the strongest in small air-cooled models. Another type is Krypton lasers.

Advantages: Their ability to produce c.w. output of a few milliwatts to tens of watts in the visible and up to 10 W in the ultraviolet.

Helium–cadmium lasers

Emit on lines of ionized metal vapours. Electric discharge pumping. C.w. output up to about 150 mW at 442 nm or powers to about 50 mW at 325 nm.

Advantages: Offer short wavelength at moderate output power which can be focused to a small spot.

Carbon dioxide lasers

Pumping: Electric discharge, RF or gas-dynamic. Transitions between vibrational levels. Infrared radiation between 9 and 11 μm. Several distinct types. Can produce c.w. output powers from under 1 W for scientific applications to many kilowatts for materials working. Can generate pulses from the nanosecond to millisecond regimes. Custom-made CO_2 lasers have produced c.w. beams of hundreds of kilowatts for military weapon research or nanosecond pulses of 40 kJ for research in laser-induced nuclear fusion. Advantages: No other commercial laser can generate as intense c.w. output.

Chemical lasers

Pumping by means of chemical reaction. Three most important media: hydrogen fluoride, deuterium fluoride and iodine. Emits at wavelengths between 1.3 μm and 4.2 μm. Military research has demonstrated building-sized lasers that have generated nominally c.w. outputs to a couple of megawatts. Commercial devices produce much lower powers.

Copper and gold vapour lasers

Emit in or near the visible region on lines of neutral metal vapor. Pumping: Electric discharge. Operate as pulsed lasers only. Commercial copper vapour lasers can emit over 100 W in the green and yellow, gold vapour lasers can generate several watts in the red. Advantages: High power and high efficiency in the actual wavelength region with repetition rates of several kilohertz.

Eximer lasers

Eximer is a contraction of 'excited dimer', a description of a molecule consisting of two identical atoms which exists only in an excited state, e.g. He_2 and Xe_2. Since the ground state does not exist, population inversion is obtained as long as there are molecules in the excited state. Pumping: Electric discharge transverse to the gas flow. Most important media: rare gas halides such as: argon fluoride, krypton fluoride, xenon fluoride and xenon chloride. Emit powerful pulses (average power of up to 100 W) lasting nanoseconds or tens of nanoseconds in or near the ultraviolet.

Advantages: Very high gain. No other commercial laser can generate such high average power at such a short wavelength.

Nitrogen lasers

Pumping: Electric discharge. Can produce nanosecond or subnanosecond pulses (average power of a few hundred milliwatts) of wavelength 337 nm.

Advantages: Low-cost. So simple to build that it was once featured in the 'Amateur Scientist' column of *Scientific American*.

5.4.3 Liquid Lasers

Dye lasers

The discussion of liquid lasers almost invariably starts and ends with the dye laser. The active medium is a fluorescent organic dye dissolved in a liquid solvent.

Pumping: Optical, with a flashlamp or (more often) with an external laser. The output wavelength can be tuned from the near-ultraviolet into the near-infrared. Dye lasers can be adjusted to operate over an extremely narrow spectral bandwidth and can also produce ultrashort pulses, much shorter than a picosecond.

Disadvantages: Very complex. Tuning wavelength across the visible spectrum requires several changes of dye. Complex optics are needed to produce either ultra-narrow linewidth or picosecond pulses.

5.4.4 Semiconductor Diode Lasers. Light Emitting Diodes

As mentioned in Appendix E, light can be emitted from a semiconductor material as a result of electron-hole recombination. A light-emitting diode (LED) is a forward-biased p-n junction where electrons and holes are injected into the same region of space. The resulting recombination radiation is called injection electroluminescence (see Figure 5.11a). If the forward voltage is increased beyond a certain value, the number

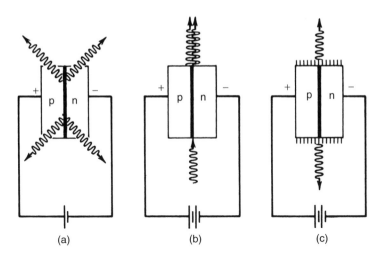

Figure 5.11   A forward-biased semiconductor p-n junction diode operated as (a) an LED; (b) a semiconductor optical amplifier; and (c) a semiconductor injection laser. From Saleh, B.E.A., and Teich, M.C. (1991) *Fundamentals of Photonics*. Reproduced by permission of John Wiley & Sons, New York

of electrons and holes in the junction region can become sufficiently large so that population inversion is achieved. We then can have stimulated emission and the junction may be used as a diode laser amplifier (Figure 5.11(b)) or, with appropriate feedback, as an injection laser diode (Figure 5.11(c)). Both LEDs and injection lasers are highly efficient electronic-to-photonic transducers and are readily modulated by the injected current. Their successful applications include lamp indicators, display devices, scanning, reading and printing systems, fibre-optic communication systems and optical data storage systems such as CD players.

LEDs. The photon flux generated in the junction is radiated uniformly in all directions. However, because of the high refractive index of many semiconductor materials (for GaAs, $n = 3.6$) most of the light suffers total internal reflection (see Section 9.5). Thus, for $n = 3.6$, only 3.9% of the total generated photon flux can be transmitted. A technique to increase the output flux is to encapsulate the junction in a plastic material with a refractive index of about 1.5. LEDs may be constructed in either surface or edge-emitting configuration: see Figure 5.12. Figure 5.13 shows the observed wavelength spectral densities for a number of LEDs that operate in the visible and near-infrared regions.

In a semiconductor injection laser (or **laser diode, LD**) the feedback is usually obtained by cleaving the semiconductor material along its crystal planes. The sharp refractive index difference between the crystal and the surrounding air causes the cleaved surfaces to act as reflectors. In comparison with other types of lasers, the laser diode has a number of advantages: small size, high efficiency, integrability with electronic components, and ease of pumping and modulation by electric current injection. However, the spectral linewidth of LDs is typically larger than that of other lasers.

If the thickness of the active region (the junction) could be reduced, the optical gain would be the same with a far lower current density. This is a problem, however, because

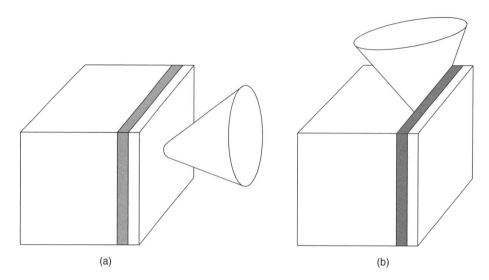

Figure 5.12   (a) Surface-emitting LED and (b) edge-emitting LED

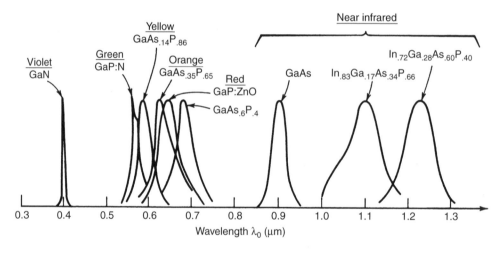

Figure 5.13   Spectral densities versus wavelength for semiconductor LEDs with different band-gaps. The peak intensities are normalized to the same value. From Saleh, B.E.A., and Teich, M.C. (1991) *Fundamentals of Photonics*. Reproduced by permission of John Wiley & Sons, New York

the carriers tend to diffuse out of the region. The solution to this problem is to use a heterostructure device which confines the light within the active medium which acts as an optical waveguide. By comparing LEDs and LDs we note that LDs produce light even below threshold. When operated below threshold, the LD acts as an edge-emitting LED. In fact, most LEDs are simply edge-emitting double-heterostructure devices. LDs with sufficiently strong injection so that stimulated emission is much greater than spontaneous emission, but with little feedback so that the lasing threshold is high, are called superluminescent LEDs.

Table 5.2   Diode laser wavelengths under 1 μm and power levels available commercially. (From Hecht, J. (1992), The Laser Guidebook (2nd edn), McGraw-Hill, New York). (Reproduced by permission of McGraw-Hill, Inc)

Nominal wavelength (nm)	Compound	Maximum continuous-wave power (single element)
635	InGaAlP	3 mW
660	InGaAlP	3 mW
670	$Ga_{0.5}In_{0.5}P$	10 mW
750	GaAlAs	8 mW
780	GaAlAs	35 mW
810	GaAlAs	100 mW single
		10 W linear array
		60 W quasi-continuous-wave (pulsed) array
		1500 W quasi-continuous-wave stacked array
830	GaAlAs	150 mW
850	GaAlAs	100 mW
880 or 895	GaAlAs	Pulsed only
905	GaAs (nominal)	Pulsed only
910	InGaAs	Pulsed only
980	InGaAs	50 mW

LDs can be divided into short-wavelength (below about 1.1 μm) and long-wavelength lasers. The lasers belonging to the first group are listed in Table 5.2. LDs with $\lambda > 1.1$ μm are used primarily for fibre-optic communication. Work has been concentrated on 1.31 μm where silica step-index single-mode fibres have zero wavelength dispersion and loss about 0.5 dB per kilometre, and on 1.55 μm where silica fibres have their lowest loss, about 0.15 dB/km: see Section 13.3.

5.4.5 Solid-State Lasers

A solid-state laser is one in which the active medium is a non-conductive solid, a crystalline material, or glass doped with a species that can emit laser light. In crystalline or glass solid-state lasers, the active species is an ion embedded in a matrix of another material, generally called the 'host'. It is excited by light from an external source.

Neodymium lasers

The active medium is triply ionized neodymium in a crystal or glass matrix. The most common host materials are: yttrium lithium fluoride (YLE), phosphate glass, gadolinium scandium gallium garnet (GSGG), silicate glass and yttrium aluminum garnet (YAG) with wavelengths ranging from 1047 to 1064 nm (Nd-YAG). Neodymium lasers can generate c.w. beams of a few milliwatts to over a kilowatt, short pulses with peak powers in the gigawatt range, or pulsed beams with average powers in the kilowatt range.

Advantages: A very versatile laser that can be doubled, tripled and quadrupled by means of harmonic generation and generate short pulses with high power by means of Q switching and modelocking, see Section 5.4.7.

Ruby lasers

The first laser demonstrated (Maiman 1960). Ruby laser rods are grown from sapphire (Al_2O_3) doped with about 0.01 to 0.5 percent chromium. Emits at 694.3 nm. C.w operation has been demonstrated in the laboratory, but is difficult to achieve. Oscillators can produce millisecond pulses of 50–100 J. Require active cooling.

Tunable vibronic solid-state lasers

Tunable wavelength due to operation on 'vibronic' transitions in which the active medium changes both electronic and vibrational states. Commercial lasers made from: alexandrite (chromium-doped $BeAl_2O_4$) which can be tuned between 701 and 826 nm. Titanium-doped sapphire (Al_2O_3), tunable from 660 to 1180 nm, and cobalt doped magnesium fluoride between 1750 and 2500 nm (wavelength ranges given at room temperature). Can be operated both c.w. and pulsed. Output power depends on wavelength. Commercial pulsed alexandrite lasers can generate average powers to 20 W, Ti sapphire reaches several watts c.w.

Advantages: Ti sapphire has the broadest tuning range of any single conventional laser medium. (Dye lasers can be tuned across a broader range only by switching dyes.)

Fibre lasers and amplifiers

The fibre laser is a miniaturization of solid-state lasers. Interest has concentrated on fibre amplifiers to replace conventional electro-optic repeaters used in fibre-optic systems. Such repeaters detect a weak optical signal, convert it into electronic form, amplify and process the electronic signal, and use it to drive a laser transmitter.

The basic concept is shown in Figure 5.14. A fibre is made from a solid-state material (typically a glass) doped with an ion which emits at the desired wavelength λ_1. It is illuminated from one end by a weak signal at λ_1 and a stronger steady beam at a second wavelength λ_2 which excites the ion in the fibre to the upper laser level. As the weak

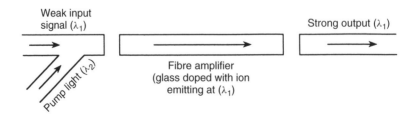

Figure 5.14 Operating principle of a fibre amplifier

signal passes through the fibre, it stimulates emission from excited ions at λ_1. Interest has centred on $\lambda = 1.3$ µm with neodymium as the laser ion and $\lambda = 1.54$ µm with erbium. For practical applications, diode lasers are used for pumping.

Other solid-state lasers

A lot of solid-state laser materials have been demonstrated in the laboratory. Here we mention the erbium–glass laser and the crystalline erbium laser (Er-YAG). The first emits at 1.54 µm and is therefore a candidate for eye-safe laser range finders (The 1.06 µm wavelength of Nd-YAG poses a serious eye hazard.) The most important line of Er-YAG is at 2.94 µm which is absorbed strongly by water, so it leaves a thinner damaged layer between healthy tissue and the zone removed by surgery. The absorption is so strong that it can be used to cut bone.

5.4.6 Other Lasers

Here we mention two types:

The free electron laser

The central idea is to extract light energy from electrons passing through a magnetic field with periodic variations in intensity and directions. It is therefore not based on stimulated emission but promises extremely high powers and exceptionally broad tunability, from microwaves to X-rays.

X-ray lasers

Visible and near-ultraviolet lasers operate on electronic transitions in the outer or valence shells of atoms. Transitions from outer to inner shells involve much more energy, thus producing X-rays. However, conditions for producing population inversion on such inner-shell transitions are extremely difficult to obtain. Two methods have been demonstrated by the Lawrence Livermore National Laboratory. One used a nuclear bomb explosion, the other used short, intense pulses from high-energy lasers built for fusion research.

5.4.7 Enhancements of Laser Operation

A description of lasers is not complete without mentioning some techniques that can enhance their operation. Here we give a short introduction to methods for wavelength enhancements, i.e. laser line narrowing and alteration of the laser wavelength, and changing of the pulse length.

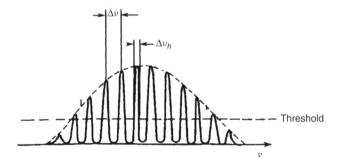

Figure 5.15   Spectral distribution of a He–Ne laser $\Delta\nu$ = resonator mode spacing, $\Delta\nu_h$ = resonator linewidth

Line narrowing

As indicated in Figure 5.15, the gain profile of even nominally monochromatic lasers normally spans several longitudinal modes, each with slightly different wavelength. The basic idea of line narrowing is to insert into the laser cavity optical elements, which restricts oscillations to a range of wavelengths so narrow that it includes only a single longitudinal mode. The commonest line-narrowing component is the Fabry–Perot etalon, typically a transparent plate with two reflective surfaces forming a short resonator that can be inserted within the laser cavity. In this way, frequency bandwidths as low as 500 kHz are obtainable from commercial lasers.

Wavelength alteration

Techniques for changing the wavelength from a laser include harmonic generation, parametric oscillation and Raman shifting. The method in most practical use is harmonic generation. This is based on the nonlinear interactions between light and matter (usually a non-linear crystal) which can generate harmonics at multiples of the light-wave frequency. The magnitude of the non-linear effect is proportional to the square of the incident power. Therefore, for most practical applications only the second, third and sometimes the fourth harmonics are produced. Conventionally, the laser beam makes a single pass through the crystal (usually potassium dihydrogen phosphate, KDP) which is placed outside the laser cavity. The commonest use of harmonic generation is with the 1064 nm Nd-YAG laser producing the 532 nm second, the 355 nm third and the 266 nm fourth harmonic. Dye laser output is often frequency doubled to obtain tunable ultraviolet light and also GaAlAs semiconductor lasers to give blue light.

Three techniques which operate by interacting with light inside the laser cavity for producing short pulses are in widespread use. These are Q-switching, cavity dumping and modelocking.

Q-switching

Like any oscillator, a laser cavity has a quality factor Q, defined as

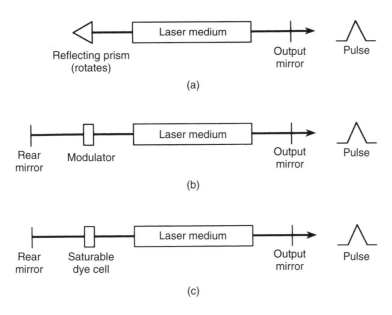

Figure 5.16 Types of Q switching: (a) rotating mirror or prism; (b) active modulator; and (c) passive

$$Q = \frac{\text{energy stored per pass}}{\text{energy dissipated per pass}}$$

Normally, the Q-factor of a laser cavity is constant, but if the Q-factor is kept artificially low, energy will gradually accumulate in the laser medium because the Q-factor is too low for laser oscillation to occur and dissipate the energy. If the Q-factor is increased abruptly, the result is a large population inversion in a high-Q cavity, producing a high-power burst of light, a few nanoseconds to several hundred nanoseconds long, in which the energy is emitted. This rapid change is called Q-switching. Figure 5.16 shows the three basic variations on Q-switching. The first uses a rotating mirror or prism as the rear cavity mirror. Periodically the rotating mirror passes through the point where it is properly aligned with the output mirror, causing cavity Q to increase abruptly and producing an intense Q-switched pulse. The second is insertion of a modulator (usually electro-optic or acousto-optic devices) into the cavity, blocking off one of the cavity mirrors. The third variation is insertion of a non-linear lossy element into the cavity that becomes transparent once intra-cavity power exceeds a certain level.

Cavity dumping

The basic idea of cavity dumping is to couple laser energy directly out of the cavity without having it pass through an output mirror. In this case, both cavity mirrors are totally reflective and sustain a high circulating power within the laser cavity. The concept is illustrated in Figure 5.17 where a mirror pops up into the cavity and deflects a pulse with length close to the cavity round-trip time. In practice, cavity dumping is done with

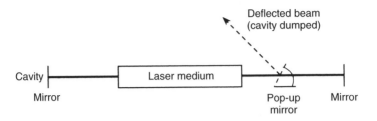

Figure 5.17   The basic concept of cavity dumping

other arrangements. Cavity dumping can be used with c.w. lasers which do not store energy in the upper excited levels and hence cannot be Q-switched. Cavity dumping of a c.w. laser can generate pulses of 10–50 ns at repetition rates of 0.5–5 MHz.

Modelocking

Modelocked pulses can be visualized as a group of photons clumped together and aligned in phase as they oscillate through the laser cavity. Each time they hit the partially transparent output mirror, part of the light escapes as an ultra-short pulse. The clump of photons then makes another round trip before the next pulse is emitted. Thus the pulses are separated by the cavity round-trip time $2L/c$, where L is the cavity length and c the speed of light. The physics is much more complex and well beyond the scope of this book. Modelocking can generate pulses in the picosecond regime. It requires a laser that oscillates in many longitudinal modes and will therefore not work for many gas lasers with narrow emission lines. However, it can be used with argon or krypton ion lasers, semiconductor lasers and dye lasers. The pulse length in inversely proportional to the laser's oscillating bandwidth, so dye lasers can generate very short pulses because of their broad gain bandwidths.

Q-switching, cavity dumping and modelocking may be used in combination.

5.4.8 Applications

Since its invention in 1960, the laser has found numerous applications. With regard to optical metrology, the He–Ne laser and also the argon ion laser have long been a standard choice for holography and interferometry. In recent years also single narrow-stripe semiconductor diode lasers (GaAlAs) operating in a single longitudinal mode have been applied in holography. For pulsed holography, the ruby laser and the neodymium laser are most common.

Most of the laser types described above have been used in numerous different types of applications. Below we mention some of the more successful application areas.

- *Compact-disc players.* Designed around the 780 nm beam of a few milliwatts from GaAlAs semiconductor diode laser. Nearly 20 million CD audio players are sold each year. The most successful commercial single applications of lasers.

- *Writing and recording.* In laser printers, the beam scans a photoconductive drum, discharging the electrostatic charge held by the surface at points where the beam is 'on'.

The resulting pattern is then printed on paper by a copierlike process. Lasers also can encode data as a series of dots on light-sensitive discs for computer data storage. He−Ne lasers initially were used for these applications, but today all but the most expensive high-speed systems use semiconductor lasers. For high-speed systems, He-Cd are the most common lasers.

- *Reading and scanning*. The biggest single application for He−Ne lasers has been in bar-code readers for supermarket checkout. The beam repetively scans a well-defined pattern. This application requires red light and good beam quality, so He−Ne lasers are likely to remain dominant in the near future.

- *Medicine*. Surgery is performed with the 10 μm line from CO_2 lasers with output power in the 50 W range. Neodymium lasers are standard tools for cataract surgery (The membrane inside the eye becomes cloudy). Unlike the CO_2, light from Nd lasers can be carried through optical fibres allowing use in endoscopes for e.g. gall-bladder surgery. Er-YAG is a promising laser for surgery.

- *Materials working*. The CO_2 laser is used for materials working, primarily cutting and welding of metals and non-metals and heat treating of metals. It is complemented by Nd-YAG which is better for drilling, spot-welding and laser marking.

- *Range finding*. The largest single use of Nd-YAG lasers probably is as military range finders and target designators. Their pulses pose a serious eye hazard to friendly troops and Erbium glass is an alternative.

- *Spectroscopy and analytical chemistry*. The ability to tune the dye laser wavelength and light emission to a narrow spectral bandwidth and the generation of ultra-short picosecond and femtosecond pulses makes dye lasers extremely useful in this research area. The tuneable Ti-sapphire laser is an alternative in the near-infrared regime.

- *Communications*. The InGaAsP semiconductor diode laser is used for fibre-optic communication systems, see Section 13.3. The 511 nm line of the copper vapour laser is suitable for underwater transmission.

5.4.9 The Coherence Length of a Laser

Although the laser light has a well-defined wavelength (or frequency), it has nevertheless a certain frequency spread. By spectral analysis of the light, it turns out that it consists of one or more distinct frequencies called resonator modes, separated by a frequency difference equal to

$$\Delta v = c/2L \tag{5.38}$$

where $c = $ the speed of light and L is the distance between the laser mirrors, i.e. the resonator length. Thus, the spectral distribution of the light from a multimode He−Ne laser is typically as given in Figure 5.15.

Now, assume that we apply a laser with two resonator modes as the light source in the Michelson interferometer in Figure 3.15. We then have two wave fields u_1 and u_2 with frequencies

$$v_1 = \frac{c}{\lambda_1} \tag{5.39a}$$

$$v_2 = \frac{c}{\lambda_2} = v_1 + \Delta v \tag{5.39b}$$

u_1 will interfere with itself but not with u_2, and vice versa, and the total intensity thus becomes (see Equation (3.36)):

$$I(l) = 2I\left(1 + \cos\frac{2\pi v_1 l}{c}\right) + 2I\left(1 + \cos\frac{2\pi v_2 l}{c}\right) \tag{5.40}$$

where l is the path length difference and where we have assumed u_1 and u_2 to have equal intensity I. Equation (5.40) can be rearranged to give

$$I(l) = 4I\left[1 + \cos\frac{2\pi(v_1 - v_2)l}{2c}\cos\frac{2\pi(v_1 + v_2)l}{2c}\right] \tag{5.41}$$

We see that the interference term is the same as that obtained with a light source with the mean frequency $(v_1 + v_2)/2$, but multiplied (modulated) by the factor

$$\cos\frac{2\pi\Delta v l}{2c}$$

This means that each time

$$\frac{\Delta v l}{2c} = \frac{n}{2} \quad \text{for} \quad n = 0, 1, 2, \ldots$$

i.e.

$$l = n\frac{c}{\Delta v} = n2L \tag{5.42}$$

the contrast or visibility of the interference pattern will have a maximum. The visibility and therefore the temporal degree of coherence (see Section 3.3) in this case is therefore equal to

$$|\gamma(\tau)| = \left|\cos\frac{\pi\Delta v l}{c}\right| = \left|\cos\frac{\pi l}{2L}\right| \tag{5.43}$$

The path-length difference corresponding to the first minimum in the visibility function Equation (5.43), is called the coherence length (see Section 3.3). This is illustrated in Figure 5.18. If more than two resonator modes had been taken into account, the result would have been essentially the same, i.e. the same locations of the minima, but with a more steeply varying visibility function $|\gamma(\tau)|$. From this we conclude that when applying a laser in interferometry, the path length difference should be nearly zero or an integer number of twice the resonator length.

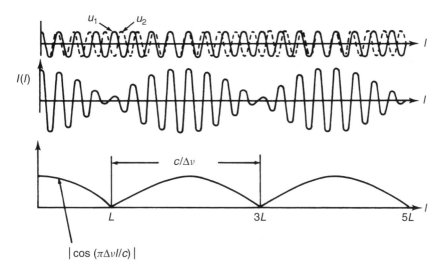

Figure 5.18 (a) Field amplitudes u_1 and u_2; (b) intensity distribution; and (c) temporal degree of coherence (visibility function) resulting from two resonator modes of a laser

5.5 HOLOGRAM RECORDING MEDIA

5.5.1 Silver Halide Emulsions

When making a hologram recording such as that illustrated in Figure 6.1, the photosensitive material must resolve the complicated intensity distribution resulting from the interference between the object and reference waves. The mean frequency in this pattern is given by the mean angle between these two waves. In practice this angle can be about $20°$ and greater. From Figure 3.5 we see that this angle corresponds to a spatial frequency of about 500 lines/mm. Normally a resolution of 1000 to 2000 lines/mm is desirable. This criterion is met by several silver halide emulsions, having a resolution of up to 5000 lines/mm. They also have a high sensitivity from about 1 to 10 μJ/cm^2.

In the description of hologram recording in Section 6.2, we assume a linear relation between the amplitude transmittance t and the exposure E of a hologram, where E is the intensity times the exposure time, i.e. the energy density per unit area. This assumption is not strictly true. A typical t–E curve for a film emulsion is shown in Figure 5.19. Another common transmission characteristic of film is the Hurter-Driffield curve, which is a density versus $\log E$ curve. The density D is defined as

$$D = \log \frac{1}{|t|^2}$$

Density is a common parameter for ordinary photography since the eye detects brightness differences on an approximately logarithmic scale. Photographic films are often characterized by the slope γ of the linear portion of the D–$\log E$ curve.

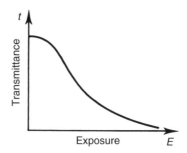

Figure 5.19 $t-E$ curve for holographic emulsion

In Section 6.5 we shall see that to obtain maximum diffraction efficiency of an amplitude hologram, the bias transmittance t_b should be equal to 0.5 which corresponds to $D = 0.6$. Because of the non-linearity of the $t-E$ curve it is advantageous to have t_b slightly lower (i.e. D slightly higher) than this value. In ordinary photography the bias density is centred at the linear portion of the $D-\log E$ curve which does not coincide with the linear portion of the $t-E$ curve. A properly recorded amplitude hologram therefore looks underexposed compared to a photograph.

Since the previous edition of this book, the use of silver halide emulsions in holography has gone down drastically. The holographic emulsions from Kodak, Agfa and Ilford are no longer on the market.

5.5.2 Thermoplastic Film

A thermoplastic, e.g. Staybelite Ester 10, is not photosensitive and must therefore be combined with a photoconductor in a film structure (Urbach and Meier 1966). The system is usually built up from a substrate of glass upon which is coated a conducting layer of, for example, tin oxide. On this is deposited a photoconductor such as polyvinylcarbazole sensitized with trinitro-9-fluorenone, and on top of this is deposited a thermoplastic layer. The recording technique consists of a number of steps, beginning with establishing a uniform electrostatic charge on the surface of the thermoplastic with a corona discharge assembly. This charge is capacitively divided between the photoconductor and the thermoplastic layers, and upon subsequent exposure the interference pattern causes the photoconductor to discharge in a spatial pattern dependent on the exposure. However, this does not cause any variation in the charge on the thermoplastic. This is accomplished by recharging the surface uniformly, which increases the charge in the illuminated areas. The thermoplastic is then heated to the softening temperature, allowing electrostatic forces to deform it so that it becomes thinner at illuminated areas and thicker elsewhere. Cooling quickly to room temperature, the deformations are frozen in, resulting in a hologram with thickness variations, i.e. a phase hologram. Reheating the thermoplastic to a higher temperature tends to restore it to its original state. Thus the material has a write–erase recycling capability. This is a quite complicated procedure, but complete camera units are commercially available, giving a hologram ready for reconstruction within 5 s of the exposure.

A peculiar feature of the thermoplastic film is that it has a band-limited spatial frequency response centred at about 1000–2000 lines/mm. The sensitivity is between 10 and 100 $\mu J/cm^2$.

5.5.3 Photopolymer Materials

A photopolymer recording material consists of three parts: a photopolymerizable mono-mer, an initiator system and a polymer. When exposed, a part of the monomer is polymer-ized. This gives rise to diffusion of monomer molecules from the regions of high intensity to the regions of low intensity. The material is then exposed to light of uniform intensity until the remaining monomer is polymerized. A difference in the refractive index within the material is then obtained.

Photoplymer materials can be used for recording phase holograms, where applications in mass-production of display holograms and optical elements are of main interest. Com-panies such as AT&T Bell Laboratories, du Pont and Hughes have produced photopolymer materials for recording holograms. Advantages are a low noise level and its suitability for applying dry processing techniques. The sensitivity is about 10 mJ/cm^2.

Of the other materials for hologram recording, we mention dichromated gelatin, pho-toresist, photochromic materials and ferroelectric crystals.

5.6 PHOTOELECTRIC DETECTORS

Optical detectors can be classified as in the block diagram of Figure 5.20. Here we classify photographic film, photopolymers, etc. as chemical detectors and they are described in Section 5.5. They do not give a signal output in the usual sense as do the other types, termed electronic detectors, which are divided into two branches: thermal and photon detectors. In thermal detectors, the absorption of light raises the temperature of the device and this in turn results in changes in some temperature-dependent parameter (e.g. electrical conductivity). Most thermal detectors are rather inefficient and quite slow, and because

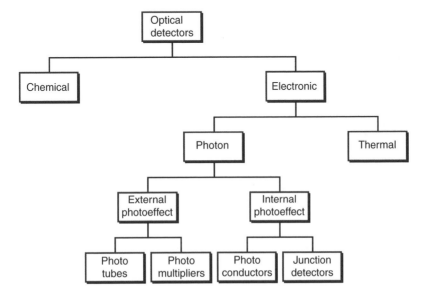

Figure 5.20 Optical detectors

of their relative unimportance in the field of optical metrology we name only some of the better known types – the thermocouple, the bolometer and pyroelectric detectors. The last type can be made with response times in the nanosecond region and with a wavelength response up to 100 μm. They have proved very useful as low cost, robust infrared detectors for fire detection and intruder alarms, for example.

The operation of photon (photoelectric) detectors is based on the photoeffect, in which the absorption of photons by some materials results directly in an electronic transition to a higher energy level and the generation of mobile charge carriers. Since the energy of a single photon is $E = h\nu = hc/\lambda$, photon detectors have a maximum wavelength beyond which they do not operate. A problem common to all photon detectors operated in the infrared is that the photon energy $h\nu$ becomes comparable with the average thermal energy ($\approx kT$) of the atoms in the detector itself. Therefore, most photon detectors operating above a wavelength of about 3 μm must be cooled to liquid nitrogen temperatures (77 K) or below.

The photoeffect takes two forms: external and internal. The former process involves photoelectric emission, in which the photo-generated electrons escape from the material (the photocathode) as free electrons with a maximum kinetic energy given by Einstein's photoelectric equation:

$$E_{\max} = h\nu - W$$

where the work function W is the energy difference between the vacuum and the Fermi levels of the material. Pure metals are rarely used as photocathodes since they have low quantum efficiencies ($\approx 0.1\%$) and high work functions ($W = 2.1$ eV for Cs) which makes them useful only in the visible and ultraviolet regions of the spectrum. However, semiconductors can operate with higher quantum efficiencies and lower work functions corresponding to wavelengths up to about 1.1 μm. Photoemissive devices usually take the form of vacuum tubes called phototubes. Electrons emitted from the photocathode travel to an electrode (the anode) which is kept at a higher electric potential. As a result, an electric current proportional to the photon flux incident on the cathode is created in the circuit. In a photomultiplier, the electrons are accelerated towards a series of electrodes (called dynodes) maintained at successively higher potentials. From the dynodes a cascade of electrons are emitted by secondary emission, resulting in an amplification by a factor as high as 10^7.

A microchannel plate consists of an array of channels (of internal diameter ≈ 10 μm) in a slab of insulating material (≈ 0.5 mm thick). Both faces of the plate are coated with thin metal films that act as electrodes, and the interior walls of each channel are made slightly conducting. Each channel thus acts like a miniature photomultiplier tube. On emerging from the channels, the electrons can generate light (and thereby an optical image) by striking a phosphor screen. The latter combination is called an image intensifier.

In the internal photoeffect, the photoexcited carriers (electrons and holes) remain within the material.

5.6.1 Photoconductors

Photoconductor detectors rely directly on the light-induced increase in the conductivity, an effect exhibited by almost all semiconductors (see Appendix E). The absorption of a photon results in the generation of a free electron excited from the valence band to

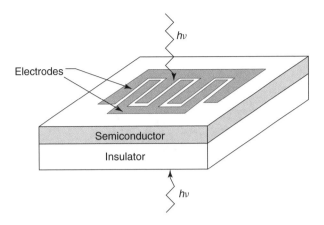

Figure 5.21 Photoconductor detector design

the conduction band, and a hole is generated in the valence band. An external voltage source connected to the material causes the electrons and holes to move, resulting in a detectable electric current. The detector operates by registering either the current (which is proportional to the photon flux) or the voltage drop across a series resistor. Unlike the quantum efficiency for the photoelectric effect, for example, the gain in a photoconductor may be larger than unity. The semiconducting material may take the form of a slab or a thin film. The contact electrodes are often placed on the same surface of the material in a geometry such as in Figure 5.21 to maximize the light transmission while minimizing the transit time. CdS and CdSe are both used for low cost visible radiation sensors in, for example, light meters for cameras. They usually have high gains ($10^3 - 10^4$) but poor response times (≈ 50 ms). Other photoconductor materials for infrared detectors are PbS, InSb and HgCdTe.

5.6.2 Photodiodes

The photodiode detector is a p-n junction structure where photons absorbed in the depletion layer generate electrons and holes which are subject to the local electric field within that layer. Because of this field, the two carriers drift in opposite directions and an electric current is induced in the external circuit. Photodiodes have been fabricated from many of the semiconductor materials listed in Table E.2, as well as from ternary and quarternary compound semiconductors such as InGaAs and InGaAsP. Devices are often constructed in such a way that the light impinges normally on the p-n junction instead of parallel to it. A typical construction is seen in Figure 5.22. There are three classical modes of photodiode operation: open circuit (photovoltaic), short-circuit, and reverse biased (photoconductive). The usual i–V characteristic is seen in Figure E.2 (Appendix E). With increasing photon flux, the i–V characteristics move downwards as in Figure 5.23. In the photovoltaic mode, a voltage V_p is produced across the device that increases as a logarithmic function of the incident light irradiance. This mode is used, for example, in solar cells. In the photoconductive mode, a relatively large reverse bias (≈ 10 V or more) is applied across the diode. Here the circuit current is directly proportional to the incident

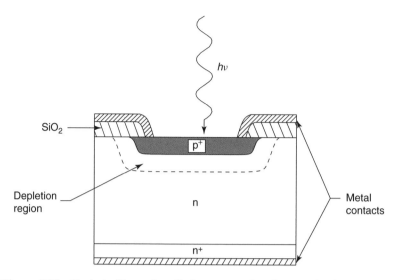

Figure 5.22   Typical silicon photodiode structure for photoconductive operation

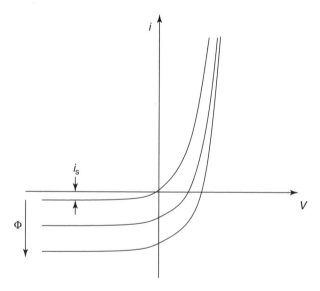

Figure 5.23   Current–voltage characteristics of a p-n junction under various levels of illumination

light irradiance. Other advantages of the photoconductive mode are faster response, better stability and greater dynamic range.

A strong reverse bias increases the width of the depletion layer, resulting in a larger photosensitive area, reduced junction capacitance and improved response time. A structure that results in a good long-wavelength response with modest reverse bias levels is the so-called pin (or PIN) structure. This is a p-n junction with an intrinsic layer sandwiched between the p and n layers. Here only a few volts of reverse bias are needed to cause the depletion layer to extend all the way through the n region.

By replacing the p-type (or n-type) layer in the p-n junction by a thin metallic film, we get a metal–semiconductor photodiode (also called a Schottky-barrier photodiode). There are a number of reasons why Schottky-barrier photodiodes are useful:

(1) Not all semiconductors can be prepared in both p-type and n-type forms.

(2) In p-n junctions one gets a substantial surface recombination and thereby a reduced efficiency. The metal–semiconductor junction has a depletion layer present immediately at the surface, thus eliminating this effect.

(3) The low resistance of the metal decreases the RC time constant thereby increasing the speed. Response times in the picosecond regime ($\approx$100 GHz bandwidths) are readily available.

Of particular interest is the Schottky-barrier photodiode of PtSi on p-type Si which is sensitive to wavelengths from the near ultraviolet to about 6 μm in the infrared (it must be cooled to 77 K). When used as elements in a CCD (see Section 5.7) one gets a device with multispectral imaging capabilities.

Finally it should be mentioned that with sufficiently large reverse bias, the electrons and holes may acquire sufficient energy to liberate more electrons and holes. Devices in which this internal amplification process occurs are known as avalanche photodiodes (APDs).

5.7 THE CCD CAMERA

5.7.1 Operating Principles

Until the mid 1960s, electronic devices for the pick-up of optical images were in the form of vacuum-type camera tubes. During the 1960s solid-state arrays with individual photoconductor elements connected to X–Y conductors and sequentially activated by voltages from thin-film shift registers were developed. The resulting images were however severely limited by response non-uniformities and other form of spatial noise associated with the X–Y readout techniques.

Workers at the Bell Telephone Laboratories (Boyle and Smith 1970) presented a new semiconductor device concept based on the manipulation of charge packets rather than the modulation of electric currents. Below we give a brief description of this concept. For a more thorough description, the article by Barbe and Campana (1977) is recommended.

A CCD is essentially a series of metal oxide semiconductor (MOS) capacitors. Figure 5.24 shows a simplified sketch of one of the capacitors. A semiconductor substrate of p-type silicon is covered with a thin layer of insulating silicon oxide which insulates the Si substrate from the metal electrode. When a positive voltage is applied between the electrode and the Si substrate, the minority carriers (holes in p-type Si) will be repelled from the interface between semiconductor and insulator, creating a region free of mobile carriers directly underneath the electrode. This region is known as the depletion region and has a thickness of a few micrometres. The metal electrodes (usually made of polycrystalline silicon) are transparent for wavelengths larger than about 400 nm.

If an incident photon has an energy larger than the bandgap in Si, it can create an electron-hole pair in the semiconductor. When this creation occurs in or near the depletion region, the photon generated electron is attracted towards the potential well which is

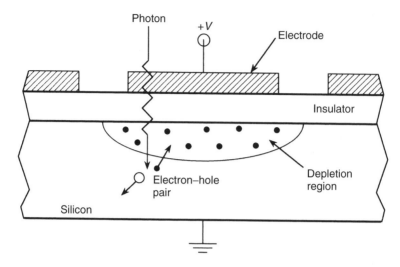

Figure 5.24   Simplified sketch of an MOS capacitor

formed under the positive charged electrode. In this way, a charge packet is formed consisting of photon-electrons which were created in the vicinity of a specific electrode.

Figure 5.25 shows how the charge packets are transferred from one electrode to the next by proper clocking of the potentials of the electrodes. In a device having a planar oxide and uniform substrate doping, at least three phases are required for unidirectional charge transfer, i.e. a barrier is maintained behind the charge packet while a deeper well is formed in front of the packet. The clocking diagrams are shown in Figure 5.25(b). At $t = t_1$, charge resides in the wells under the ϕ_1 electrodes. At $t = t_2$, the potential on ϕ_2 is made positive forming wells under the ϕ_2 electrodes. Charge will then flow from the ϕ_1 wells into the ϕ_2 wells. At $t = t_3$, the potential on the ϕ_1 electrodes is reduced to a low value, and the remaining charge in the ϕ_1 wells will be pushed into the ϕ_2 wells. This sequence repeats with the result that the charge configuration moves from one cell to the next every clock period.

To allow the device to be clocked with two phases, potential barriers between electrodes must be built in. This is done either by forming alternate thin and thick oxide insulators, the so-called stepped oxide barrier, or the implanted barrier technique by non-uniform doping of the substrate.

There are two different ways in which CCDs are organized when applied as an imaging sensor. In the following description standard TV rates are assumed, i.e. 1/25 s, European standard (CCIR) or 1/30 s, American standard (RS-170) frame time. We also assume two-phase clocking of the electrodes. Figure 5.26(a) shows the organization of the so-called frame-transfer structure. This sensor is divided into two identical areas, the image section and a masked storage section. A TV-frame is divided into two fields A and B, see Section 5.9. Field A is formed by collecting photoelectrons under the odd rows of electrodes for 1/50 s (1/60 s for RS-170). This charge configuration is shifted into the shielded storage register in a time that is short (several MHz) compared with the integration (exposure) time. Field A is then read out a line at a time while field B is being formed by collecting photoelectrons under the even electrodes.

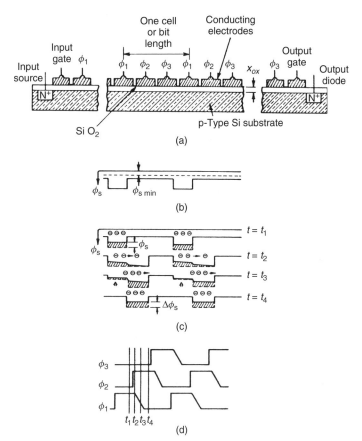

Figure 5.25 Three-phase CCD: (a) cross-sectional view showing input section, transfer section, and output section; a primitive electrode structure having unprotected gaps is shown for simplicity; (b) surface-potential profile showing potential wells under the ϕ_1 electrodes; (c) surface-potential profiles showing progression of charge transfer during one clock period; and (d) clocking waveforms used to drive the CCD during transfer. (From Barbe 1975. Reproduced by permission of IEEE)

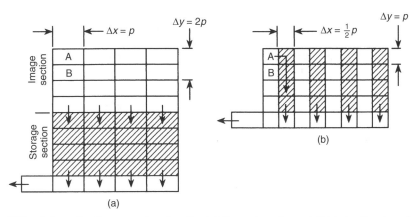

Figure 5.26 Schematic diagrams showing the (a) frame-transfer and (b) inter-line transfer array organizations

Figure 5.26(b) shows the organization of the so-called interline transfer structure. Here the shielded vertical readout registers are interdigitated with the photosensitive column. Potential wells are formed in the photosensitive regions by applying voltages to the vertical polycrystalline silicon (polysilicon) stripes. The horizontal polysilicon stripes are used to clock the vertical shielded register. Because the integrating cells and shift-out cells are separate, the effective integration time for both fields A and B is 1/25 s. The operation is as follows. After collecting photoelectrons in field A for 1/25 s, the charge configuration is shifted into the shielded registers and down, a line at a time, into the horizontal output register. When field A has been completely read out (1/50 s), field B is shifted into the shielded registers and out. Note that the effective integration time for the interline transfer structure is twice that of the frame-transfer structure because the integration in the interline transfer structure is performed in sites separate from the transport registers.

5.7.2 Responsivity

A very important parameter for an imaging sensor used for optical metrology is its responsitivity R, particularly the responsitivity as a function of the spatial frequency $R(f)$ of the imaged scene. This can be formulated as

$$R(f) = R(0) \prod_i (MTF)_i \tag{5.44}$$

where $R(0)$ is the responsitivity at zero spatial frequency and the other term is the product of all of the modulation transfer function (MTF, see Section 4.6.2) factors that affect the frequency response of the chip. These factors are: (1) the loss of frequency response due to the geometry of the integrating cell ($\text{MTF}_{\text{integ}}$), (2) the loss of frequency response due to transfer inefficiency ($\text{MTF}_{\text{transfer}}$), and (3) the loss of frequency response due to the diffusion of charge between photon absorption and photoelectron collection (MTF_{diff}). We here will consider three chip designs: The front-illuminated interline transfer CCD (FIIT), the front illuminated frame transfer CCD (FIFT) and the back-illuminated frame transfer CCD (BIFT). For FIIT approximately one-half of the chip area is photosensitive because the other half is occupied by the vertical transport registers. When used in standard TV interlaced mode, however, the integration time for FIFT and BIFT is half of that for FIIT. The efficiency with which photons are absorbed and the resulting photoelectrons are collected in the integrating cells are twice as high for BIFT than for both FIIT and FIFT when averaged over the 0.4–1.0 μm wavelength band. In conclusion therefore, R(0) is twice as high for BIFT than for FIIT and FIFT.

In modern CCD chips, $\text{MTF}_{\text{transfer}}$ and MTF_{diff} have negligible effects compared to $\text{MTF}_{\text{integ}}$. The latter is, however, a fundamental effect on the chip responsitivity and is related to the finite size of the integrating cells.

To calculate $\text{MTF}_{\text{integ}}$ it is sufficient to find the response to a sinusoidal grating of frequency f

$$H_1 = H_0[1 + m \cos 2\pi f x] \tag{5.45}$$

When sensed by an array of cells of width Δx and inter-cell distance p, the output charge pattern becomes

$$H_2 = \frac{1}{p} \int_{x-\Delta x/2}^{x+\Delta x/2} H_1 \, dx = \frac{H_0}{p} \int_{x-\Delta x/2}^{x+\Delta x/2} [1 + m \cos 2\pi f x] \, dx$$

$$= \frac{H_0 \Delta x}{p} \left[1 + m \frac{\sin \pi \Delta x}{\pi f \Delta x} \cos 2\pi f x \right] \tag{5.46}$$

Therefore the MTF of the integrating process is (cf. Section 4.6.2, Equation (4.56))

$$MTF_{\text{integ}} = \left| \frac{\sin \pi f \Delta x}{\pi f \Delta x} \right| \tag{5.47}$$

By introducing the so-called Nyquist frequency $f_{\text{n}} = 1/2p$, this can be written

$$MTF_{\text{integ}} = \left| \frac{\sin \dfrac{\pi}{2} \dfrac{f}{f_{\text{n}}} \dfrac{\Delta x}{p}}{\dfrac{\pi}{2} \dfrac{f}{f_{\text{n}}} \dfrac{\Delta x}{p}} \right| \tag{5.48}$$

Figure 5.27 shows the responsivity as a function of spatial frequency for FIIT, FIFT and BIFT in the horizontal and vertical directions.

5.8 SAMPLING

5.8.1 Ideal Sampling

Consider a one-dimensional function $f(x)$ which might represent e.g. the irradiance distribution along a TV-line on a CCD camera. To sample this function means to find the values of f at regular intervals, i.e. $f(np)$ where $n = 0, 1, 2, \ldots$ and p is a constant called the sampling period (Goodman 1968). This is equivalent to multiplying f by a comb function (Equation (B.16), Appendix B.2) to get the sampled function f_{s},

$$f_{\text{s}} = \text{comb}\left(\frac{x}{p}\right) f(x) \tag{5.49}$$

The spectrum F_{s} of f_{s} is given by its Fourier transform

$$F_{\text{s}}(f_x) = \mathscr{F}\left\{ \text{comb}\left(\frac{x}{p}\right) f(x) \right\} = \mathscr{F}\left\{ \text{comb}\left(\frac{x}{p}\right) \right\} \otimes F(f_x) \tag{5.50}$$

where the last equality follows from the convolution theorem and $F(f_x)$ is the spectrum of $f(x)$. Now we have that

$$\mathscr{F}\left\{ \text{comb}\left(\frac{x}{p}\right) \right\} = \text{pcomb}(pf_x) = \sum_{n=-\infty}^{\infty} \delta\left(f_x - \frac{n}{p}\right) \tag{5.51}$$

It follows that the spectrum of the sampled function is given by

$$F_{\text{s}}(f_x) = \sum_{n=-\infty}^{\infty} F\left(f_x - \frac{n}{p}\right) \tag{5.52}$$

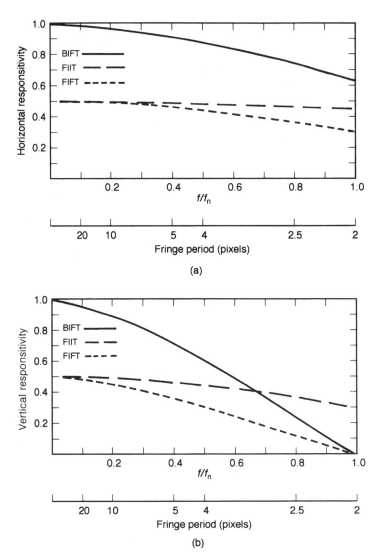

Figure 5.27   Responsivity for back-illuminated FT (BIFT), front-illuminated FT (FIFT) and IT (FIIT) arrays: (a) versus normalized horizontal spatial frequency and fringe period and (b) versus normalized vertical spatial frequency and fringe period. The relations between Δx, Δy and p are as in Figure 5.26, i.e. standard video signal transfer is assumed

Evidently the spectrum of f_s can be found simply by erecting the spectrum of f about each point n/p along the f_x-axis as shown in Figure 5.28c).

Now assume (as in Figure 5.28(a)) that the spectrum F of f vanishes outside some interval $[-W, W]$. A function whose transform has this property for any finite value of W is called a band-limited function. From Figure 5.28(c) we see that if

$$\frac{1}{p} \geq 2W \tag{5.53}$$

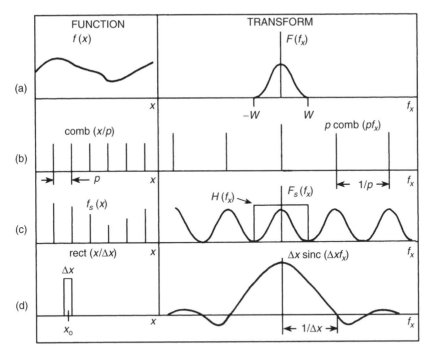

Figure 5.28 Sampling process in the x-domain and the spatial frequency domain

the spectra $F(f_x - n/p)$ constituting the spectrum of f_s do not overlap. If inequality (5.53) is fulfilled, we therefore can separate the $n = 0$ term of F_s from all the other terms by multiplying it by a filter function

$$H(f_x) = \text{rect}\left(\frac{f_x}{2W}\right) \qquad (5.54)$$

We see therefore that F is recovered from F_s in that

$$F_s(f_x)\,\text{rect}\left(\frac{f_x}{2W}\right) \equiv F(f_x) \qquad (5.55)$$

The inverse Fourier transform of Equation (5.55) yields

$$\mathscr{F}^{-1}\{F(f_x)\} = f(x) = \mathscr{F}^{-1}\left\{F_s(f_x)\,\text{rect}\left(\frac{f_x}{2W}\right)\right\}$$

$$= \mathscr{F}^{-1}\{F_s(f_x)\} \otimes \mathscr{F}^{-1}\left\{\text{rect}\left(\frac{f_x}{2W}\right)\right\}$$

$$= f_s(x) \otimes 2W \text{ sinc } (2Wx) = f(x)\text{comb}\left(\frac{x}{p}\right) \otimes 2W \text{ sinc } (2Wx) \quad (5.56)$$

Noting that

$$f(x)\text{comb}\left(\frac{x}{p}\right) = p \sum_{n=-\infty}^{\infty} f(np)\delta(x - np) \tag{5.57}$$

Equation (5.56) becomes

$$f(x) = 2pW \sum_{n=-\infty}^{\infty} f(np) \text{ sinc } [2W(x - np)] \tag{5.58}$$

Finally, when the sampling interval p is taken to have its maximum allowable value $1/2W$, we have that

$$f(x) = \sum_{n=-\infty}^{\infty} f\left(\frac{n}{2W}\right) \text{sinc} \left[2W\left(x - \frac{n}{2W}\right)\right] \tag{5.59}$$

Equation (5.59) represents a fundamental result which we refer to as the Whittaker–Shannon sampling theorem. It implies that exact recovery of a bandlimited function can be achieved from an appropriately spaced array of its sampled values, the recovery is accomplished by injecting, at each sample point, an interpolation function consisting of a sinc function.

It should be noted that other choices of the filter function $H(f_x)$ than that given in Equation (5.54) is possible as long as $H(f_x)$ passes the $n = 0$ term of F_s and excludes other terms. In fact it is a multitude of choices which will result in alternative sampling theorems.

5.8.2 Non-Ideal Sampling

The sampling of a function by discrete points is an idealized situation. In reality the sampling pulses always have finite width.

Consider Figure 5.28(d) with such a sampling pulse of width Δx centred at x_0. If this pulse represents the cell in a CCD chip, the irradiance $f(x)$ will be integrated over this cell. The charge at x_0 will therefore be given by

$$f_i(x_0) = \frac{1}{\Delta x} \int_{x_0-\Delta x/2}^{x_0+\Delta x/2} f(x) \, dx \tag{5.60}$$

If we introduce the rectangle function (Equation (B.13), Appendix B.2), this integral can be written as

$$f_i(x_0) = \frac{1}{\Delta x} \int_{-\infty}^{\infty} \text{rect} \left(\frac{x - x_0}{\Delta x}\right) f(x) \, dx \tag{5.61}$$

Since rect $(-x) = $ rect (x) (it is symmetric), we have

$$f_i(x) = \frac{1}{\Delta x} \int_{-\infty}^{\infty} \text{rect} \left(\frac{x - \xi}{\Delta x}\right) f(\xi) \, d\xi \tag{5.62}$$

where we also have changed variables. Equation (5.62) is recognized as a convolution integral, i.e.

$$f_i(x) = \frac{1}{\Delta x} \text{rect} \left(\frac{x}{\Delta x} \right) \otimes f(x) \tag{5.63}$$

This function is then sampled in the same way as for ideal sampling:

$$f_{sN} = f_i(x)\text{comb} \left(\frac{x}{p} \right) = \left[\frac{1}{\Delta x} \text{rect} \left(\frac{x}{\Delta x} \right) \otimes f(x) \right] \text{comb} \left(\frac{x}{p} \right) \tag{5.64}$$

The spectrum now becomes

$$F_{sN} = \mathscr{F} \left\{ \frac{1}{\Delta x} \text{rect} \left(\frac{x}{\Delta x} \right) \otimes f(x) \right\} \otimes \mathscr{F} \left\{ \text{comb} \left(\frac{x}{p} \right) \right\}$$

$$= \text{sinc} \, (\Delta x f_x) F(f_x) \otimes p\text{comb}(pf_x) = \sum_{n=-\infty}^{\infty} F_N \left(f_x - \frac{n}{p} \right) \tag{5.65}$$

where

$$F_N = \text{sinc} \, (\Delta x f_x) F(f_x) \tag{5.66}$$

Apart from $F(f_x)$ being multiplied by a sinc function, this is the same result as for ideal sampling. This does not matter so much as long as f_x is well below $1/\Delta x$, the first zero of the sinc function: see Figure 5.28(d).

5.8.3 Aliasing

If inequality (5.53) is not fulfilled, the repeated spectra F_N will overlap each other as seen in Figure 5.29. Since natural scenes are not band limited, the spectra will always overlap unless $F(f_x)$ is prefiltered. Overlapping of the spectra causes frequencies higher than the Nyquist limit ($f_n = 2/p$) to appear in the passband ($-f_n \leq f_x \leq f_n$) as lower-frequency components – thus the term 'aliasing'. Thus, for example the frequency $1.5 f_n$ in F_N for $n = 1$ would give a response in F_N for $n = 0$ at $0.5 f_n$.

An example of aliasing is shown in Figure 5.30. Here vertical bars of different spacings are imaged onto a 100×100 element interline transfer CCD chip. The Nyquist frequency in this case was 12.3 cycles/mm. Thus only the top row of Figure 5.30 represents the true imagery. The remaining six views are moirè patterns produced by the interaction of the CCD structure and the bars of the test chart.

5.9 SIGNAL TRANSFER

Most electronic cameras are equipped with a video output signal. This is an analog signal containing the image data (Grob 1984). To guide the scanning beam of the TV-monitor, this video signal also contains some timing information, see Figure 5.31. The timing information is transmitted between each horizontal scan line and is called the horizontal

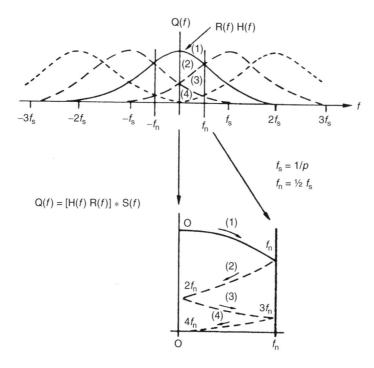

$$Q(f) = [H(f)\,R(f)] * S(f)$$

$$f_s = 1/p$$
$$f_n = \tfrac{1}{2}\,f_s$$

Figure 5.29 Spatial frequency domain analysis of the sampling process. Branch (1) is the desired response, and branches (2), (3) and (4) represent response to $f > f_n$ aliased into the passband. (From Barbe, D.F. and Campana, S.B., (1976) Aliasing and MTF effects in photosensor arrays, in P.G. Jespers, F.v.de Wiele and M.H. White (eds), Solid State Imaging, Noordhoff, Leiden. Reproduced by permission of D.F. Barbe)

blanking period. In this period the electron beam intensity is set to zero to avoid disturbing the picture during flyback to the next TV line. The horizontal blanking period consists of a horizontal (hsync) pulse which is used for synchronization and a back porch. In the AD converter of the frame grabber (see Section 10.2), this porch is used for adjusting the grey level to a known value (black). All video signals use interlaced raster technique. This means that the horizontal scan lines of a TV frame are divided into two fields. The even field consists of all the even-numbered lines in the frame, starting with line zero. The odd fields consists of the odd-numbered scan lines. This is done to avoid flicker on the TV screen. The even and odd fields are separated by a vertical blanking period.

A complete scanning pattern is shown in Figure 5.32, where the corresponding horizontal and vertical sawtooth waveforms illustrate odd-line interlaced scanning. A total of 21 lines in the frame is used for simplicity, instead of 525 (American standard). The 21 lines are interlaced with two fields per frame. Of the 10.5 lines in a field, we can assume that 1 line is scanned during vertical retrace to have a convenient vertical flyback time. So 9.5 lines are scanned during vertical trace in each field. Therefore in the entire frame, 19 lines are scanned during vertical trace, so 2 lines are lost in the vertical retrace lines. Starting at point A in the upper left corner, the beam scans the first line from left to right and retraces to the left to begin scanning the third line in the frame. Then the beam scans all succeeding odd lines until it reaches point B at the bottom, when vertical flyback

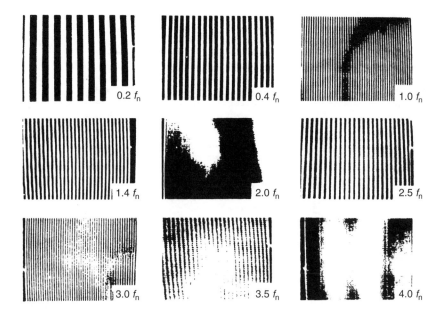

Figure 5.30 Bar-pattern imagery produced by the 100×100 element CCD array. The fundamental frequencies are given relative to the Nyquist frequency, f_n. (From Barbe, D.F. and Campana, S.B., (1976) Aliasing and MTF effects in photosensor arrays, in P.G. Jespers, F.v.de Wiele and M.H. White (eds), Solid State Imaging, Noordhoff, Leiden. Reproduced by permission of D.F. Barbe)

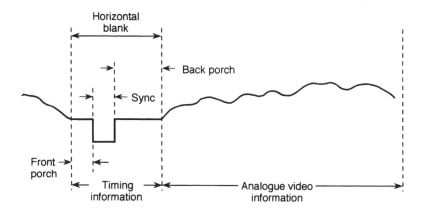

Figure 5.31 The analog video signal with timing information

begins. Note that this vertical retrace begins in the middle of a horizontal line. During this vertical retrace the scanning beam is brought to point C, which is separated from A by exactly one half-line so that the scanning of the second field can begin. In commercial TV broadcasting, some of the 'lost' lines in vertical retrace are digitally encoded to carry data for reproduction of full pages of alphanumeric characters for videotext. This requires a decoder at the receiver to gate out the specific lines and process the digital signal.

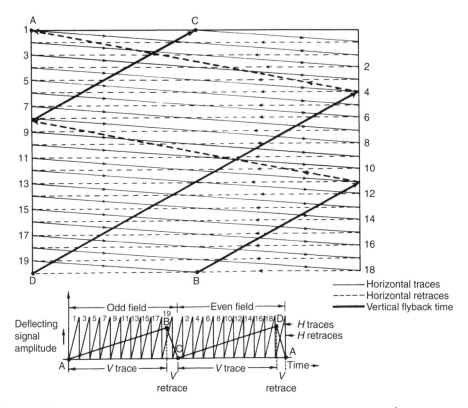

Figure 5.32 A sample scanning pattern for 21 interlaced lines per frame and $10\frac{1}{2}$ lines per field. The corresponding H and V sawtooth deflection waveforms are shown below pattern. Starting at point A, the scanning motion continues through B, C and D and back to A again. ((From Gro, B. (1984) Basic Television and Video Systems. (5th edn), McGraw-Hill, N.Y. Reproduced by permission of McGraw-Hill Inc., N.Y)

Table 5.3 Three different television standard video systems distinguished in acquisition speed, frame size and grey-value resolution

Pixel frequency	Slow scan <5 MHz	Standard video 7.5–15 MHz	HDTV >15 MHz
Frame size in pixels	1024 × 1024 4096 × 4096	256 × 256 512 × 512 780 × 540	1280 × 1024
Grey-value resolution	6–16 bit	6–10 bit	6–8 bit

Unfortunately, the video signal has several standards. Analogue input comes in three different forms, essentially distinguished by the corresponding 8-bit conversion frequency: slow-scan, standard video and high definition video (HDTV). Table 5.3 shows the different frequencies, frame grabber sizes and grey value resolutions.

The standard video norms for black and white are RS-170 (used in North and South America and Japan) and CCIR (used in Europe). A detailed overview of these systems is found in Table 5.4.

Table 5.4 Comparison of the black and white video systems RS-170 and CCIR

	RS-170	CCIR
Frame rate/field rate	30/60 Hz	25/50 Hz
Number of lines	525	625
Number of active lines	480	576
Field time	16 2/3 ms	20 ms
Time per line	63.49 µs	64 µs
Active line period	52.5 µs	52 µs
Nominal video bandwidth	4.5 MHz	5.5 MHz
Resolution	472	572
Aspect ratio	4:3	4:3

Standard video uses interlacing to avoid flicker in human perception of video images. This is not demanded in machine vision and video metrology, but will in many cases be a drawback, especially when monitoring high-speed phenomena. When using non-interlaced video, the resolution in the vertical direction of the frame transfer CCD cameras is also doubled, see Section 5.7.2. Specialized camera manufacturers such as EG & G, Fairchild and Dalsa offer matrix cameras with capabilities of up to 6000×6000 pixels for non-standard video transfer. The scan rate is driven by an external clock and can be selected by the operator. Such cameras are of course superior to standard video cameras with regard to resolution, but the amount of data and therefore the processing time increases dramatically, from $512 \times 512 \times 8 = 200$ kbyte to 288 Mbyte for a single image.

As mentioned in Section 10.2, for frame grabbers receiving composite video, line jitter is a problem especially when making measurements with sub-pixel accuracy. This problem is avoided when the signal is transferred digitally from the camera to the frame grabber and not via an analogue video signal. This is achieved by placing the A/D converter inside the camera and transferring the signal via a data cable. Such cameras are manufactured by Kodak (Videk Megaplus) and Cohu. Most frame grabbers today accept such signals.

PROBLEMS

5.1 Assume the Sun to be a 6000 K blackbody source and that its diameter subtends an angle $\alpha = 9.3$ mrad at the Earth.

(a) Find the wavelength λ_m corresponding to the maximum solar spectral radiant exitance M_λ.

(b) Find M_λ of the Sun's surface at this wavelength.

(c) What is the spectral radiance L?
The area of the solar disc dA is given by

$$dA = \pi \alpha^2 s^2 / 4$$

where s is the Earth-Sun distance.

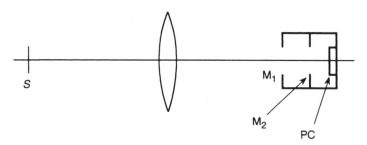

Figure P5.1

(d) Calculate the radiant flux Φ on an area of 1 m^2 (i.e. the irradiance) on the top of the earth's atmosphere at the wavelength λ_m.

5.2 A Lambert source S with radiance $L_e = 10^3$ W/m^2, 1 cm in diameter, is located 45 cm from a thin lens of 30 cm focal length and 10 cm diameter, see Figure P5.1. A detector is located 120 cm to the right of the lens. It consists of two masks M_1 and M_2 each with 0.1 mm diameter pinholes on axis 1 cm apart, followed closely by a large photocell PC with a sensitivity of 1 A/W. Neglect losses in the lens and find the value of the photocurrent coming from the cell.

5.3 The responsitivity of the integrating process, cf. Equations (5.44), (5.48), we write as

$$R(f) = \frac{\Delta x}{p} \frac{\sin\left(\dfrac{\pi}{2}\dfrac{f}{f_n}\dfrac{\Delta x}{p}\right)}{\dfrac{\pi}{2}\dfrac{f}{f_n}\dfrac{\Delta x}{p}} = r\,\frac{\sin\left(\dfrac{\pi}{2}qr\right)}{\dfrac{\pi}{2}qr} = R(q,r)$$

where $q = f/f_n$, $r = \Delta x/p$. Assume that r can be varied between 0 and 1. The lowest possible relative frequency q for which $R(q,r) = R(q,1)$, we denote q_e.

(a) Find q_e expressed by r. Can $R(q,r) > R(q,1)$ for $q < 1$?

(b) Find q_e for $r = 0.5$.

5.4 Light sources producing uniform beams is an idealization. In reality the intensity varies across the transverse plane. A particularly important transverse pattern is the Gaussian distribution, radiated by most gas lasers and some specially designed laser diodes and also single-mode fibres. The Gaussian field distribution is given as

$$U = U_o \exp\left[-\frac{r^2}{w^2}\right]$$

Given that

$$\int_{-\infty}^{\infty} e^{-u^2}\, du = \sqrt{\pi}$$

(a) Calculate the one-dimensional Fraunhofer diffraction pattern of this distribution. An accepted definition of the radius of the spot size of this beam is the distance at which the intensity has dropped to $1/e^2 = 0.135$.

(b) Calculate the spot size in the focal plane of a lens (with diameter $D > w$) of focal length f. Compare the resulting pattern with the Airy pattern.

5.5 Consider a laser oscillating in three resonator modes of finite width Δv_h. The translated spectral distribution function $S(v)$ then can be written as

$$S(v) = G(v) \otimes \tfrac{1}{2}[\delta(v) + \tfrac{1}{2}\delta(v - \Delta v) + \tfrac{1}{2}\delta(v + \Delta v)]$$

where each mode has a Gaussian distribution

$$G(v) = \sqrt{\left(\frac{2}{\pi}\right)} \frac{1}{\Delta v_h} \exp\left[-2\left(\frac{v}{\Delta v_h}\right)^2\right]$$

and where Δv is the resonator frequency spacing.

Calculate the temporal degree of coherence $\gamma(\tau)$. Assume $\Delta v = 3\Delta v_h$ and compare the result with that found in Section 5.4.9.

5.6 To look more closely into the sampling process, consider the simple input signal

$$f(x) = \tfrac{1}{2}(1 + \cos 2\pi f_o x)$$

(a) What is the bandwidth $2W$ of this signal?

This signal is ideally sampled at a sampling interval of p (i.e a sampling frequency $f_s = 1/p$).

(b) Find the spectrum of the sampled signal. Assume that we apply a filter function $H(f_x) = \text{rect } (f_x/2W)$ with the value of W found in (a).

Find the resulting spectrum and the output signal in the following cases: (i) $f_s > 2f_o$, (ii) $f_s = 1.5f_o$, (iii) $f_s = 0.4f_o$.

6
Holography

6.1 INTRODUCTION

Holography is the synthesis of interference and diffraction. In recording a hologram, two waves interfere to form an interference pattern on the recording medium. When reconstructing the hologram, the reconstructing wave is diffracted by the hologram. When looking at the reconstruction of a 3-D object, it is like looking at the real object. It is therefore said that: 'A photograph tells more than a thousand words and a hologram tells more than a thousand photographs'.

Although holography requires coherent light, it was invented by Gabor back in 1948, more than a decade before the invention of the laser. By means of holography an original wave field can be reconstructed at a later time at a different location. This technique therefore has many potential applications. In this book we concentrate on the technique of holographic interferometry. Because of the above-mentioned properties, we shall see that holographic interferometry has many advantages compared to standard interferometry.

6.2 THE HOLOGRAPHIC PROCESS

Figure 6.1(a) shows a typical holography set-up. Here the light beam from a laser is split in two by means of a beamsplitter. One of the partial waves is directed onto the object by a mirror and spread to illuminate the whole object by means of a microscope objective. The object scatters the light in all directions, and some of it impinges onto the hologram plate. This wave is called the object wave. The other partial wave is reflected directly onto the hologram plate. This wave is called the reference wave. In the figure this wave is collimated by means of a microscope objective and a lens. This is not essential, but it is important that the reference wave constitutes a uniform illumination of the hologram plate. The hologram plate must be a light-sensitive medium, e.g. a silver halide film plate with high resolution. We now consider the mathematical description of this process in more detail. For more comprehensive treatments, see Collier *et al.* (1971), Smith (1969), Caulfield (1979) and Hariharan (1984).

Let the object and reference waves in the plane of the hologram be described by the field amplitudes u_o and u respectively. These two waves will interfere, resulting in an intensity distribution in the hologram plane given by

$$I = |u + u_o|^2 = |u|^2 + |u_o|^2 + u_o^* u + u_o u^* \tag{6.1}$$

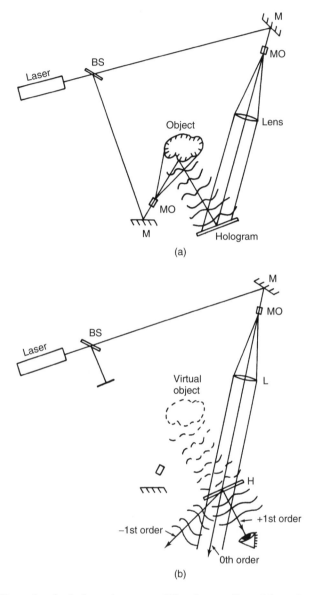

Figure 6.1   (a) Example of a holography set-up. BS = beamsplitter, M = mirrors, MO = microscope objectives and (b) Reconstruction of the hologram

We now expose the film plate to this intensity distribution until it gets a suitable blackening. Then it is removed from the plate holder and developed. We now have a hologram. The process so far is called a hologram recording.

This hologram has an amplitude transmittance t which is proportional to the intensity distribution given by Equation (6.1). This means that

$$t = \alpha I = \alpha |u|^2 + \alpha |u_o|^2 + \alpha u_o^* + \alpha u_o u^* = t_1 + t_2 + t_3 + t_4 \qquad (6.2)$$

We then replace the hologram back in the holder in the same position as in the recording. We block the object wave and illuminate the hologram with the reference wave which is now termed the reconstruction wave (see Figure 6.1(b)). The amplitude distribution u_a just behind the hologram then becomes equal to the field amplitude of the reconstruction wave multiplied by the amplitude transmittance of the hologram, i.e.

$$u_a = t \cdot u = \alpha \lfloor |u|^2 + |u_0|^2 \rfloor u + \alpha (uu)u_o^* + \alpha |u|^2 u_o \qquad (6.3)$$

As mentioned above, the reference (reconstruction) wave is a wave of uniform intensity. The quantity $|u|^2$ is therefore a constant and the last term of Equation (6.3) thus becomes (apart from a constant) identical to the original object wave u_o. We therefore have been able to reconstruct the object wave, maintaining its original phase and relative amplitude distribution. The consequence is that, by looking through the hologram in the direction of the object, we will observe the object in its three-dimensional nature even though the physical object has been removed. Therefore this reconstructed wave is also called the virtual wave.

The other two terms of Equation (6.3) represent waves propagating in the directions indicated in Figure 6.1(b). In fact, a hologram can be regarded as a very complicated grating where the first term of Equation (6.3) represents the zeroth order and the second and third terms represent the ±first side orders diffracted from the hologram. If we could use u^*, the conjugate of u, as the reconstruction wave, we see that the second term of Equation (6.3) would have become proportional to $|u|^2 u_o^*$, i.e. the conjugate of the object wave would have been reconstructed. The physical meaning of this deserves some explanation. Complex conjugation of a field amplitude means changing the sign of its phase. It thus gives a wave field returning back on its own path. u_o^* therefore represents a wave propagating from the hologram back to the object forming an image of the object. It is therefore termed the real wave. To reconstruct the hologram with u^* in the case of a pure plane wave, the reconstruction wave can be reflected back through the hologram by means of a plane mirror. An easier way, which also applies for a general reference (reconstruction) wave, is to turn the hologram 180° around the vertical axis. By placing a screen in the real wave, we can observe the image of the object on the screen.

In Figure 6.2 another possible realization of a holography set-up is sketched. Here the expanded laser beam is wavefront-divided by means of a mirror which reflects the

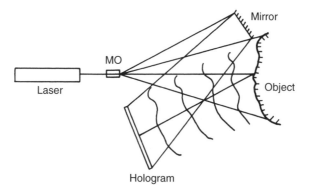

Figure 6.2

reference wave onto the hologram. This set-up is normally more stable than in Figure 6.1 since fewer components are involved.

6.3 AN ALTERNATIVE DESCRIPTION

An alternative and more physical description of the holographic process has already been touched on in Section 4.3.1. Let the point source P in Figure 4.7(a) represent the light from a point on the object, and let the plane wave represent the reference wave. The resulting zone plate pattern is recorded on a film. In Figure 4.7(b) this developed film (the hologram) is illuminated by a plane wave (the reconstruction wave). When viewed through the film, the diffracted, diverging spherical wave looks as if it is coming from P. This argument can be repeated for all points on the object and give us the virtual reconstructed object wave. The spherical wave converging to point P′ represents the real wave.

The circular zone plate is therefore also termed a unit hologram. In the general case when the object- and reference waves are not normally incident on the hologram, the pattern changes from circular to elliptical zone plate patterns, and the diffracted virtual and real waves propagate in different directions in the reconstruction process.

6.4 UNCOLLIMATED REFERENCE AND RECONSTRUCTION WAVES

We now consider in more detail the locations of the virtual and real images for the most general recording and reconstructing geometries. To do this, it suffices to consider a single object point source with coordinates (x_o, y_o, z_o): see Figure 6.3. Here the hologram film is placed in the xy-plane and the reference wave is coming from a point source with coordinates (x_r, y_r, z_r). Using quadratic (Fresnel) approximations to the spherical waves, the object and reference fields of wavelength λ_1 incident on the xy-plane may be written

$$u_o = U_o \exp\left\{i\frac{\pi}{\lambda_1 z_o}[(x - x_o)^2 + (y - y_o)^2]\right\} \tag{6.4}$$

$$u = U \exp\left\{i\frac{\pi}{\lambda_1 z_r}[(x - x_r)^2 + (y - y_r)^2]\right\} \tag{6.5}$$

The transmittance of the resulting hologram we write as

$$t \propto |u_o + u|^2 = t_1 + t_2 + t_3 + t_4 \tag{6.6}$$

where the interesting terms (cf. Equation (6.2)) are

$$t_3 = \alpha U U_o \exp\left\{i\frac{\pi}{\lambda_1 z_r}[(x - x_r)^2 + (y - y_r)^2] - i\frac{\pi}{\lambda_1 z_o}[(x - x_o)^2 + (y - y_o)^2]\right\} \tag{6.7}$$

$$t_4 = \alpha U U_o \exp\left\{-i\frac{\pi}{\lambda_1 z_r}[(x - x_r)^2 + (y - y_r)^2] + i\frac{\pi}{\lambda_1 z_o}[(x - x_o)^2 + (y - y_o)^2]\right\} \tag{6.8}$$

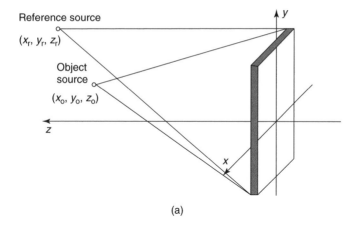

(a)

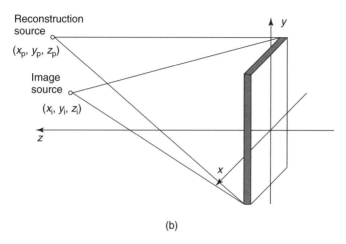

(b)

Figure 6.3   (a) Recording and (b) reconstruction geometries of point sources

In reconstruction, the hologram is illuminated by the spherical wave

$$u_p = U_p \exp\left\{i\frac{\pi}{\lambda_2 z_p}[(x - x_p)^2 + (y - y_p)^2]\right\} \tag{6.9}$$

where we have allowed for both a displaced (relative to the reference wave) point source and a different wavelength λ_2. The two reconstructed waves of interest are $u_3 = t_3 u_p$ and $u_4 = t_4 u_p$ which gives (writing out the x-dependence only)

$$u_3 = t_3 u_p \propto \exp\left\{i\frac{\pi}{\lambda_1 z_r}(x^2 + x_r^2 - 2x_r x) - i\frac{\pi}{\lambda_1 z_o}(x^2 + x_o^2 - 2x_o x) + i\frac{\pi}{\lambda_2 z_p}\right.$$
$$\left. \times (x^2 + x_p^2 - 2x_p x)\right\}$$

$$= \exp\left\{i\pi\left[\frac{x_r^2}{\lambda_1 z_r} - \frac{x_o^2}{\lambda_1 z_o} + \frac{x_p^2}{\lambda_2 z_p}\right]\right\} \exp\left\{i\pi\left(\frac{1}{\lambda_1 z_r} - \frac{1}{\lambda_1 z_o} + \frac{1}{\lambda_2 z_p}\right)x^2\right\}$$

$$\times \exp\left\{-2i\pi\left(\frac{x_r}{\lambda_1 z_r} - \frac{x_o}{\lambda_1 z_o} + \frac{x_p}{\lambda_2 z_p}\right)x\right\} \tag{6.10}$$

By performing the same calculations for the wave u_4, we get for the phase terms depending on x^2 and x

$$u_4 \propto \exp\left\{i\pi\left(-\frac{1}{\lambda_1 z_r} + \frac{1}{\lambda_1 z_o} + \frac{1}{\lambda_2 z_p}\right)x^2\right\} \exp\left\{-2i\pi\left(-\frac{x_r}{\lambda_1 z_r} + \frac{x_o}{\lambda_1 z_o} + \frac{x_p}{\lambda_2 z_p}\right)x\right\} \tag{6.11}$$

A spherical wave diverging from a point (x_i, y_i, z_i) (writing out only the x-dependence) is given as:

$$u_i = U_i \exp\left\{i\frac{\pi}{\lambda_2 z_i}(x - x_i)^2\right\} = U_i \exp\left\{i\frac{\pi}{\lambda_2 z_i}(x^2 + x_i^2 - 2x_i x)\right\}$$

$$= U_i \exp\left\{i\frac{\pi}{\lambda_2 z_i}x_i^2\right\} \exp\left\{i\frac{\pi}{\lambda_2 z_i}x^2\right\} \exp\left\{-2i\pi\frac{x_i}{\lambda_2 z_i}x\right\} \tag{6.12}$$

By comparing this with the above expressions for u_3 and u_4, we get

$$\frac{1}{\lambda_2 z_i} = \pm\frac{1}{\lambda_1 z_r} \mp \frac{1}{\lambda_1 z_o} + \frac{1}{\lambda_2 z_p}, \quad \text{i.e.} \quad z_i = \left(\frac{1}{z_p} \pm \frac{\lambda_2}{\lambda_1 z_r} \mp \frac{\lambda_2}{\lambda_1 z_o}\right)^{-1} \tag{6.13}$$

and

$$\frac{x_i}{\lambda_2 z_i} = \pm\frac{x_r}{\lambda_1 z_r} \mp \frac{x_o}{\lambda_1 z_o} + \frac{x_p}{\lambda_2 z_p}, \quad \text{i.e.} \quad x_i = \mp\frac{\lambda_2 z_i}{\lambda_1 z_o}x_o \pm \frac{\lambda_2 z_i}{\lambda_1 z_r}x_r + \frac{z_i}{z_p}x_p \tag{6.14}$$

and with a completely analogous expression for y_i:

$$y_i = \mp\frac{\lambda_2 z_i}{\lambda_1 z_o}y_o \pm \frac{\lambda_2 z_i}{\lambda_1 z_r}y_r + \frac{z_i}{z_p}y_p \tag{6.15}$$

Here the upper set of signs applies for u_3, the real reconstructed wave, and the lower set for u_4, the virtual wave. What we have done is to find the coordinates (x_i, y_i, z_i) of the image point expressed by the coordinates of the object point, the source point of the reference and the reconstruction waves. We see that when $\lambda_2 = \lambda_1$ and $z_p = z_r$, we get for the virtual wave $z_i = z_o$. When, in addition, $z_r = \infty$ (collimated reference and reconstruction waves), $z_i = -z_o$ for the real wave.

From our calculations, we can associate a transversal magnification

$$m = \left|\frac{x_i}{x_o}\right| = \left|\frac{y_i}{y_o}\right| = \left|\frac{\lambda_2 z_i}{\lambda_1 z_o}\right| = \left|1 - \frac{z_o}{z_r} \mp \frac{\lambda_1 z_o}{\lambda_2 z_p}\right|^{-1} \tag{6.16}$$

6.5 DIFFRACTION EFFICIENCY. THE PHASE HOLOGRAM

Assume the object- and reference waves to be described by

$$u_o = U_o e^{i\phi_o} \tag{6.17a}$$

and

$$u = U e^{i\phi} \tag{6.17b}$$

respectively. The resulting amplitude transmittance then becomes

$$t = \alpha[U^2 + U_o^2 + UU_o e^{i(\phi-\phi_o)} + UU_o e^{-i(\phi-\phi_o)}]$$
$$= \alpha(I + I_0)[1 + V \cos(\phi - \phi_0)] \tag{6.18}$$

which can be written as

$$t = t_b \left[1 + \frac{V}{2} e^{i(\phi-\phi_o)} + \frac{V}{2} e^{-i(\phi-\phi_o)} \right] \tag{6.19}$$

where $I = U^2$, $I_0 = U_0^2$ and where we have introduced the visibility V (see eq. (3.29)) and the bias transmittance $t_b = \alpha(I + I_o)$. Since the transmittance t never can exceed unity and $0 \le V \le 1$, we see from Equation (6.18) that $t_b \le 1/2$.

The reconstructed object wave u_r is found by multiplying the last term of Equation (6.19) by the reconstruction wave u:

$$u_r = t_b \frac{V}{2} U e^{i\phi_o} \tag{6.20}$$

and the intensity

$$I_r = |u_r|^2 = \tfrac{1}{4} U^2 t_b^2 V^2 \tag{6.21}$$

The diffraction efficiency η of such a hologram we define as the ratio of the intensities of the reconstructed wave and the reconstruction wave, i.e.

$$\eta = I_r/I = \tfrac{1}{4} t_b^2 V^2 \tag{6.22}$$

From this expression we see that the diffraction efficiency is proportional to the square of the visibility. η therefore reaches its maximum when $V = 1$, i.e. when $I_o = I$, which means that the diffraction efficiency is highest when the object and reference waves are of equal intensity.

Maximum possible diffraction efficiency is obtained for $V = 1$ and $t_b = \frac{1}{2}$, which gives

$$\eta_{max} = \frac{1}{16} = 6.25\%$$

This type of hologram is called an amplitude hologram because its transmittance is a pure amplitude variation. A hologram with a pure phase transmittance is called a phase hologram. Such holograms can be produced in different ways. A commonly applied method consists of bleaching the exposed silver grains in the film emulsion of a standard amplitude hologram. The recorded amplitude variation then changes to a corresponding variation in emulsion thickness. The transmittance t_p of a phase hologram formed by bleaching of an amplitude hologram can be written as

$$t_p = e^{iM\cos(\phi_0 - \phi)} = \sum_{n=-\infty}^{\infty} i^n J_n(M) e^{in(\phi_0 - \phi)} \tag{6.23}$$

where J_n is the nth-order Bessel function. Here M is the amplitude of the phase delay. From this expression we see that a sinusoidal phase grating will diffract light into n orders in contrast to a sinusoidal amplitude grating which has only $\pm$1st orders. The amplitude of the first-order reconstructed object wave is found by multiplying Equation (6.23) by the reconstruction wave u for $n = 1$, i.e.

$$u_r = J_1(M)U e^{i\phi_0} \tag{6.24}$$

and the intensity

$$I_r = U^2 J_1^2(M) \tag{6.25}$$

The diffraction efficiency becomes

$$\eta_p = I_r/I = J_1^2(M) \tag{6.26}$$

Since $J_{1\max}(M) = 0.582$ for $M = 1.8$, the maximum possible diffraction efficiency of a phase hologram is

$$\eta_{p,\max} = 0.339 = 34\%$$

6.6 VOLUME HOLOGRAMS

Up to now we have regarded the hologram film emulsion as having negligible thickness.

For emulsions of non-negligible thickness, however, volume effects, hitherto not considered, must be taken into account. For example, a thick phase hologram can reach a theoretical diffraction efficiency of 100 per cent.

Consider Figure 6.4(a) where two plane waves are symmetrically incident at the angles $\theta/2$ to the normal on a thick emulsion. These waves will form interference planes parallel to the yz-plane with spacings (cf. eq. (3.21)).

$$d = \frac{\lambda}{2\sin(\theta/2)} \tag{6.27}$$

After development of this hologram, the exposed silver grains along these interference planes will form silver layers that can be regarded as partially reflecting plane mirrors. In Figure 6.4(b) this hologram is reconstructed with a plane wave incident at an angle ψ. This wave will be reflected on each 'mirror' at an angle ψ.

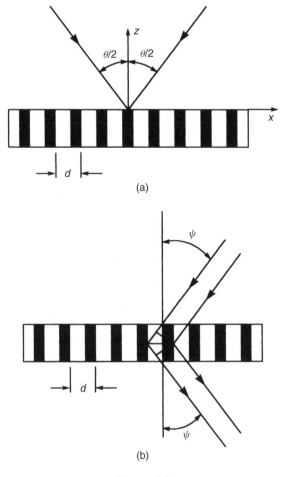

Figure 6.4

To obtain maximum intensity of the reflected, reconstructed wave, the path length difference between light reflected from successive planes must be equal to λ. From the triangles in Figure 6.4(b) this gives

$$2d \ \sin \psi = \lambda \qquad (6.28)$$

which, by substitution of Equation (6.27), gives

$$\sin \psi = \sin \theta / 2 \qquad (6.29)$$

i.e. the angles of incidence of the reconstruction and reference waves must be equal. It can be shown that for a thick hologram, the intensity of the reconstructed wave will decrease rapidly as ψ deviates from $\theta/2$; see Section 13.6. This is referred to as the Bragg effect and Equation (6.29) is termed the Bragg law.

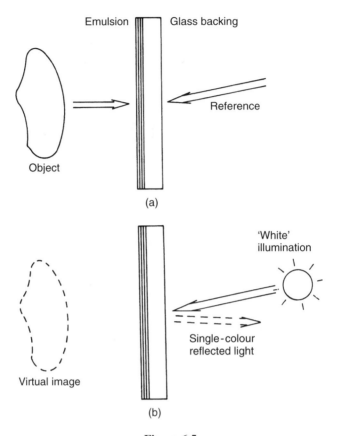

Figure 6.5

A special type of volume hologram, called a reflection hologram, is obtained by sending the object and reference waves from opposite sides of the emulsion, as shown in Figure 6.5(a). Then $\theta = 180°$ and the stratified layers of metallic silver of the developed hologram run nearly parallel to the surface of the emulsion with a spacing equal to $\lambda/2$ (see Equation (6.27)). Owing to the Bragg condition, the reconstruction wave must be a duplication of the reference wave with the same wavelength, i.e. the hologram acts as a colour filter in reflection. Therefore a reflection hologram can be reconstructed in white light giving a reconstructed wave of the same wavelength as in the recording (see Figure 6.5(b)). In practice the wavelength of the reflected light is shorter than that of the exposing light, the reason being that the emulsion shrinks during the development process and the silver layers become more closely spaced.

6.7 STABILITY REQUIREMENTS

In the description of the holographic recording process we assumed the spatial phases of both the object- and reference waves to be time independent during exposure. It is clear, however, that relative movements between the different optical components (like mirrors, beamsplitters, the hologram, etc.) in the hologram set-up will introduce such phase changes. If, for instance, a mirror makes vibrations of amplitude greater than $\lambda/4$ during the exposure time, adjacent dark and bright interference fringes interchange their positions randomly, which can lead to a uniform blackening of the hologram film and therefore ruin the experiment. The exposure time using a 5 mW H-Ne laser is typically of the order of seconds. This poses stringent requirements on the stability of the set-up. Therefore the standard methods of holography are normally performed on vibration-isolated heavy tables with the optical components mounted in massive holders. There are, however, special techniques by which unwanted movements can to a certain extent be compensated for or subtracted from. Thus, successful holographic experiments executed on the factory floor using continuous wave lasers have been reported.

By using pulse lasers, exposure times down to the order of nanoseconds can be achieved. In such cases, unwanted movements become less important. The application of pulse lasers therefore substantially reduces the stringent stability requirements.

6.8 HOLOGRAPHIC INTERFEROMETRY

In Section 3.5 we mentioned an imaginary experiment where two waves reflected from two identical objects could interfere. With the method of holography now at hand, we are able to realize this type of experiment by storing the wavefront scattered from an object in a hologram. We then can recreate this wavefront by hologram reconstruction, where and when we choose. For instance, we can let it interfere with the wave scattered from the object in a deformed state. This technique belongs to the field of holographic interferometry (Vest 1979; Erf 1974; Jones and Wykes 1989). In the case of static deformations, the methods can be grouped into two procedures, double-exposure and real-time interferometry.

6.8.1 Double-Exposure Interferometry

In this method, two exposures of the object are made on the same hologram. This might be recordings before and after the object has been subject to load by, for instance, external forces or two other object states that are to be compared. By reconstructing the hologram, the two waves scattered from the object in its two states will be reconstructed simultaneously and interfere. This double-exposed hologram can be stored and later reconstructed for analysis of the registered deformation at the time appropriate for the investigator. If a lot of different states of the object (e.g. different load levels) are to be investigated, many holograms have to be recorded, which makes the method time-consuming and elaborate.

6.8.2 Real-Time Interferometry

In this method, a single recording of the object in its reference state is made. Then the hologram is processed and replaced in the same position as in the recording. By looking through the hologram we are now able to observe the interference between the reconstructed object wave and the wave from the real object in its original position. Thus we are able to follow the deformation as it develops in real time by observing the changes in the interference pattern. These changes might be recorded on film for later playback and analysis. A disadvantage of the method is that the hologram must be replaced in its original position with very high accuracy. This can be overcome by developing the hologram *in situ* in a transparent cuvette or using a thermoplastic film, see Section 5.5.2. Also the contrast of the interference fringes is not as good as in the double-exposure method.

6.8.3 Analysis of Interferograms

As we have seen, holographic interferometry enables the wave scattered from the object in its reference state described by the field amplitude $u_1 = U_1 e^{i\phi_1}$ and the wave scattered from the object in a deformed state described by the field amplitude $u_2 = U_2 e^{i\phi_2}$ to occur simultaneously. The actual deformations will be so small that we can put $U_1 = U_2 = U$. These two waves will form an interference pattern in the usual way given by

$$I = 2U^2[1 + \cos \Delta\phi] \tag{6.30}$$

where

$$\Delta\phi = \phi_1 - \phi_2 \tag{6.31}$$

The problem is then to find the relation between $\Delta\phi$ and the deformation.

Consider Figure 6.6 where a point O on the object is moved along the displacement vector **d** to the point O′ due to a deformation of the object. The object is illuminated by a plane wave (point source placed at infinity) which propagation direction **n**$_1$ makes an angle θ_1 with the displacement vector **d**. Assume that we are looking through the hologram from infinity along the direction **n**$_2$ making an angle θ_2 with **d**. We realize that the geometrical path length from the light source via the object point to the point of observation will be different before and after the deformation has taken place. In our case this difference is equal to the path length $AO + OB$ which by applying simple trigonometry becomes equal to

$$d(\cos \theta_1 + \cos \theta_2) \tag{6.32}$$

From Section 1.4 we know that the phase difference $\Delta\phi$ is equal to the path length multiplied by the wave number k:

$$\Delta\phi = kd(\cos \theta_1 + \cos \theta_2) \tag{6.33}$$

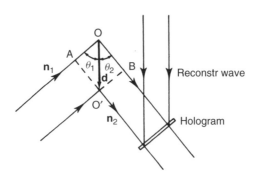

Figure 6.6

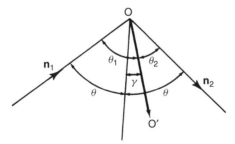

Figure 6.7

In Figure 6.7 a portion of Figure 6.6 is redrawn and the line bisecting the angle 2θ between $\mathbf{n}_1$ and $\mathbf{n}_2$ is introduced. This bisector is inclined at an angle γ to d. By using trigonometric formulas we find the geometry factor g to be

$$g = \cos\theta_1 + \cos\theta_2 = 2\cos\gamma\cos\theta \tag{6.34}$$

which yields

$$\Delta\phi = (2\pi/\lambda)2d(\cos\gamma)\cos\theta \tag{6.35}$$

By inserting Equation (6.35) into Equation (6.30) we find that the interference pattern has a maximum (bright fringe) whenever

$$\Delta\phi = (2\pi/\lambda)2d(\cos\gamma)\cos\theta = n2\pi \quad \text{for} \quad n = 0, 1, 2, \ldots$$

i.e. when

$$d\cos\gamma = \frac{n\lambda}{2\cos\theta} \tag{6.36}$$

and a minimum (dark fringe) whenever

$$\Delta\phi = (2\pi/\lambda)2d(\cos\gamma)\cos\theta = (2n+1)\pi \quad \text{for} \quad n = 0, 1, 2, \ldots$$

i.e. when

$$d\cos\gamma = \frac{2n+1}{2}\frac{\lambda}{2\cos\theta} \tag{6.37}$$

Here $d\cos\gamma$ is the component of the displacement vector onto the line bisecting the angle between the illumination and observation directions. This applies also when $\mathbf{d}$ does not lie in the plane defined by $\mathbf{n}_1$ and $\mathbf{n}_2$ as in Figure 6.7.

When interpreting interference patterns (also called interferograms) due to deformations of extended objects, we therefore can imagine the space to be filled with equispaced parallel planes which are normal to the bisector of $\mathbf{n}_1$ and $\mathbf{n}_2$ with a spacing equal to $\lambda/(2\cos\theta)$. Each time the surface of the deformed object intersects one of these planes we get a bright (or dark) fringe. To measure the deformation at a given point, one therefore simply has to count the number of fringes and multiply it by $\lambda/(2\cos\theta)$. This is illustrated

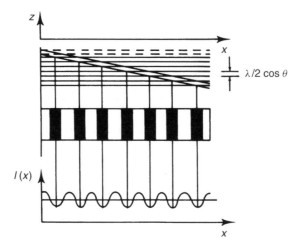

Figure 6.8

in Figure 6.8 which shows the interference pattern seen on a beam that is rotated as a rigid body about an axis (the y-axis).

Highest sensitivity is obtained when both the illumination and observation directions are parallel to the displacement vector, i.e. when $\gamma = \theta_1 = \theta_2 = \theta = 0$. The displacement corresponding to one fringe spacing is then

$$d = \lambda/2$$

from Equation (6.37) we see that the first-order dark fringe ($n = 0$) then occurs at

$$d_{min} = \lambda/4$$

For the He–Ne laser wavelength $\lambda = 0.6328$ μm, this corresponds to a displacement equal to 0.15 μm which therefore gives a representative figure for the sensitivity of standard holographic interferometry.

To obtain the best measurement accuracy when analysing holographic interferograms, the fringe positions should be determined with the highest possible accuracy (or fractional fringe width). This can be achieved in many different ways as will be discussed in chapter 11. Here we mention one method based on the heterodyne principle (see Section 3.6.4) called heterodyne holographic interferometry (HHI) (Dandliker and Thalmann 1985). In this method the two wavefronts corresponding to the object in its two states have different frequencies ν_1 and ν_2. Equation (6.30) then becomes

$$I = 2U^2\{1 + \cos[2\pi(\nu_1 - \nu_2)t + \Delta\phi]\} \tag{6.38}$$

where $\nu_1 - \nu_2$ corresponds to the intermediate frequency of Equation (3.41) and the analysis to recover $\Delta\phi$ follows the principle described in Section 3.6.4. Such a measurement can be made to an accuracy of typically 10^{-1} radians or $3 \cdot 10^{-3}$ of a cycle and hence enables very small deformations to be resolved. Moreover the measurement is independent of U which affects only the amplitude of the signal. Heterodyning may be used in both real-time and double exposure holography. In real-time HHI, the 'live' object beam and the reconstructing beam have different frequencies, whilst in double exposure HHI, a single frequency is used to make both exposures but two separate reference beams are employed. At the reconstruction stage these two reference beams are set at different frequencies. In both methods two detectors are required which are used in one of two ways. For the first of these, one of the detectors is tracked across the fringe pattern whilst the other is held static and hence generates the fixed reference frequency. Alternatively the two detectors may be maintained at a fixed distance with respect to one another and both tracked across the fringe field. This enables a differential measurement to be made and hence the variations in the gradient of $\Delta\phi$ is found. In practice the two frequencies are obtained by splitting a single continuous laser beam and passing the two beams through separate optoacoustical modulators.

When the point source and the point of observation are placed at finite distances from the object, the illumination and observation directions ($\mathbf{n}_1$ and $\mathbf{n}_2$) will vary across the object surface. This will turn the above-mentioned plane equispaced surfaces into equidistant ellipsoids with the point source and the point of observation as foci. The sensitivity (displacement per fringe) then varies across the object surface but for small objects and reasonably long distances to the source point and the observation point this variation will be quite small. As can be seen from Equation (6.35) and Figure 6.8 the method is insensitive to displacements parallel to the equispaced planes or along the ellipsoids in the general case.

This should be kept in mind when making a set-up for holographic interferometry. The concept called the holodiagram (which is essentially the ellipsoids mentioned above) developed by Abramson (1970, 1972) could be helpful in this respect.

Figure 6.9 shows some typical examples of interferograms obtained by means of holographic interferometry on different objects. It should be noted that holographic interferometry is incapable of measuring surface contour deviations between two different objects as pointed out in Section 3.5. It is therefore also impossible to measure deformations if the microstructure of the object changes drastically, as for example in plastic deformations.

6.8.4 Localization of Interference Fringes

Another difficulty in evaluating holographic interferograms comes from the phenomenon of fringe localization. It is an annoying fact that the fringes in holographic interferometry only in special cases are localized on the object surface.

This means that when imaging an interferogram by focusing on the object, the fringes may be completely lost because they focus (localize) in a plane (which might be curved) which lies far away from the object surface. That the fringes are localized in a certain plane means that they have maximum contrast or visibility in that plane. Loss of fringes in the plane of the object or other planes therefore means that their contrast is too low to be detected in that plane.

To see this, we must remember that the interferogram is formed by interference between light scattered from pairs of corresponding points on the object surface in the first and second exposure. Interference between non-identical points on the two displaced surfaces will contribute a random noisy background to the interferogram, thereby reducing its contrast.

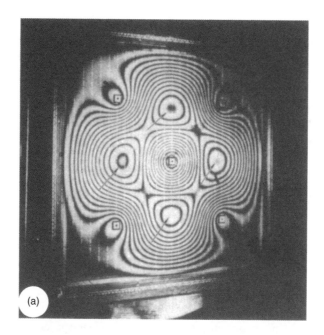

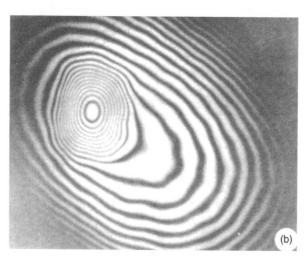

Figure 6.9 Examples of holographic interferograms. (a) Deflection of a rectangular plate fastened with five struts and subjected to a uniform pressure. From Wilson *et al.* 1971. (Photograph courtesy of Dr A. D. Wilson, Thomas J. Watson Research Center, Yorktown Heights, New York. Reproduced by permission of SEM.); (b) Detection of debonded region of a honeycomb construction panel. From Vest 1979. (Reproduced by permission of John Wiley & Sons Inc.); (c) A bullet in flight observed through a doubly-exposed hologram. From Collier *et al.* 1971. (Reproduced by permission of Dr R. F. Wuerker, TRW Inc.); and (d) Holographic reconstruction of a solid turbine blade illustrating (A) the first flexural resonance at 981 Hz, (B) a second-order flexural resonance at 4624 hz, and the 2nd and 3rd torsional resonances at (C) 6406 Hz. From Erf 1974. (Photograph courtesy at Dr R. K. Erf, United Technologies Research Center, East Hartiord, Connecticut. Reproduced by permission of Academic Press, New York)

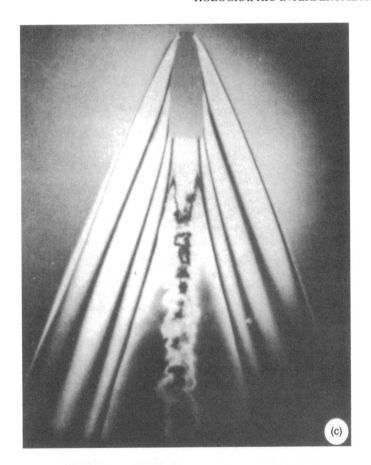

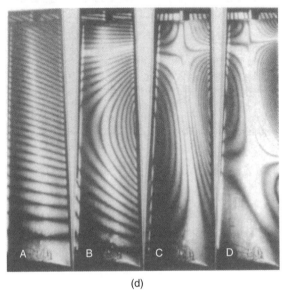

(d)

Figure 6.9   (*continued*)

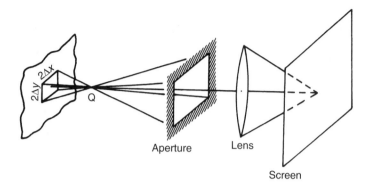

Figure 6.10

In the analysis below, we shall show that this problem might be overcome by stopping down the aperture of the imaging system, thereby increasing the depth of focus to obtain simultaneously an image of sufficient quality of both the fringe pattern and the object surface. For more details, see Vest (1979), Walles (1969) and Molin and Stetson (1971).

Consider Figure 6.10 where the holographic interferogram due to displacement of the object surface (drawn as a single surface in the figure) is imaged onto a viewing screen by a lens through a rectangular aperture. Assume that the lens images a plane containing the point Q onto the viewing screen. The central ray passing through Q emanates from a point $P(x_0, y_0)$ on the object surface. The intensity at Q is the integral of the intensity of all the ray pairs within the ray cone defined by the aperture. This cone subtends a rectangular area of dimension $2\Delta x$, $2\Delta y$ on the object surface, hence the intensity at Q is

$$I(Q) = \int_{x_0-\Delta x}^{x_0+\Delta x} \int_{y_0-\Delta y}^{y_0+\Delta y} [1 + \cos \delta(x, y)] \, dx \, dy \tag{6.39}$$

where

$$\delta(x, y) = kd(\cos \theta_1 + \cos \theta_2) \tag{6.40}$$

is the phase difference due to the displacement of each point on the object between the two exposures (see Equation (6.33)). For small Δx, Δy, $\delta(x, y)$ can be approximated by the initial terms of a Taylor series expansion about x_0, y_0:

$$\delta(x, y) = \delta(x_0, y_0) + (x - x_0)\frac{\partial \delta}{\partial x}\bigg|_{x_0, y_0} + (y - y_0)\frac{\partial \delta}{\partial y}\bigg|_{x_0, y_0}$$

$$= \delta_0 + (x - x_0)\delta_0^x + (y - y_0)\delta_0^y \tag{6.41}$$

where δ_0^x and δ_0^y denote partial derivatives of δ at x_0, y_0 with respect to x and y respectively.

Substituting Equation (6.41) into Equation (6.39) and evaluating the integral yields

$$I(Q) = 4\Delta x \Delta y \left[1 + \frac{\sin(\delta_0^x \Delta x)}{\delta_0^x \Delta x} \frac{\sin(\delta_0^y \Delta y)}{\delta_0^y \Delta y} \cos \delta_0\right] \tag{6.42}$$

The contrast of this intensity distribution is found from the definition of contrast or visibility (cf. Equation (3.8)):

$$V = \frac{I_{\text{max}} - I_{\text{min}}}{I_{\text{max}} + I_{\text{min}}} \tag{6.43}$$

which by putting $\cos \delta_0 = 1$ for I_{max} and $\cos \delta_0 = -1$ for I_{min} in Equation (6.42) gives

$$V = \left| \frac{\sin(\delta_0^x \Delta x)}{\delta_0^x \Delta x} \frac{\sin(\delta_0^y \Delta y)}{\delta_0^y \Delta y} \right| \tag{6.44}$$

From this expression we see that the visibility is equal to 1 for δ_0^x or $\delta_0^y = 0$, i.e. when the variations in phase over the viewing cone are minimized. This defines the area of fringe localization. If the value of $\delta_0^x \Delta x$ or $\delta_0^y \Delta y$ increases rapidly as the distance z is changed from the localization distance, the region of fringe localization is sharply defined. If $\delta_0^x \Delta x$ or $\delta_0^y \Delta y$ increases slowly as z is varied, the region of localization is broad. Since δ_0^x and δ_0^y are determined by the system geometry and object displacement field, we see from Equation (6.44) that the sharpness of localization can be controlled by the viewing aperture $\Delta x \Delta y$. Therefore, by decreasing the aperture, relatively distinct fringes can be observed over an extended region and fringes can be seen in the plane of the object even though it is at some distance from the region of localization.

We also see from Equation (6.44) that the contrast is a periodic function of $\delta_0^x \Delta x$, $\delta_0^y \Delta y$. As we move away from the region of localization, the contrast therefore can assume periodic maxima but with a much lower value due to the sinc-function dependence (see Figure B.1.c).

From Equation (6.40) we see that δ is dependent on the illumination and observation directions as well as the displacement vector d. The determination of the regions of fringe localization must therefore be calculated in each separate case by maximizing Equation (6.44). Here we shall not go into such detailed calculations but merely quote two general results which are:

(1) For a rigid body translation the fringes localize at infinity.

(2) For a rigid body rotation about an axis inside the object's surface and normal to the illumination and observation directions the fringes localize on the object.

In conclusion, fringe localization rarely poses any problem to the experimentalist since it can be solved by decreasing the viewing aperture. On the other hand, the phenomenon can be taken advantage of since it conveys additional information about the deformation under investigation.

6.9 HOLOGRAPHIC VIBRATION ANALYSIS

Up to now we have been dealing with holographic interferometry applied on static deformations. In this section we shall show that this method can also be applied to vibrating objects.

Assume that the object point in Figure 6.6 executes harmonic vibrations given by

$$d(x, t) = D(x) \cos \omega t \tag{6.45}$$

where $D(x)$ is the amplitude, x represents the spatial coordinates of the object point and ω is the vibration frequency. Light scattered from this point can be described by a field amplitude in the hologram plane given by

$$u_o(x, t) = U_o(x)e^{i\phi} \qquad (6.46)$$

where

$$\phi = kgd(x, t). \qquad (6.47)$$

Here $g = \cos\theta_1 + \cos\theta_2$ is the geometry factor determined by the illumination and observation directions in the same way as in the static case.

Let u_o of Equation (6.46) be the object wave in a hologram recording and u the reference wave. Just as in the static case, the reconstructed wave will be given by the last term of Equation (6.3). However, since u_o is time-varying during the exposure, the recorded intensity distribution (Equation (6.1)) and hence the hologram amplitude transmittance (Equation (6.2)) will be averaged over the exposure time. The reconstructed object wave therefore becomes equal to

$$u_a = \alpha|u|^2\bar{u}_o \qquad (6.48)$$

where the bar denotes time averaging. For exposure times much longer than the vibrating period of the object (Løkberg 1979) this time averaging is equivalent to averaging over one vibration period T. Therefore

$$u_a = \alpha|u|^2\frac{1}{T}\int_0^T u_o(x, t)\, dt$$

$$= \alpha|u|^2 U_o \frac{1}{T}\int_0^T e^{ikg D(x)\cos\omega t}\, dt = \alpha|u|^2 U_o J_o[kgD(x)] \qquad (6.49)$$

where J_o is the zeroth-order Bessel function. Here we have applied the relation

$$\frac{1}{2\pi}\int_0^{2\pi} e^{i\eta\cos\xi}\, d\xi = J_o(\eta) \qquad (6.50)$$

The observable intensity distribution of the reconstructed wave becomes

$$I_a = |u_a|^2 = K J_0^2[kgD(x)] \qquad (6.51)$$

where all constants are gathered into a common constant K.

As an illustrative example, consider Figure 6.11(a) where a bar is vibrating as a rigid body about an axis. In Figure 6.11(b) we have drawn the Bessel function squared which represents the intensity distribution that will be observed in the reconstruction of a hologram recording of the vibrating bar. We see that the region around the axis which is at rest (the nodal point) will show up as a bright zero-order fringe of much higher intensity than the higher-order bright fringes farther along the bar. This is in contrast to the cos-fringes obtained in the case of a static deformation where all bright fringes have equal intensity.

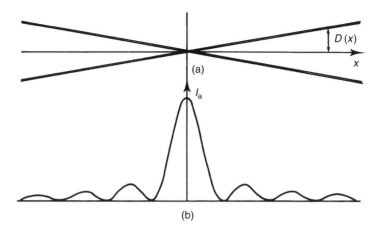

Figure 6.11   (a) Vibration of a bar about an axis and (b) Intensity distribution of the resulting time-average holographic recording

To find the vibration amplitude at the higher-order fringes we consult a table of Bessel function values from which we find that

$$J_0^2(\eta) = \max \text{ for } \eta = 0, 3.83, 7.02, 10.17, 13.32, 16.47, \ldots$$

$$J_0^2(\eta) = 0 \text{ for } \eta = 2.40, 5.52, 8.65, 11.79, 14.93, \ldots$$

For $g = 2$ (illumination and observation directions parallel to the displacement) and $\lambda = 632.8$ nm (wavelength of the He–Ne laser) this gives:

Bright fringes when $D(x) = 0, 0.19, 0.35, 0.51, 0.67, 0.83, \ldots [\mu m]$

Dark fringes when $D(x) = 0.12, 0.28, 0.44, 0.59, 0.75, \ldots [\mu m]$

By means of standard holographic techniques one has been able to observe fringes up to the 50th order by this method. A very detailed map of the amplitude distribution is therefore obtained. The frequency range of holographic vibration measurements is only limited by the method of object excitation. Using piezoelectric transducers for excitation, values of hundreds of kilohertz are easily obtained.

The method described above is called the time-average method. Another method similar to the real-time method for static displacements can also be applied. It consists of first recording a hologram while the object is at rest, then replacing the hologram in its original position and observing the resulting fringe pattern when the object vibrates. The contrast of this pattern is very low due to the resulting $(1 - J_0)$-dependence (Vest 1979). While observing this real-time pattern the laser beam can be chopped with the same frequency. This is equivalent to the real-time method for static displacements. Here one observes fringes due to the displacement of the object, when at rest and when illuminated by the light pulse. In that way, by slowly varying the phase between the light pulse and the object vibration, one can observe the vibration of the object in slow motion. Thus it is also possible to observe the phase of the displacements on the different parts of the object surface.

Another method is to phase modulate the reference wave (Aleksoff 1971, 1974). This can be done by placing a vibration mirror in the light path of the reference wave. The argument of the Bessel function of Equation (6.51) then changes from $(4\pi/\lambda)D(x)$ (for $g = 2$) to

$$\frac{4\pi}{\lambda}[D_o^2(x) + R^2 - 2D_o(x)R\cos(\psi_o(x) - \psi_r)]^{1/2} \qquad (6.52)$$

where

$\quad\quad D_o(x) = $ the vibrating amplitude of the object,

$\quad\quad\quad R = $ the vibrating amplitude of the reference mirror,

$\psi_o(x) - \psi_r = $ the phase difference between the vibration of the object and reference mirror.

The result of this reference wave modulation is that the centre of the Bessel function is moved from the nodal points of the object to points that vibrate at the same amplitude and phase as the modulating mirror. These points then show up as zero-order bright fringe areas. In that way it is possible to extend the measurable amplitude range considerably, in practice up to about 10 μm. By varying the phase of the reference mirror it is also possible to trace out the areas of the object vibrating in the same phase as the reference mirror, thereby mapping the phase distribution of the object.

Reference mirror modulation can also be used to measure very small vibration amplitudes by moving the steepest part of the central maximum of the Bessel function to coincide with zero object vibration amplitude (Metherell *et al.* 1969; Hogmoen and Løkberg 1976).

Application of a TV camera as the recording medium gives a very versatile instrument for vibration analysis using the methods described above. This will be treated in more detail in Section 12.2.

6.10 HOLOGRAPHIC INTERFEROMETRY OF TRANSPARENT OBJECTS

Up to now we have mainly considered the application of holographic interferometry to opaque objects. The method can, however, equally well be used for the analysis of transparent objects (Vest 1979). In fact, the set-up becomes slightly simpler, for example that shown in Figure 6.12.

The quantity actually measured by this method is the change in refractive index due to some change in the object volume. For three-dimensional objects, the corresponding phase difference is an integrated value through the object volume along the ray path. In the same way as for opaque objects, the resulting interferogram is given as (cf. Equation (6.30))

$$I = 2U^2(1 + \cos\Delta\phi) \qquad (6.53)$$

where $\Delta\phi$ is the phase difference in the two recordings. In most applications the refractive index during one exposure, say the first, is uniform and can be denoted by n_0. Then the difference in the general case is given as

$$\Delta\phi = k\int[n(x, y, z) - n_0]\,dz \qquad (6.54)$$

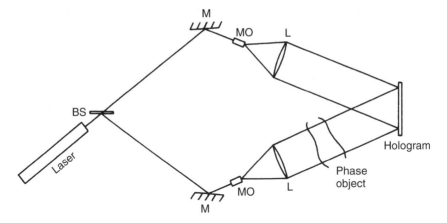

Figure 6.12   Holography set-up for transparent phase objects

where $n(x, y, z)$ is the refractive index distribution during the second exposure and where we have assumed plane wave object illumination along the z-axis. From this equation we see that it is impossible to determine the phase contribution from a single specific point inside the object volume. There are, however, special cases, for example refractive index variations only in the x- and y-directions, radially symmetric distributions, etc. for which Equation (6.54) can be explicitly solved.

Holographic interferometry of transparent phase objects has nevertheless become a versatile tool in research areas such as aerodynamics, heat transfer, plasma diagnostics and stress analysis of transparent models. The latter will be treated in more detail in Section 9.7.

Figure 6.9(c) shows a typical example from the field of aerodynamics. It is a two-exposure holographic interferogram of high speed air flow past a cone. The change in refractive index due to air compression can be found by counting the number of fringes starting from the tip of the cone.

We shall not here go into details about the various techniques used in this field. In fact, a lot of different methods have been applied. For reference, we merely quote some relations between the refractive index and some physical quantities appropriate to the different measurement problems.

In aerodynamics and flow visualization, interferometry is used to determine the distribution of density (the mass per unit volume) in a gas. Density, denoted by ρ, is related to the refractive index of the gas by the Gladstone-Dale equation:

$$n - 1 = K\rho \tag{6.55}$$

where K, the Gladstone-Dale constant, is a property of the gas. K is nearly independent of the wavelength of light and of temperature and pressure under moderate physical conditions.

In heat and mass transfer, interferometry is used to determine the spatial distribution of temperature or concentration of chemical species. For gases, the relation between temperature and refractive index is found by combining the ideal gas equation of state

$$\rho = \frac{MP}{RT} \qquad (6.56)$$

and the Gladstone-Dale equation (Equation (6.55)) to yield

$$n - 1 = \frac{KMP}{RT} \qquad (6.57)$$

Where P is the pressure, M is the molecular weight of the gas, T is the absolute temperature and R is the universal gas constant. The slope of a curve of refractive index versus temperature is therefore

$$\frac{dn}{dT} = -\frac{KMP}{RT^2} \qquad (6.58)$$

For small temperature changes, the right-hand side of Equation (6.58) is approximately constant, giving a linear relation for the rate of change of refractive index with temperature. For liquids, the density is not a simple function of temperature so empirical relations must be found between n and T.

In plasma diagnostics, interferometry is used to determine the spatial distribution of densities of the plasma which is a collection of atoms, ions and electrons created by very high temperatures. The refractive index of a plasma is the sum of the refractive indices of these particles weighted by their number densities. For atoms and ions, the refractive index is given by the Gladstone-Dale relation, Equation (6.55). For electrons, the refractive index n_e is given as

$$n_e = \left(1 - \frac{N_e e^2 \lambda^2}{2\pi m_e c^2} \right)^{1/2} \qquad (6.59)$$

where e is the charge and m_e the mass of an electron, c is the speed of light and N_e is the number density of electrons. We see that this relation is strongly dependent on the wavelength λ.

PROBLEMS

6.1 In Section 3.4 we found an expression for the interfringe distance d_x measured along the x-axis when two plane waves are incident on the xy-plane, see Figure 3.2 and Equation (3.23).

Calculate the interfringe distance along the x-axis when the two plane waves are refracted into a medium (a hologram plate) of refractive index n.

6.2 Imagine that you cut a small piece out of a recorded hologram plate (you might have dropped the hologram glass plate on the floor!) and use this piece in the reconstruction. What do you see? Explain.

6.3 In Section 6.5 we stated that maximum diffraction efficiency for a thin amplitude hologram is obtained when the visibility (or contrast) of the intensity distribution is unity, which means that the ratio $R = I_o/I_r$ between the object and reference intensities is equal to 1.

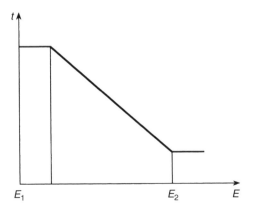

Figure P6.1

The transmittance versus exposure $(t - E)$ curve for a holographic film typically looks like that sketched in Figure P6.1 with a linear portion between the exposures E_1 and E_2. To get a linear response, it is therefore advantageous to have the exposure lying between these values. Calculate R in terms of E_1 and E_2 when the whole linear portion of the $t - E$ curve is utilized.

6.4 A thin hologram is recorded with the object- and reference waves being unit amplitude, plane waves with angles of incidence θ_o and θ_r respectively, and with wavelength λ_1.

(a) Calculate the intensity distribution in the hologram plane.
The hologram is reconstructed with a plane wave with angle of incidence θ_i and wavelength λ_2.

(b) Use the grating equation (Equation (4.21)) to find an expression for the angle θ_s of the reconstructed wave and the ray angle magnification $M_\alpha \simeq \sin\theta_s / \sin\theta_o$.

6.5 As mentioned in Section 5.5.2, a thermoplastic film has a bandlimited spatial frequency response centred at about 1500 lines/mm. What should preferably be the angle between the object- and reference waves when using this film?

6.6 The transmittance of a phase hologram is given by Equation (6.23). Do you get a single reconstructed wave? What happens to the zero-order wave as M approaches 2.4?

6.7 (a) Consider the interferogram in Figure 6.9(a). If we have $\theta_1 = 10°$, $\theta_2 = 20°$ and an He−Ne laser is used, what is the maximum deflection of the plate?

(b) Assuming $g = 2$ and $\lambda = 632.8$ nm, what is the maximum vibration amplitude on the left and right side of the turbine blade in B of Figure 6.9(d) (Consult a table of Bessel functions).

6.8 Consider the vibrating bar in Figure 6.11(a) and let the angular vibrating amplitude be α_m. Assume that we record a time-average hologram with a modulated reference wave with a vibrating amplitude of the reference mirror equal to R and the phase $\psi_r = \pi$. Sketch (qualitatively) the intensity distribution of the resulting interferogram.

7
Moiré Methods. Triangulation

7.1 INTRODUCTION

Figure 3.2 is an illustration of two interfering plane waves. Let us look at the figure for what it really is, namely two gratings that lie in contact, with a small angle between the grating lines. As a result, we see a fringe pattern of much lower frequency than the individual gratings. This is an example of the moiré effect and the resulting fringes are called moiré fringes or a moiré pattern. Figures 3.4, 3.8 and 3.9 are examples of the same effect. The mathematical description of moiré patterns resulting from the superposition of sinusoidal gratings is the same as for interference patterns formed by electromagnetic waves. The moiré effect is therefore often termed mechanical interference. The main difference lies in the difference in wavelength which constitutes a factor of about 10^2 and greater.

The moiré effect can be observed in our everyday surroundings. Examples are folded fine-meshed curtains (moiré means watered silk), rails on each side of a bridge or staircase, nettings, etc.

Moiré as a measurement technique can be traced many years back. Today there is little left of the moiré effect, but techniques applying gratings and other type of fringes are widely used. In this chapter we go through the theory for superposition of gratings with special emphasis on the fringe projection technique. The chapter ends with a look at a triangulation probe.

7.2 SINUSOIDAL GRATINGS

Often, gratings applied in moiré methods are transparencies with transmittances given by a square-wave function. Instead of square-wave functions, we describe linear gratings by sinusoidal transmittances (reflectances) bearing in mind that all types of periodic gratings can be described as a sum of sinusoidal gratings. A sinusoidal grating of constant frequency is given by

$$t_1(x, y) = a + a \cos\left(\frac{2\pi}{p}x\right) \qquad (7.1)$$

where p is the grating period and where $0 < a < \frac{1}{2}$. The principle behind measurement applications of gratings is that they in some way become phase modulated (see

Section 4.7). This means that the grating given by Equation (7.1) can be expressed as

$$t_2(x, y) = a + a \cos 2\pi \left(\frac{x}{p} + \psi(x) \right) \tag{7.2}$$

$\psi(x)$ is the modulation function and is equal to the displacement of the grating lines from its original position divided by the grating period

$$\psi(x) = \frac{u(x)}{p} \tag{7.3}$$

where $u(x)$ is the displacement.

When the two gratings given by Equations (7.1) and (7.2) are laid in contact, the resulting transmittance t becomes the product of the individual transmittances, viz.

$$\begin{aligned}
t(x, y) &= t_1 t_2 \\
&= a^2 \left\{ 1 + \cos \frac{2\pi}{p} x + \cos 2\pi \left[\frac{x}{p} + \psi(x) \right] \right. \\
&\left. + \frac{1}{2} \cos 2\pi \left[\frac{2x}{p} + \psi(x) \right] + \frac{1}{2} \cos 2\pi \psi(x) \right\}
\end{aligned} \tag{7.4}$$

The first three terms represent the original gratings, the fourth term the second grating with doubled frequency, while the fifth term depends on the modulation function only. It is this term which describes the moiré pattern.

Another way of combining gratings is by addition (or subtraction). This is achieved by e.g. imaging the two gratings by double exposure onto the same negative. By addition we get

$$t(x, y) = t_1 + t_2 = 2a \left\{ 1 + \cos \pi \psi(x) \cos 2\pi \left[\frac{x}{p} + \frac{1}{2} \psi(x) \right] \right\} \tag{7.5}$$

Here we see that the term $\cos \pi \psi(x)$ describing the moiré fringes are amplitude modulating the original grating.

Both Equations (7.4) and (7.5) have a maximum resulting in a bright fringe whenever

$$\psi(x) = n, \quad \text{for} \quad n = 0, \pm 1, \pm 2, \pm 3, \ldots \tag{7.6}$$

and minima (dark fringes) whenever

$$\psi(x) = n + \tfrac{1}{2}, \quad \text{for} \quad n = 0, \pm 1, \pm 2, \pm 3, \ldots \tag{7.7}$$

Both grating t_1 and t_2 could be phase-modulated by modulation functions ψ_1 and ψ_2 respectively. Then $\psi(x)$ in Equations (7.6) and (7.7) has to be replaced by

$$\psi(x) = \psi_1(x) - \psi_2(x) \tag{7.8}$$

In both multiplication and addition (subtraction), the grating becomes demodulated (see Section 3.6.4) thereby getting a term depending solely on $\psi(x)$, describing the moiré fringes. By using square wave (or other types) of gratings, the result will be completely analogous.

Below we shall find the relations between $\psi(x)$ (and $u(x)$) and the measuring parameters for the different applications.

7.3 MOIRÉ BETWEEN TWO ANGULARLY DISPLACED GRATINGS

The mathematical description of this case is the same as for two plane waves interfering under an angle α (see Section 3.4). When two gratings of transmittances t_1 and t_2 are laid in contact, the resulting transmittance is not equal to the sum $t_1 + t_2$ as in Section 3.4, but the product $t_1 \cdot t_2$. The result is, however, essentially the same, i.e. the gratings form a moiré pattern with interfringe distance (cf. Equation (3.21))

$$d = \frac{p}{2 \sin \frac{\alpha}{2}} \qquad (7.9)$$

This can be applied for measuring α by measurement of d.

7.4 MEASUREMENT OF IN-PLANE DEFORMATION AND STRAINS

When measuring in-plane deformations a grating is attached to the test surface. When the surface is deformed, the grating will follow the deformation and will therefore be given by Equation (7.2). The deformation $u(x)$ will be given directly from Equation (7.3):

$$u(x) = p\psi(x) \qquad (7.10)$$

To obtain the moiré pattern, one may apply one of several methods (Post 1982; Sciammarella 1972, 1982):

(1) Place the reference grating with transmittance t_1 in contact with the model grating with transmittance t_2. The resulting intensity distribution then becomes proportional to the product $t_1 \cdot t_2$.

(2) Image the reference grating t_1 onto the model grating t_2. The resulting intensity then becomes proportional to the sum $t_1 + t_2$. This can also be done by forming the reference grating by means of interference between two plane coherent waves.

(3) Image the model grating t_2, and place the reference grating t_1 in the image plane. t_1 then of course has to be scaled according to the image magnification. The resulting intensity becomes proportional to $t_1 \cdot t_2$.

(4) Image the reference grating given by t_1 onto a photographic film and thereafter image the model grating given by t_2 after deformation onto another film. Then the two films are laid in contact. The result is $t_1 \cdot t_2$.

(5) Do the same as under (4) except that t_1 and t_2 are imaged onto the same negative by double exposure. The result is $t_1 + t_2$.

Other arrangements might also be possible. In applying methods (1), (3) and (4), the resulting intensity distribution is proportional to $t_1 \cdot t_2$ and therefore given by Equation (7.4) which can be written

$$I(x) = I_0 + I_1 \cos 2\pi \psi(x) + \text{terms of higher frequencies} \tag{7.11}$$

By using methods (2) and (5), the intensity distribution becomes equal to $t_1 + t_2$ and therefore given by Equation (7.5), which can be written

$$I(x) = I_0 + I_1 \cos \pi \psi(x) \cos \frac{2\pi x}{p} + \text{other terms} \tag{7.12}$$

We see that by using methods (1), (3) and (4) we essentially get a DC-term I_0, plus a term containing the modulation function. In methods (2) and (5) this last term amplitude modulates the original reference grating. When applying low-frequency gratings, all these methods may be sufficient for direct observation of the modulation function, i.e. the moiré fringes. When using high-frequency gratings, however, direct observation might be impossible due to the low contrast of the moiré fringes. This essentially means that the ratio I_1/I_0 in Equations (7.11) and (7.12) is very small. We then have the possibility of applying optical filtering (see Section 4.5). For methods (4) and (5), this can be accomplished by placing the negative into a standard optical filtering set-up. Optical filtering techniques can be incorporated directly into the set-up of methods (1) and (2) by using coherent light illumination and observing the moiré patterns in the first diffracted side orders. A particularly interesting method (belonging to method (2)) devised by Post (1971) is shown in Figure 7.1. Here the reference grating is formed by interference between a plane wave and its mirror image. The angle of incidence and grating period are adjusted so that the direction of the first diffracted side order coincides with the object surface normal. Experiments using model gratings of frequencies as high as 600 lines/mm have been reported by application of this method. To get sufficient amount of light into the first diffraction order one has to use phase-type gratings as the model grating. For the description of how to replicate fine diffraction gratings onto the object surface the reader is referred to the work of Post.

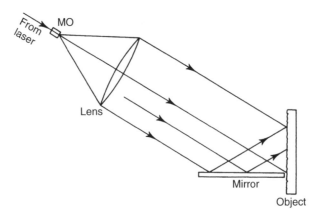

Figure 7.1

By using methods (3), (4) and (5) the grating frequency (i.e. the measuring sensitivity) is limited by the resolving power of the imaging lens. For curved surfaces, the model grating will be modulated due to the curvature, which can lead to false information about the deformation when using methods (1), (2) and (3). This is not the case for methods (4) and (5) because this modulation is incorporated in the reference grating (the first exposure). Surface curvature might also be a problem when using methods (3), (4) and (5) because of the limited depth of focus of the imaging lens. If we neglect the above-mentioned drawbacks, methods (1), (2) and (3) have the advantage of measuring the deformation in real time.

By using one of these methods, we will, either directly or by means of optical filtering, obtain an intensity distribution of the same form as given in the two first terms in Equation (7.11) or (7.12). This distribution has a

$$\text{maximum whenever } \psi(x) = n, \quad \text{for} \quad n = 0, 1, 2, \dots$$

$$\text{minimum whenever } \psi(x) = n + \tfrac{1}{2}, \quad \text{for} \quad n = 0, 1, 2, \dots$$

According to Equation (7.10) this corresponds to a displacement equal to

$$u(x) = np \quad \text{for maxima} \tag{7.13a}$$

$$u(x) = (n + \tfrac{1}{2})p \quad \text{for minima} \tag{7.13b}$$

Figure 7.2(a) shows an example of such an intensity distribution with the corresponding displacement and strain in Figures 7.2(b) and (c).

By orienting the model grating and the reference grating along the y-axis, we can in the same manner find the modulation function $\psi_y(y)$ and the displacement $v(y)$ in the y-direction. $\psi_x(x)$ and $\psi_y(y)$ can be detected simultaneously by applying crossed gratings, i.e. gratings of orthogonal lines in the x- and y-directions. Thus we also are able to calculate the strains

$$\varepsilon_x = p \frac{\partial \psi_x}{\partial x} \tag{7.14a}$$

$$\varepsilon_y = p \frac{\partial \psi_y}{\partial y} \tag{7.14b}$$

$$\gamma_{xy} = p \left[\frac{\partial \psi_x}{\partial y} + \frac{\partial \psi_y}{\partial x} \right] \tag{7.14c}$$

7.4.1 Methods for Increasing the Sensitivity

In many cases the sensitivity, i.e. the displacement per moiré fringe, may be too small. A lot of effort has therefore been put into increasing the sensitivity of the different moiré techniques (Gåsvik and Fourney 1986). The various amendments made to the solution of this problem can be grouped into three methods: fringe multiplication, fringe interpolation and mismatch.

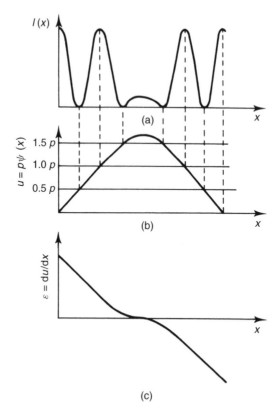

Figure 7.2 (a) Example of the intensity distribution of a moiré pattern with the corresponding; (b) displacement; and (c) strain

Fringe multiplication

In moiré methods one usually employs square-wave or phase gratings as model gratings. An analysis of such gratings would have resulted in expressions for the intensity distribution equivalent to Equations (7.11) and (7.12), but with an infinite number of terms containing frequencies which are integral multiples of the basic frequency. When using such gratings it is therefore possible to filter out one of the higher-order terms by means of optical filtering. By filtering out the Nth order, one obtains N times as many fringes and therefore an N-fold increase of the sensitivity compared to the standard technique. This is the concept of fringe multiplication. However, the intensity distribution of the harmonic terms generally decreases with increasing orders which therefore sets an upper bound to the multiplication process. Although in some special cases multiplications up to 30 have been reported, practical multiplications can rarely exceed 10.

Fringe interpolation

This method consists of determining fractional fringe orders. It can be done by scanning the fringe pattern with a slit detector or taking microdensitometer readings from a photograph of the fringes. It can also be done by digitizing the video signal from a TV picture. These methods are limited by the unavoidable noise in the moiré patterns. When forming the reference grating by interference between two plane waves, interpolation can be achieved by moving the phase of one of the plane waves. This is easily obtained by means of e.g. a quarterwave plate and a rotatable polarizer in the beam of the plane wave.

For more details of such methods, see Chapter 11.

Mismatch

This is a term concerning many techniques. It consists of forming an initial moiré pattern between the model and reference grating before deformation. Instead of counting fringe orders due to the deformation, one measures the deviation or curvature of the initial pattern. The initial pattern can be produced by gratings having different frequencies, by a small rotation between the model and reference grating or by a small gap between them. In this way one can increase the sensitivity by at least a factor of 10.

This is equivalent to the spatial carrier method described in Section 11.4.3.

7.5 MEASUREMENT OF OUT-OF-PLANE DEFORMATIONS. CONTOURING

7.5.1 Shadow Moiré

We shall now describe an effect where moiré fringes are formed between a grating and its own shadow: so-called shadow moiré. The principle of the method is shown in Figure 7.3.

The grating lying over the curved surface is illuminated under the angle of incidence θ_1 (measured from the grating normal) and viewed under an angle θ_2. From the figure we see that a point P_0 on the grating is projected to a point P_1 on the surface which by viewing is projected to the point P_2 on the grating. This is equivalent to a displacement of the grating relative to its shadow equal to

$$u = u_1 + u_2 = h(x, y)(\tan \theta_1 + \tan \theta_2) \tag{7.15}$$

where $h(x, y)$ is the height difference between the grating and the point P_1 on the surface. In accordance with Equation (7.3), this corresponds to a modulation function

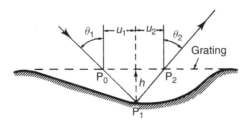

Figure 7.3   Shadow moiré

equal to

$$\psi(x) = \frac{u}{p} = \frac{h(x, y)}{p}(\tan\theta_1 + \tan\theta_2) \tag{7.16}$$

A bright fringe is obtained whenever $\psi(x) = n$, for $n = 0, 1, 2, \ldots$, which gives

$$h(x, y) = \frac{np}{\tan\theta_1 + \tan\theta_2} \tag{7.17a}$$

and

$$h(x, y) = \frac{(n + \frac{1}{2})p}{\tan\theta_1 + \tan\theta_2} \tag{7.17b}$$

for dark fringes. In this way, a topographic map is formed over the surface.

In the case of plane wave illumination and observation from infinity, θ_1 and θ_2 will remain constant across the surface and Equation (7.17) describes a contour map with a constant, fixed contour interval. With the point source and the viewing point at finite distances, θ_1 and θ_2 will vary across the surface resulting in a contour interval which is dependent on the surface coordinates. This is of course an unsatisfactory condition. However, if the point source and the viewing point are placed at equal heights z_p above the surface and if the surface height variations are negligible compared to z_p, then $\tan\theta_1 + \tan\theta_2$ will be constant across the surface resulting in a constant contour interval. This is a good solution, especially for large surface areas which are impossible to cover with a plane wave because of the limited aperture of the collimating lens.

If the surface height variations are large compared to the grating period, diffraction effects will occur, prohibiting a mere shadow of the grating to be cast on the surface. Shadow moiré is therefore best suited for rather coarse measurements on large surfaces. It is relatively simple to apply and the necessary equipment is quite inexpensive. It is a valuable tool in experimental mechanics and for measuring and controlling shapes.

Perhaps the most successful application of the shadow moiré method is in the area of medicine, such as the detection of scoliosis, a spinal disease which can be diagnozed by means of the asymmetry of the moiré fringes on the back of the body. Takasaki (1973, 1982) has worked extensively with shadow moiré for the measurement of the human body. He devised a grating made by stretching acrylic monofilament fibre on a frame using screws or pins as the pitch guide. According to him, the grating period should be 1.5–2.0 mm, and the diameter should be half the grating period. The grating should be sprayed black with high-quality dead back paint. Figure 7.4 shows an example of contouring of a mannequin of real size using shadow moiré.

7.5.2 Projected Fringes

We now describe a method where fringes are projected onto the test surface. Figure 7.5 shows fringes with an inter-fringe distance d projected onto the xy-plane under an angle θ_1 to the z-axis. The fringe period along the x-axis then becomes

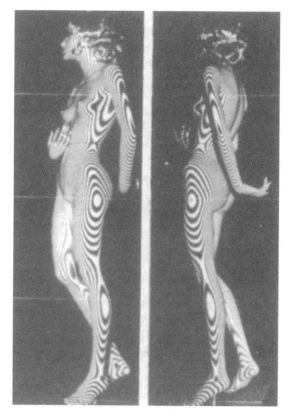

Figure 7.4   Shadow moiré contouring. (Reproduced from Takasaki 1973 by permission of Optical Society of America.)

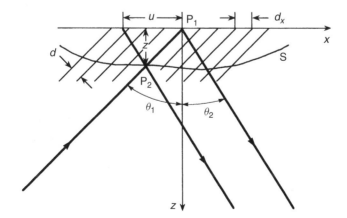

Figure 7.5   Fringe projection geometry. θ_1 = projection angle. θ_2 = viewing angle

$$d_x = \frac{d}{\cos \theta_1} \tag{7.18}$$

Also in the figure is drawn a curve S representing a surface to be contoured. From the figure we see that a fringe originally positioned at P_1 will be displaced to P_2. This displacement is given by

$$u = z(\tan \theta_1 + \tan \theta_2) \tag{7.19}$$

where z is the height of P_2 above the xy-plane and θ_2 is the viewing angle. From Equation (7.3) this gives a modulation function equal to

$$\psi(x) = \frac{u}{d_x} = \frac{z(\tan \theta_1 + \tan \theta_2)}{d/\cos \theta_1} = \frac{z}{d}(\sin \theta_1 + \cos \theta_1 \tan \theta_2) = \frac{z}{d}\frac{\sin(\theta_1 + \theta_2)}{\cos \theta_2} = \frac{z}{d}G \tag{7.20}$$

where we have introduced the geometry factor

$$G = G(\theta_1, \theta_2) = \sin \theta_1 + \cos \theta_1 \tan \theta_2 = \frac{\sin(\theta_1 + \theta_2)}{\cos \theta_2} \tag{7.21}$$

One method of projecting fringes on a surface is by means of interference between two plane waves inclined at a small angle α to each other. This can be achieved by means of a Twyman-Green interferometer with a small tilt of one of the mirrors (Figure 7.6(a)). The distance between the interference fringes is then equal to

$$d = \frac{\lambda}{2 \sin(\alpha/2)} \tag{7.22}$$

where λ is the wavelength and α is the angle between the two plane waves.

From Equation (7.2) and Equations (7.18)–(7.21), the intensity distribution across the surface can then be written as

$$I = 2\left\{1 + \cos 2\pi \left[\frac{x}{d_x} + \psi(x)\right]\right\} = 2\left\{1 + \cos \frac{2\pi}{d}[x \cos \theta_1 + zG]\right\} \tag{7.23}$$

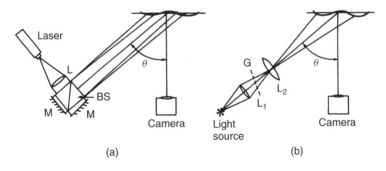

Figure 7.6 Fringe projection by means of (a) interference and (b) grating imaging. L = lenses, M = mirrors, BS = beamsplitter, G = grating

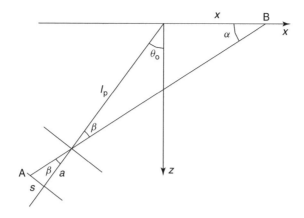

Figure 7.7 Grating projection

Figure 7.6(b) shows another method for projecting a fringe pattern onto the surface. Here, a grating is imaged onto the surface by means of a lens L$_2$. This situation can be analysed more closely from Figure 7.7 where a light ray through the centre of the projection lens goes from point A on the grating to point B on the xy-plane. A lies a distance s from the optical axis of the projection lens and B is a distance x from the origin of the coordinate system. From the figure we see that

$$\frac{x}{\sin \beta} = \frac{l_p}{\sin \alpha} = \frac{l_p}{\cos(\theta_o + \beta)} = \frac{l_p}{\cos \theta_o \cos \beta - \sin \theta_o \sin \beta} \tag{7.24}$$

which gives

$$\frac{x}{\tan \beta} = \frac{l_p}{\cos \theta_o - \tan \beta \sin \theta_o} \tag{7.25}$$

where l_p is the distance from the lens to the origin of the coordinate system. By inserting

$$\tan \beta = s/a \tag{7.26}$$

where a is the grating–lens distance, we get

$$x = x(s) = \frac{s l_p}{a \cos \theta_o - s \sin \theta_o} \tag{7.27}$$

Equation (7.27) gives the position $x = x(s)$ as a function of the position s on the grating. By increasing s by d_g, the grating period, we get for the fringe period d_x along the x-axis

$$d_x = x(s + d_g) - x(s) = \frac{d_g a l_p \cos \theta_o}{(a \cos \theta_o - s \sin \theta_o)^2 - d_g \sin \theta_o (a \cos \theta_o - s \sin \theta_o)} \tag{7.28}$$

In the following we approximate d_x/d_g by dx/ds, the derivative of x with respect to s:

$$\frac{d_x}{d_g} \approx \frac{dx}{ds} = \frac{a l_p \cos \theta_o}{(a \cos \theta_o - s \sin \theta_o)^2} \tag{7.29}$$

which we see is equal to Equation (7.28) when putting $d_g = 0$ in the denominator. This is a good approximation since d_g will be small compared to a. From Equation (7.27) we solve for s:

$$s = \frac{ax \cos \theta_o}{l_p + x \sin \theta_o} \tag{7.30}$$

which put into Equation (7.29) finally gives

$$d_x = \frac{d_g(l_p + x \sin \theta_o)^2}{al_p \cos \theta_o} = \frac{d_g l_p}{a \cos \theta_o}\left[1 + \frac{x \sin \theta_o}{l_p}\right]^2 = d_{xo}\left[1 + \frac{x \sin \theta_o}{l_p}\right]^2 \tag{7.31}$$

In the last equation we have introduced the quantity

$$d_{xo} = \frac{d_g}{\cos \theta_o}\frac{l_p}{a} \tag{7.32}$$

the fringe period for $x = 0$. When the grating is focused at $x = 0$, the projection magnification is

$$m_p = l_p/a \tag{7.33}$$

which gives

$$d_{xo} = m_p \frac{d_g}{\cos \theta_o} \tag{7.34}$$

For the instantaneous frequency we get

$$f_x = \frac{1}{d_x} = f_o\left[1 + \frac{x \sin \theta_o}{l_p}\right]^{-2} \tag{7.35}$$

where

$$f_o = \frac{1}{d_{xo}} = \frac{\cos \theta_o}{m_p d_g} \tag{7.36}$$

The phase along a line normal to the grating lines is given as s/d_g. We find the phase φ in the xy-plane from Equation (7.30):

$$\varphi = \frac{s}{d_g} = \frac{ax \cos \theta_0}{d_g(l_p + x \sin \theta_o)} = \frac{f_o x}{\left(1 + \dfrac{\sin \theta_o}{l_p}x\right)} \tag{7.37}$$

where f_o is given by Equation (7.36). The intensity in the xy-plane therefore can be written in the 'usual' way as

$$I(x, y) = A + B \cos 2\pi \varphi \tag{7.38}$$

with φ given from Equation (7.37). From the definition of the instantaneous frequency, we get

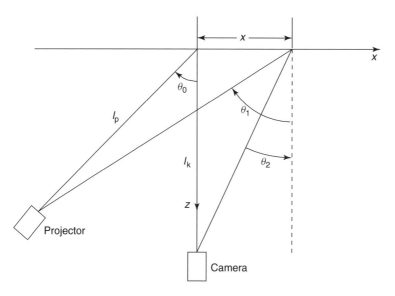

Figure 7.8 Fringe projection geometry

$$f_x = \frac{d\phi}{dx} = f_o \left[1 + \frac{\sin \theta_o}{l_p} x \right]^{-2}$$ (7.39)

which agrees with Equation (7.35).

When the camera is pointing along the z-axis, we see from Figure 7.8 that

$$\tan \theta_1 = \frac{l_p \sin \theta_o + x}{l_p \cos \theta_o}$$ (7.40)

$$\tan \theta_2 = \frac{-x}{l_k}$$ (7.41)

where θ_o is the projection angle measured from the z-axis and l_p and l_k are the projection and camera distances respectively. This gives for the displacement u and the phase ψ:

$$u(x) = z(\tan \theta_1 + \tan \theta_2) = \frac{z}{\cos \theta_o} \left[\sin \theta_o + \frac{(l_k - l_p \cos \theta_o)x}{l_p l_k} \right]$$ (7.42)

$$\psi(x) = \frac{u(x)}{d_x} = \frac{z}{d_{xo} \cos \theta_o} \left[\sin \theta_o + \frac{(l_k - l_p \cos \theta_o)x}{l_p l_k} \right] \left[1 + \frac{x \sin \theta_o}{l_p} \right]^{-2}$$ (7.43)

From Equation (7.42) we see that the displacement u becomes dependent on x only through z (i.e. the contour interval becomes independent of the position on the surface) if the projection lens and the camera lens are placed at equal heights above the xy-plane ($l_k - l_p \cos \theta_o = 0$).

Note that in general (from Equation (7.31)) the fringe period d_x is not constant but depends on x. The phase $\psi(x)$ given by Equation (7.43) therefore becomes more and more prone to error as the displacement $u(x)$ exceeds d_x with a factor much greater than 1. This can, however, be solved by dividing by a sum of d_x in the right direction from the evaluation point. By inverting Equations (7.42) and (7.43) we get

$$z(x) = \cos\theta_o \left[\sin\theta_o + \frac{(l_k - l_p \cos\theta_o)x}{l_p l_k} \right]^{-1} u(x) \tag{7.44}$$

$$z(x) = d_{xo} \cos\theta_o \left[\sin\theta_o + \frac{(l_k - l_p \cos\theta_o)x}{l_p l_k} \right]^{-1} \left[1 + \frac{x \sin\theta_o}{l_p} \right]^2 \psi(x) \tag{7.45}$$

7.5.3 Vibration Analysis

In the same manner as for holographic interferometry, moiré technique using projected fringes (or shadow moiré) can be applied for vibration analysis of surfaces (Hazell and Niven 1968; Vest and Sweeney 1972; Harding and Harris 1983). Let us analyse this method more closely.

Assume that a point on the surface in Figure 7.5 executes harmonic out-of-plane vibrations given by

$$z = z_0 + a \cos\omega t \tag{7.46}$$

where z_0 is the equilibrium position, a is the amplitude and ω the frequency.

The intensity distribution of the projected pattern (cf. Equation (7.23)) now becomes

$$I(x, t) = 2 \left[1 + \cos\frac{2\pi}{d}(x \cos\theta + (z_0 + a \cos\omega t) \sin\theta) \right] \tag{7.47}$$

where we for simplicity have assumed that the camera is imaging from infinity along the z-axis, i.e. $G = \sin\theta$ where θ is the projection angle.

The expression can be written as

$$I(x, t) = 2[1 + \cos(\phi_c + \phi_t)] = 2 + e^{i(\phi_c + \phi_t)} + e^{-i(\phi_c + \phi_t)} \tag{7.48}$$

where

$$\phi_c = (2\pi/d)(x \cos\theta + z_0 \sin\theta) \tag{7.49a}$$

$$\phi_t = (2\pi/d) \sin\theta a \cos\omega t \tag{7.49b}$$

By photographing this pattern with an exposure time much longer than the vibration period T, the resulting transmittance t of the film becomes proportional to $I(x, t)$ averaged over the vibration period. This is analogous to time-average holography (see Section 6.9) and gives for the transmittance

$$t = \frac{1}{T} \int_0^T I(x, y) \, dt = 2 + \frac{e^{i\phi_c}}{T} \int_0^T e^{i\phi_t} \, dt + \frac{e^{-i\phi_c}}{T} \int_0^T e^{-i\phi_t} \, dt \tag{7.50}$$

Now we have

$$\frac{1}{T} \int_0^T e^{\pm i\phi_t} \, dt = \frac{1}{T} \int_0^T \exp[\pm i\,(2\pi/d)\sin\theta a \cos\omega t]\,dt$$
$$= J_0 \left(\frac{2\pi}{d} a \sin\theta\right) \tag{7.51}$$

which inserted into Equation (7.50) gives

$$t = 2\left[1 + J_0\left(\frac{2\pi}{d} a \sin\theta\right)\cos\phi_c\right] \tag{7.52}$$

where J_0 is the zeroth-order Bessel function. From this expression we see that the Bessel function amplitude modulates the fringe pattern on the static surface given by z_0. This is illustrated in Figure 7.9.

The negative with the transmittance given from Equation (7.52) can be placed in an optical filtering system in the same way as in the static case, resulting in an amplitude distribution in the image plane equal to

$$u = J_0\left(\frac{2\pi}{d} a \sin\theta\right) e^{i\phi_c} \tag{7.53}$$

and hence an intensity distribution given by

$$I = J_0^2\left(\frac{2\pi}{d} a \sin\theta\right) \tag{7.54}$$

From the values of the arguments of the Bessel function corresponding to its maximum and zeros given on page 167, we find that light fringes occur when

$$a = \frac{d}{2\pi \sin\theta} \times [0, 3.83, 7.02, 10.17, \ldots] \tag{7.55a}$$

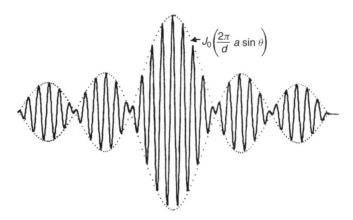

Figure 7.9

and dark fringes occur when

$$a = \frac{d}{2\pi \sin \theta} \times [2.40, 5.52, 8.65, 11.79, \ldots] \tag{7.55b}$$

The first dark fringe of this pattern thus corresponds to an amplitude a_l equal to

$$a_l = 0.38 \frac{d}{\sin \theta} \tag{7.56}$$

which is a figure representing the sensitivity of the method.

7.5.4 Moiré Technique by Means of Digital Image Processing

A very convenient way of adding (subtracting) pictures is by means of digital image processing (Gåsvik 1983). A set-up for projection moiré is shown in Figure 7.10 where a grating is projected onto the object. The surface with projected fringes are imaged by a TV camera and the video signal is sent to an image processor, see Chapter 10. In this way it is possible to subtract a stored image from the image seen by the camera in real time. Figure 7.11 shows some examples of the results obtained with such a system. Figure 7.11(a) shows a cartridge casing with a dent. The reference image stored into the memory is taken from the undefective side of the casing. Figure 7.11(b) shows the result from two recordings of a 25-litre oil can before and after filling with water. In Figure 7.11(c) the system is applied to vibration analysis. It shows a circular plate centrally clamped to a shaft and excited by a shaker at a point in the lower right edge. The picture is a time-average recording resulting in a zeroth-order Bessel fringe function displaying the amplitude distribution of the plate as described in Section 7.5.3. For time-average recordings, the image processor is not necessary.

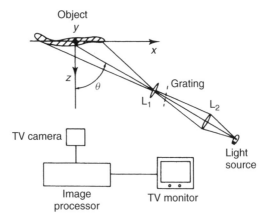

Figure 7.10   Projection moiré using TV-camera and digital image processor

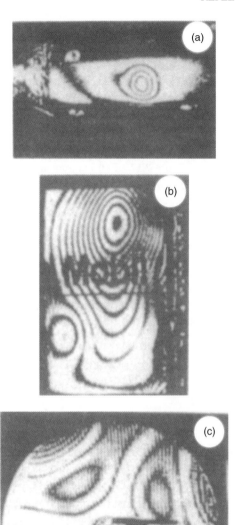

Figure 7.11 Examples of TV-moiré fringes: (a) cartridge casing with a dent. Contour interval 0.15 mm; (b) 25-litre oil can after filling with water; (c) time-average recording of a 400-mm diameter aluminium plate excited in the lower right edge. Frequency 250 Hz. Amplitude corresponding to the first dark fringe 0.16 mm

7.6 REFLECTION MOIRÉ

As we have already seen, moiré technique offers a wide variety of application methods. Most of them are, however, variations of the basic principles discussed in the preceding sections. Here we mention a method which to a certain extent differs from the previous

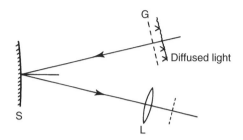

Figure 7.12   Reflection moiré

techniques. It is called Lichtenberg's method (Lichtenberg 1955) and can be applied on shiny, mirror-like surfaces and phase-objects (Liasi and North 1994).

Figure 7.12 shows the principle of the method. The smoothness of the surface S makes it possible to image the mirror image of the grating G by means of the lens L. As in previous methods, a grating can be placed in the image plane of L or the mirror image of G can be photographed before and after the deformation of S. The result is a moiré pattern defining the derivative of the height profile, i.e. the slope of the deformation.

In an analysis of the resolution of the reflection moiré method it is found that the maximum resolution that can be obtained with a viewing camera is of the order 7×10^{-3} radians.

7.7   TRIANGULATION

Shadow moiré and projected fringes are techniques based on the triangulation principle. We close this chapter by considering a simple triangulation probe. In Figure 7.13 a laser beam is incident on a diffusely scattering surface under an angle θ_1. The resulting light spot on the surface is imaged by a lens onto a detector D. The optical axis of the lens makes an angle θ_2 to the surface normal. Assume that the object moves a distance s normal to its surface. From the figure, using simple trigonometric relations, we find that

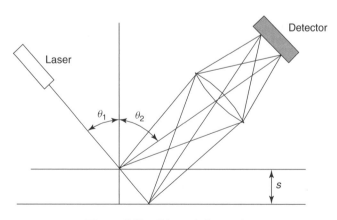

Figure 7.13   Triangulation probe

the corresponding movement of the imaged spot on the detector is given by (see Eq. 7.21)

$$s' = m\frac{s\sin(\theta_1 + \theta_2)}{\cos\theta_1} = ms(\tan\theta_1\cos\theta_2 + \sin\theta_2) \tag{7.57}$$

where m is the transversal magnification of the lens. The detector is position-sensing, i.e. it gives an output voltage proportional to the distance of the light spot from the centre of the detector. It is the centroid of the light spot that is sensed and thus the position measurement is independent of the spot diameter as long as it is inside the detector area. Therefore sharp focusing is not critical. The position of an unexpanded laser beam directly incident on such a detector can be determined to an accuracy of less than 1μm. From Equation (7.57) we see that the movement s' can be magnified by the lens, thereby increasing the sensitivity. However, the size of the light spot will also be magnified, and this must always lie inside the detector area to avoid measurement errors, thus limiting the usable magnification.

In many applications, θ_1 is set to zero. Then very precise measurements of movements along the light beam (the z-axis) can be made. Since the light spot is then always on the z-axis, it is a good idea to tilt the detector such that the spot is always focused on the detector. To make measurements on small details, the diameter of a laser beam might be too large. Then clever optics forming a thin beam through the measurement volume have to be constructed and light sources other than lasers might be a better alternative. Such probes can be used to measure the profile of screws and other small parts, for example.

PROBLEMS

7.1 Two gratings with amplitude transmittance

$$t(x, y) = a\left[1 + \cos\left(\frac{2\pi}{p}x\right)\right]$$

are laid in contact with an angle α between the grating lines. Calculate $t_1 \cdot t_2$.

7.2 A circular zone plate with centre at (x_0, y_0) has an amplitude transmittance given by

$$t(x, y) = \tfrac{1}{2}\{1 + \cos\beta[(x - x_0)^2 + (y - y_0)^2]\}$$

where β is a constant. Suppose that two zone plate transmittances are laid in contact with a displacement d between their centres.

(a) Show that the resulting moiré fringes describes a new zone-plate pattern.

(b) Find the centre and the frequency of this new zone plate pattern.

7.3 We attach a grating of period p to a bar, whereafter the bar undergoes a uniaxial tension resulting in a strain equal to ε_x.

(a) Write down the transmittances t_1 and t_2 of the grating before and after the load.

(b) Find an expression for $t_1 + t_2$ describing the moiré pattern.

(c) If $\varepsilon_x = 10^{-3}$, the grating frequency is 20 lines/mm and the length of the bar is 10 cm, how many moiré fringes are observed?

7.4 In shadow moiré, the reference grating t_1 is multiplied by the object (shadow) grating t_2. In what sense can we say that this is a sort of synchronous demodulation of t_2? What is the low-pass filter in this case?

7.5 Consider the method of projected fringes using interference between two plane waves (Figure 7.6(a)). Assume that the camera is looking normal to the surface from infinity (i.e. $\theta_2 = 0$).

(a) calculate the out-of-plane displacement Δz per fringe in this case.

(b) Based on pure geometric considerations, there is a limit to the sensitivity (displacement per fringe) in this set-up. Find this limit and the corresponding values of θ_1 and α (the angle between the two plane waves).

(c) Using $\lambda = 600$ nm, calculate the mean spatial frequency in lines/mm on the surface in this case. Why is this measuring sensitivity unattainable?

8

Speckle Methods

8.1 INTRODUCTION

When looking at the laser light scattered from a rough surface, one sees a granular pattern as in Figure 8.2. This so-called speckle pattern can be regarded as a multiple wave interference pattern with random individual phases. In the years following the advent of the laser, this pattern was considered a mere nuisance, especially in holography (and it still is!). But from the beginning of 1970 there were several reports from experiments in which speckle was exploited as a measuring tool. In this chapter the basic principles of the different techniques of speckle metrology will be described. As a spin-off from laser speckle, methods based on similar principles using incoherent light have emerged. This white-light speckle photography is included in the final section of the chapter.

8.2 THE SPECKLE EFFECT

In Figure 8.1, light is incident on, and scattered from, a rough surface of height variations greater than the wavelength λ of the light. As is shown in the figure, light is scattered in all directions. These scattered waves interfere and form an interference pattern consisting of dark and bright spots or speckles which are randomly distributed in space. In white light illumination, this effect is scarcely observable owing to lack of spatial and temporal coherence (see Section 3.3). Applying laser light, however, gives the scattered light a characteristic granular appearance as shown in the image of a speckle pattern in Figure 8.2.

It is easily realized that the light field at a specific point in a speckle pattern must be the sum of a large number N of components representing the light from all points on the scattering surface. The complex amplitude at point in a speckle pattern can therefore be written as

$$u = \frac{1}{\sqrt{N}} \sum_{k=1}^{N} u_k = \frac{1}{\sqrt{N}} \sum_{k=1}^{N} U_k e^{i\phi_k} \tag{8.1}$$

By assuming that (1) the amplitude and phase of each component are statistically independent and also independent of the amplitudes and phases of all other components, and (2) the phases ϕ_k are uniformly distributed over all values between $-\pi$ and $+\pi$, Goodman (1975) has shown that the complex amplitude u will obey Gaussian statistics. Further he

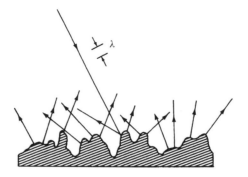

Figure 8.1   Light scattering from a rough surface

Figure 8.2   Photograph of a speckle pattern

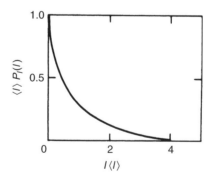

Figure 8.3

has shown that the probability density function P_I for the intensity in a speckle pattern is given as

$$P_I(I) = \frac{1}{\langle I \rangle} \exp\left(-\frac{I}{\langle I \rangle}\right) \tag{8.2}$$

where $\langle I \rangle$ is the mean intensity. The intensity of a speckle pattern thus obeys negative exponential statistics. Figure 8.3 shows a plot of $P_I(I)$. From this plot we see that the most probable intensity value is zero, that is, black.

A measure of the contrast in a speckle pattern is the ratio $C = \sigma_I/\langle I \rangle$, where σ_I is the standard deviation of the intensity given by

$$\sigma_I^2 = \langle \Delta I^2 \rangle = \langle (I - \langle I \rangle)^2 \rangle = \langle I^2 - 2\langle I \rangle I + \langle I \rangle^2 \rangle = \langle I^2 \rangle - \langle I \rangle^2 \qquad (8.3)$$

where the brackets denote mean values. By using

$$\langle I^2 \rangle = \int_0^\infty P_I(I) I^2 \mathrm{d}I = 2\langle I \rangle^2 \qquad (8.4)$$

we find the contrast C in a speckle pattern to be unity.

8.3 SPECKLE SIZE

From Figure 8.2 we see that the size of the bright and dark spots varies. To find a representative value of the speckle size, consider Figure 8.4, where a rough surface is illuminated by laser light over an area of cross-section D. The resulting so-called objective speckle pattern is observed on a screen S at a distance z from the scattering surface. For simplicity, we consider only the y-dependence of the intensity. An arbitrary point P on the screen will receive light contributions from all points on the scattering surface. Let us assume that the speckle pattern at P is a superposition of the fringe patterns formed by light scattered from all point pairs on the surface. Any two points separated by a distance l will give rise to fringes of frequency $f = l/(\lambda z)$ (see Section 3.6.1, Equation (3.35)). The fringes of highest spatial frequency f_{max} will be formed by the two edge points, for which

$$f_{max} = \frac{D}{\lambda z} \qquad (8.5)$$

The period of this pattern is a measure of the smallest objective speckle size σ_o which therefore is

$$\sigma_o = \frac{\lambda z}{D} \qquad (8.6)$$

For smaller separations l, there will be a large number of point pairs giving rise to fringes of the corresponding frequency. The number of point pairs separated by l is proportional

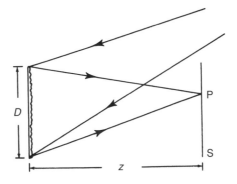

Figure 8.4   Objective speckle formation

to $(D - l)$. Since the various fringe patterns have random individual phases they will add incoherently. The contribution of each frequency to the total intensity will therefore be proportional to the corresponding number of pairs of scattering points. Since this number is proportional to $(D - l)$, which in turn is proportional to $(f_{max} - f)$, the relative number of fringes versus frequency, i.e. the spatial frequency spectrum will be linear, as shown in Figure 8.5.

Figure 8.6 shows the same situation as in Figure 8.4 except that the scattering surface now is imaged on to a screen by means of a lens L. The calculation of the size of the resulting so-called subjective speckles is analogous to the calculation of the objective speckle size. Here the cross-section of the illuminated area has to be exchanged by the diameter of the imaging lens. The subjective speckle size σ_s therefore is given as

$$\sigma_s = \frac{\lambda b}{D} \tag{8.7}$$

where b is the image distance. By introducing the aperture number

$$F = \frac{f}{D} \tag{8.8}$$

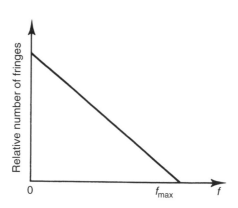

Figure 8.5

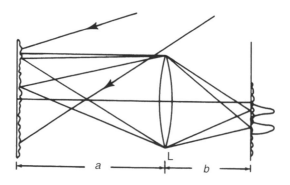

Figure 8.6 Subjective speckle formation

where f is the focal length, we get

$$\sigma_s = (1 + m)\lambda F \qquad (8.9)$$

where $m = (b - f)/f$ is the magnification of the imaging system. From this equation we see that the speckle size increases with decreasing aperture (increasing aperture number). This can be easily verified by stopping down the eye aperture when looking at a speckle pattern.

Speckle formation in imaging cannot be explained by means of geometrical optics which predicts that a point in the object is imaged to a point in the image. The field at a point in the image plane therefore should receive contributions only from the conjugate object point, thus preventing the interference with light from other points on the object surface. However, even an ideal lens will not image a point into a point but merely form a intensity distribution (the Airy disc, see Section 4.6) around the geometrical image point due to diffraction of the lens aperture. This is indicated in Figure 8.6. It is therefore possible for contributions from various points on the object to interfere so as to form a speckle pattern in the image plane.

8.4 SPECKLE PHOTOGRAPHY

Discussions on the subject of speckle photography can be found, for example, in Burch and Tokarski (1968), Dainty (1975), Erf (1978), Fourney (1978), Hung (1978) and Jones and Wykes (1989).

8.4.1 The Fourier Fringe Method

Assume that we image a speckle pattern onto a photographic film. After development this negative is placed in the object plane of a set-up for optical filtering like that in Figure 4.12. Figure 8.7 shows (Fourney 1978)

(1) a speckle pattern on the negative;

(2) the resulting diffraction pattern (the spatial frequency spectrum) in the x_f, y_f-plane; and

(3) typical form of the smoothed intensity distribution along the x_f-axis.

The dark spot in the middle of Figure 8.7(b) is due to the blocking of the strong zeroth-order component. Figure 8.7(c) displays the essential feature discussed in Section 8.2, namely that the imaged speckle pattern contains a continuum of spatial frequencies ranging from zero to $f_{max} = \pm 1/\sigma_s$, where σ_s is the smallest speckle size given from Equation (8.7). For a circular aperture the frequency distribution will not be exactly linear as in Figure 8.5, but merely as indicated in Figure 8.7(c).

Now assume that we in the diffraction plane (the filter plane) place a screen with a hole a distance x_f from the optical axis. This situation is illustrated in Figure 8.8, where Figure 8.8(a, b) shows the spectra before and after the filtering process has taken place. The same spectrum (as in Figure 8.8(b)) would have resulted by filtering out the first side

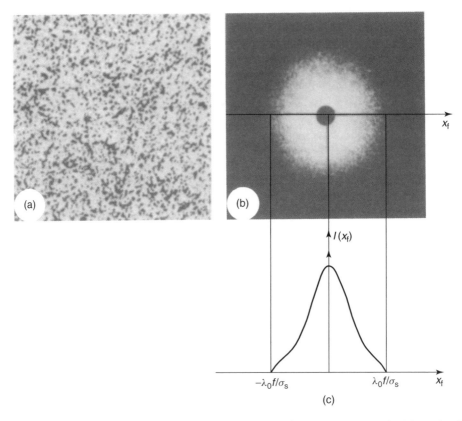

Figure 8.7   (a) Speckle pattern and its corresponding; (b) diffraction pattern; and (c) intensity distribution along the x_1-axis. ((a) and (b) reproduced from Fourney (1978) by permission of Academic Press, New York.)

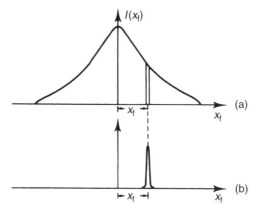

Figure 8.8   Spectrum of a speckle pattern (a) before and (b) after filtering

order from a sinusoidal grating of frequency

$$f_x = \frac{x_f}{\lambda_0 f} \tag{8.10}$$

where f is the focal length of the transforming lens and λ_0 is the wavelength of the light source applied in the filtering process. What we have actually done therefore is to select one of the numerous gratings with a continuum of directions and frequencies that constitute the laser-illuminated object.

As a consequence, we can imagine that a grating given by

$$t_1 = a(1 + \cos 2\pi f_x x) \tag{8.11}$$

is attached to the object surface. When the object undergoes an in-plane deformation, the speckle pattern will follow the displacements of the points on the object surface. Consequently, the grating will be phase-modulated and thus can be written as

$$t_2 = a[1 + \cos 2\pi (f_x x + \psi_x(x))] \tag{8.12}$$

This is closely analogous to the situation described in Section 7.4 where a model grating was attached to the object for measurement of an in-plane deformation by means of moiré technique. In the same way as described there, we can image the speckle pattern before and after the deformation onto the same film negative by double exposure. The resulting transmittance then becomes equal to $t_1 + t_2$, the sum of Equations (8.11) and (8.12). The two speckle patterns could possibly also be imaged onto two separate negatives and subsequently superposed, giving a resultant transmittance equal to $t_1 \cdot t_2$, but this is more difficult to achieve. By means of optical filtering of the double-exposed negative we get an intensity distribution dependent on the modulation function only (see Section 4.7, Equation (4.66)). This distribution will be maximum for $\psi_x(x) = n$ and minimum for $\psi_x(x) = n + (1/2)$ and correspond to displacements (cf. Equations (7.10) and (7.13)) equal to

$$u(x) = n\frac{1}{f_x} = n\frac{\lambda_0 f}{x_f} \qquad \text{for maxima} \tag{8.13a}$$

$$u(x) = \left(n + \frac{1}{2}\right)\frac{\lambda_0 f}{x_f} \qquad \text{for minima} \tag{8.13b}$$

in direct analogy with the moiré method. In deriving Equation (8.13) we have assumed unit magnification. These values therefore must be divided by the magnification m applied when recording the speckle patterns.

The speckle pattern represents gratings of all orientations in the plane of the object (cf. the spectrum in Figure 8.7(b)). Therefore, by placing the hole in the screen in the filter plane a distance y_f along the y_f-axis we obtain the modulation function $\psi_y(y)$ and the corresponding displacement $v_y(y)$ along the y-axis. Generally, by placing the hole in the filter plane at a point of coordinates x_f, y_f, the resulting intensity distribution will give a displacement equal to

$$s(x, y) = \sqrt{u^2(x) + v^2(y)} = n\lambda_0 f \sqrt{(1/x_f)^2 + (1/y_f)^2} \quad \text{for maxima} \tag{8.14a}$$

$$s(x, y) = (n + \tfrac{1}{2})\lambda_0 f \sqrt{(1/x_f)^2 + (1/y_f)^2} \quad \text{for minima} \tag{8.14b}$$

In this case we thus measure the displacement along a direction inclined an angle β to the x-axis given by

$$\tan \beta = y_f/x_f \tag{8.15}$$

In this method we can therefore vary the measuring sensitivity by varying x_f and y_f, and a remarkable property of this technique is that the sensitivity can be varied subsequent to the actual measurement, that is, after the speckle patterns have been recorded. Highest sensitivity is obtained by placing the filtering hole at the edge of the spectrum. The first-order dark fringe then corresponds to a displacement equal to

$$s_{\min} = \frac{1}{2}\sigma_s = \frac{1}{2}\frac{\lambda b}{D} = 0.5(1 + m)\lambda F \tag{8.16}$$

With a magnification $m = 1$, this gives a sensitivity equal to the laser wavelength multiplied by the aperture number of the imaging lens. The sensitivity limit by using laser speckle photography can therefore be down to about 1 μm.

Figure 8.9 shows an example of such Fourier fringes obtained by this method (Hung 1978). Note that as the filtering hole is moved away from the optical axis, the number of fringes increases, i.e. the sensitivity is increased.

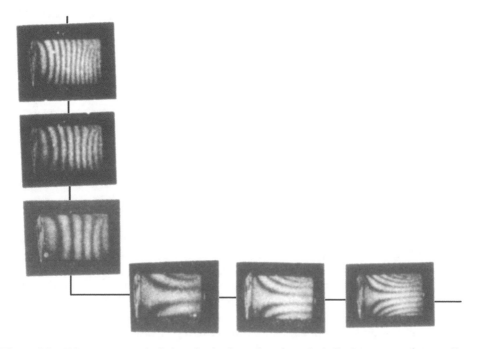

Figure 8.9 Fringe patterns depicting the horizontal and vertical displacements of a cantilever beam obtained from the various filtering positions in the Fourier filtering plane. (Reproduced from Hung 1978 by permission of Academic Press, New York.)

8.4.2 The Young Fringe Method

On photographing a speckle pattern, each speckle will form a pointlike blackening on the film. When the object undergoes an in-plane deformation, the speckle pattern will follow this deformation. A double-exposed negative of two speckle patterns resulting from a deformation therefore will consist of identical point pairs separated by a distance equal to the deformation times the magnification of the imaging system.

Assume that we illuminate this double-exposed negative with an unexpanded laser beam. When the beam covers one pair of identical points, they will act in the same way as the two holes P_1 and P_2 in the screen of the Young's interferometer (see Figure 3.13, Section 3.6.1). On a screen at a distance z from the negative we therefore will observe interference fringes which are parallel and equidistant and with a direction perpendicular to the line joining P_1 and P_2, i.e. perpendicular to the displacement.

The situation is sketched in Figure 8.10. In Section 3.6.1 we found that the distance between two adjacent fringes in this pattern is equal to

$$d = \frac{\lambda z}{D} \qquad (8.17)$$

where D is the distance between P_1 and P_2. If the displacement on the object is equal to s the separation of the corresponding speckle points on the negative is equal to $m \cdot s$ where m is the magnification of the camera. Put into Equation (8.17) this gives, for the object deformation,

$$s = \frac{\lambda z}{md} \qquad (8.18)$$

In deriving the equations for the Young fringe pattern in Section 3.6.1 we used the approximation $\sin \theta = \tan \theta$. Without this approximation Equation (8.18) becomes

$$s = \frac{\lambda}{m \sin \theta} \qquad (8.19)$$

By measuring the fringe separation d we can therefore find the object displacement at the point of the laser beam incidence using Equation (8.18). Better accuracy is obtained by measuring the distance d_n covered by n fringes on the screen. We then have $d = d_n/n$,

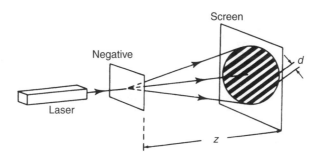

Figure 8.10   Young fringe formation

which gives

$$s = \frac{n\lambda z}{m d_n} \tag{8.20}$$

To obtain such a Young fringe pattern, the identical pairs of speckles must be separated by a distance which is at least equal to one half of the speckle size, that is $1/2\sigma_s$. This is the same sensitivity limit as found in Equation (8.16).

By placing the film negative into a slide, movable in both the horizontal and vertical directions, measurement on different points on the film, i.e. the object surface, is performed

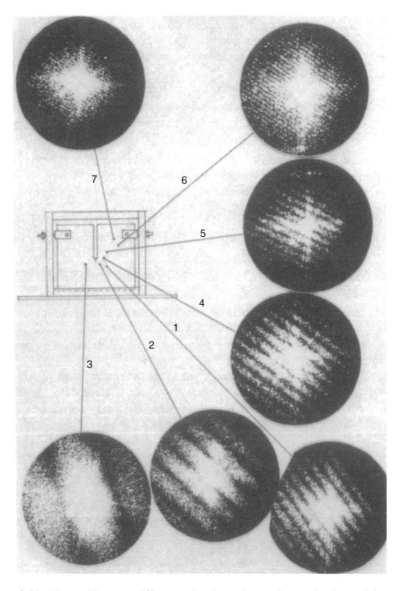

Figure 8.11   Young fringes at different points in a plate under tension in a miniature rig

quite easily and quickly. This is, however, in principle a pointwise measurement in contrast to the Fourier fringe method, which is a full field measurement. On the other hand, the Young fringe method gives better accuracy and the fringes are more easily obtained. If the Young's fringes do not appear from a double exposed specklegram it is, under normal circumstances, also impossible to obtain the Fourier fringes.

Figure 8.11 shows an example of the results obtained by this method. The object is a metal plate under tension in a miniature rig.

8.5 SPECKLE CORRELATION

As we have seen, speckle metrology is mainly concerned with the measurement of in-plane deformations of objects. When a laser-illuminated diffuse surface undergoes a displacement and/or deformation, the speckles in the scattered field or in its image show a corresponding displacement. This displacement can be represented by the peak position of the cross-correlation function c_{IX} between the intensity distributions $I_1(x, y)$ and $I_2(x, y)$ (specklegram 1 and 2) of the speckle patterns before and after the object displacement. Physically, the correlation process can be visualized as the sliding of specklegram 1 over specklegram 2 and an assessment of the similarity between I_1 and I_2 for each value of the lag. Mathematically, the cross-correlation function c_{IX} is defined by multiplying the intensity at each point on specklegram 1 by the intensity at the point on specklegram 2 displaced from it by the lag distance (components $\Delta x = x_2 - x_1$ and $\Delta y = y_2 - y_1$), averaging over the whole area, and repeating for different values of the lag:

$$c_{IX}(x_1, y_1; x_2, y_2) = \langle I_1(x_1, y_1)I_2(x_2, y_2)\rangle \tag{8.21}$$

where $\langle\ldots\rangle$ denotes the spatial averaging.

Figure 8.12(b) shows speckle patterns recorded by an electronic camera before and after in-plane translation of a piece of paper. The two-dimensional cross-correlation was computed by a digital computer. For comparison, the autocorrelation ($I_1 = I_2$) of the pattern before translation is shown in Figure 8.12(a). The peak of the autocorrelation is always located at zero. The peak of the cross-correlation corresponds to the speckle displacement and the decrease in peak height is associated with change in the structure, so-called decorrelation.

To analyse the laser speckle phenomenon further, we have to specify the statistics of the amplitudes of the speckle field. When assuming Gaussian statistics (see Section 8.2) it is an accepted fact that the autocorrelation (Goodman 1975)

$$c_I(x_1, y_1; x_2, y_2) = \langle I(x_1, y_1)I(x_2, y_2)\rangle = \langle I(x_1, y_1)\rangle\langle I(x_2, y_2)\rangle + |c_u(x_1, y_1; x_2, y_2)|^2 \tag{8.22}$$

Here, $I(x, y) = |u(x, y)|^2$ is the intensity and

$$c_u(x_1, y_1; x_2, y_2) = \langle u(x_1, y_1)u^*(x_2, y_2)\rangle \tag{8.23}$$

is the autocorrelation function of the fields, also referred to as the mutual intensity.

To calculate the field $u(x, y)$ in the xy-plane resulting from free space propagation from an illuminated rough surface in the $\xi\eta$-plane, where the field is given by $u_0(\xi, \eta)$,

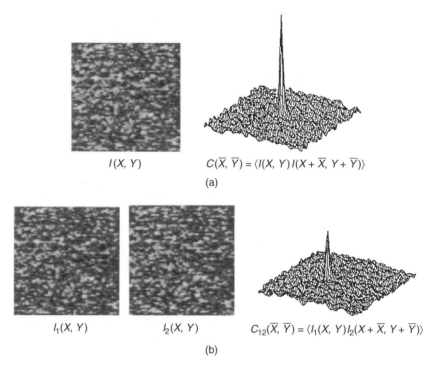

$$I(X, Y) \qquad\qquad C(\overline{X}, \overline{Y}) = \langle I(X, Y) I(X + \overline{X}, Y + \overline{Y}) \rangle$$

(a)

$$I_1(X, Y) \qquad I_2(X, Y) \qquad C_{12}(\overline{X}, \overline{Y}) = \langle I_1(X, Y) I_2(X + \overline{X}, Y + \overline{Y}) \rangle$$

(b)

Figure 8.12 Speckle patterns appearing in video images of a paper sheet before and after in-plane translation and their autocorrelation (a) and cross-correlation (b). From Yamaguchi, I. (1993) 'Theory and applications of speckle displacement and decorrelation,' in Sirohi, R.S., *Speckle Metrology*. Reproduced by permission of Marcel Dekker, Inc., New York

we apply the Fresnel approximation (writing out the x, ξ-dependence only):

$$
\begin{aligned}
u(x) &= \frac{1}{\lambda z} \int_{-\infty}^{\infty} u_0(\xi) \exp\left\{ \frac{i\pi}{\lambda z}(x - \xi)^2 \right\} d\xi \\
&= \frac{1}{\lambda z} \exp\left\{ \frac{i\pi}{\lambda z} x^2 \right\} \int_{-\infty}^{\infty} u_0(\xi) \exp\left\{ \frac{i\pi}{\lambda z} \xi^2 \right\} \exp\left\{ \frac{-i2\pi}{\lambda z} x\xi \right\} d\xi
\end{aligned}
\tag{8.24}
$$

This gives for the relation between the mutual intensity c_u in the xy-plane and c_{u0} in the $\xi\eta$-plane

$$
c_u(x_1, x_2) = \frac{1}{\lambda^2 z^2} \int_{-\infty}^{\infty} \int c_{u0}(\xi_1, \xi_2) \exp\left\{ i\frac{\pi}{\lambda z}(\xi_1^2 - \xi_2^2) \right\} \exp\left\{ -i\frac{2\pi}{\lambda z}(x_1\xi_1 - x_2\xi_2) \right\} d\xi_1 d\xi_2
\tag{8.25}
$$

where we have also omitted a phase factor dependent on $x_1^2 - x_2^2$.

Now we assume the mutual intensity in the $\xi\eta$-plane to be given as

$$
c_{u0}(\xi_1, \xi_2) \propto P(\xi_1) P^*(\xi_2) \delta(\xi_1 - \xi_2)
\tag{8.26}
$$

where $P(\xi)$ is the amplitude of the incident field. This means that we assume zero correlation except when $\xi_1 = \xi_2$. Equation (8.26) put into Equation (8.25) gives

$$c_u(x_1, y_1; x_2, y_2) = \frac{1}{\lambda^2 z^2} \int_{-\infty}^{\infty} |P(\xi_1, \eta_1)|^2 \exp\left\{i\frac{2\pi}{\lambda z}[\xi_1(x_1 - x_2) + \eta_1(y_1 - y_2)]\right\} d\xi_1 d\eta_1$$
$$(8.27)$$

where we have retained the $y\eta$-dependence. Thus the mutual intensity of the observed field depends only on the difference of the coordinates in the xy-plane. Finally, the auto-correlation function of the speckle intensity assumes the form

$$c_I(\Delta x, \Delta y) = \langle I \rangle^2 [1 + |\mu_u(\Delta x, \Delta y)|^2] \tag{8.28}$$

where

$$\mu_u(\Delta x, \Delta y) = \frac{\int_{-\infty}^{\infty}\int |P(\xi, \eta)|^2 \exp\left\{-i\frac{2\pi}{\lambda z}(\xi\Delta x + \eta\Delta y)\right\} d\xi\, d\eta}{\int_{-\infty}^{\infty}\int |P(\xi, \eta)|^2 d\xi\, d\eta} \tag{8.29}$$

which is recognized as the complex degree of spatial coherence γ_{12} (see below).

Another quantity of considerable interest is the power spectral density $W(u, v)$ of the speckle intensity distribution. According to the autocorrelation theorem this will be given by the Fourier transform of $c_I(\Delta x, \Delta y)$. Applying a Fourier transformation to Equation (8.28) gives

$$W(u, v) = \langle I \rangle^2 \left\{ \delta(u, v) + \frac{\int_{-\infty}^{\infty}\int |P(\xi, \eta)|^2 |P(\xi - \lambda z u, \eta - \lambda z v)|^2 d\xi\, d\eta}{\left[\int_{-\infty}^{\infty}\int |P(\xi, \eta)|^2 d\xi\, d\eta\right]^2} \right\} \tag{8.30}$$

When the $\xi\eta$-plane is imaged onto the xy-plane by a lens, and provided the object illumination is uniform, the size of the speckles incident on the lens pupil is very small compared to the size of the lens pupil. Then, to a good approximation, the mutual intensity of the field in the lens pupil is given by Equation (8.26) and $P(\xi, \eta)$ is the pupil function of the lens. Since free-space propagation is involved as the light passes from the pupil plane to the image plane, the results found above can be directly applied, provided the new interpretation of $P(\xi, \eta)$ is used. As an example, considering a diffraction-limited lens with (for simplicity) a square pupil of dimensions $L \times L$:

$$|P(\xi, \eta)|^2 = \text{rect}\left(\frac{\xi}{L}\right)\text{rect}\left(\frac{\eta}{L}\right) \tag{8.31}$$

we get:

$$c_I(\Delta x, \Delta y) = \langle I \rangle^2 \left[1 + \text{sinc}^2\frac{L\Delta x}{\lambda z}\,\text{sinc}^2\frac{L\Delta y}{\lambda z}\right] \tag{8.32}$$

and

$$W(u, v) = \langle I \rangle^2 \left[\delta(u, v) + \left(\frac{\lambda z}{L}\right)^2 \Lambda\left(\frac{\lambda z}{L}u\right)\Lambda\left(\frac{\lambda z}{L}v\right)\right] \tag{8.33}$$

where Λ is the triangle function. From the definition of the optical transfer function $\mathcal{H}(u, v)$ (Section 4.6, Equation (4.48)), we can write Equation (8.33) as

$$W(u, v) = \langle I \rangle^2 \left[\delta(u, v) + \left(\frac{\lambda z}{L} \right)^2 \mathscr{H}(u, v) \right] \tag{8.34}$$

Now let us go back to the Young fringe method (Section 8.4.2). A brief analysis goes as follows: assume the intensities in the first and second recording to be given by $I_1(x, y)$ and $I_2(x, y)$. Since we are illuminating the specklegram with a thin (≈ 1 mm^2) laser beam, we can to a good approximation assume the speckle displacement between the two recordings to be uniform within the illuminated area. We can then write $I_1(x, y) = I(x, y)$ and $I_2(x, y) = I(x + d, y)$ where for simplicity we have assumed the displacement to be in the x-direction. When moving the observation plane a reasonable distance from the illuminated double-exposed film with a transmittance proportional to $I_1 + I_2$, the field u_a in the observation plane is given by the Fourier transform of the transmittance, i.e.

$$u_a = \mathscr{F}\{I_1 + I_2\} = \mathscr{F}\{I_1\} + \mathscr{F}\{I_2\} \tag{8.35}$$

If we put

$$J_1(u, v) = \mathscr{F}\{I_1(x, y)\} = \mathscr{F}\{I(x, y)\} \equiv J(u, v) \tag{8.36}$$

$$J_2(u, v) = \mathscr{F}\{I_2\} = \mathscr{F}\{I(x + d, y)\} = \int_{-\infty}^{\infty} \int I(x + d, y) e^{-i2\pi(ux+vy)} \mathrm{d}x\mathrm{d}y \tag{8.37}$$

which by changing the x-variable to $x' = x + d$ becomes

$$J_2(u, v) = \mathscr{F}\{I_2\} = \int_{-\infty}^{\infty} \int I(x, y) e^{-i2\pi(u(x+d)+vy)} \mathrm{d}x\mathrm{d}y = J(u, v) \cdot e^{i2\pi ud} \tag{8.38}$$

and therefore

$$u_a = J(u, v)(1 + e^{i2\pi ud}) \tag{8.39}$$

the intensity in the observation plane then becomes

$$I_a = 2|J(u, v)|^2(1 + \cos 2\pi ud) \tag{8.40}$$

From this expression we see that $|J(u, v)|^2$ forms an envelope of the Young fringes described by $(1 + \cos 2\pi ud)$.

Now, from the autocorrelation theorem (Equation (B.2f), Appendix B)

$$|J(u, v)|^2 = \mathscr{F}\{I(x, y) \odot I(x, y)\} \tag{8.41}$$

We therefore see that in the Young fringe method we are actually making a correlation between the two displaced speckle patterns. Since $|J(u, v)|^2$ is equal to the power spectral density $W(u, v)$, we see from Equation (8.34) that we can write Equation (8.40) as

$$I_a \propto \mathscr{H}(u, v)(1 + \cos 2\pi ud) \tag{8.42}$$

We see that the Young fringes are modulated by the lens's MTF: see Figure 8.13.

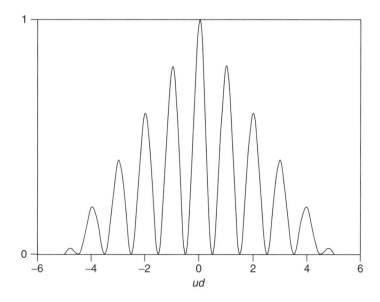

Figure 8.13 Young's fringes modulated by the lens MTF

In Section 12.4 (Digital speckle photography) we shall see that the speckle displacement is measured by detecting the position of the cross-correlation peak directly. This cannot be done optically (although in principle it can be done by means of optical filtering) but is easily performed in a computer. As we saw in Figure 8.12, lack of correlation decreases the height of the correlation peak. It is interesting to know what causes this decorrelation. In a thorough analysis by Yamaguchi (1993), it is found, among other things, that surface tilt gives decorrelation, a fact most easily realized without mathematical analysis. But also plastic deformations and other effects that change the surface microstructure will of course give decorrelation.

Since we are dealing with correlations, we might also briefly revisit the concept of coherence. The intensity in the Michelson interferometer (Section 3.6.2) we can write as

$$
\begin{aligned}
I &= \langle |u_1(t)|^2 \rangle + \langle |u_2(t)|^2 \rangle + \langle u_1^*(t)u_2(t+\tau) \rangle + \langle u_1(t)u_2^*(t+\tau) \rangle \\
&= \langle |u_1(t)|^2 \rangle + \langle |u_2(t)|^2 \rangle + 2\,\mathrm{Re}\{\langle u_1^*(t)u_2(t+\tau) \rangle\}
\end{aligned}
\tag{8.43}
$$

Here u_1 and u_2 are the field amplitudes of the light from the two interferometer arms, and the brackets mean averaging over time. By writing $u_2 = u_2(t+\tau)$ we indicate that u_2 is delayed (relative to u_1) by a time τ due to the longer path length $\Delta L = c\tau$. Assuming equal amplitudes ($u_2 = u_1 = u$), Equation (8.43) becomes

$$
I = 2\langle |u|^2 \rangle \left[1 + \mathrm{Re}\left\{ \frac{\langle u^*(t)u(t+\tau) \rangle}{\langle |u|^2 \rangle} \right\} \right] = 2I_0[1 + \mathrm{Re}\,\gamma(\tau)]
\tag{8.44}
$$

The quantity

$$
\Gamma(\tau) = \langle u^*(t)u(t+\tau) \rangle
\tag{8.45}
$$

is called the coherence function and its normalized version

$$\gamma(\tau) = \frac{\Gamma(\tau)}{\Gamma(0)} \tag{8.46}$$

is the complex temporal degree of coherence. That this really is the same γ as in Equation (3.67) (Section 3.7) we see from the following.

Averaging over time means that we have

$$\langle u^*(t)u(t+\tau)\rangle = \lim_{T\to\infty} \frac{1}{2T} \int_{-T}^{T} u^*(t)u(t+\tau)\mathrm{d}t \tag{8.47}$$

In the limit, this is equal to the autocorrelation

$$\Gamma(\tau) = u(t) \odot u(t) \tag{8.48}$$

and from the autocorrelation theorem we get

$$\mathscr{F}\{\Gamma(\tau)\} = |G(\nu)|^2 \tag{8.49}$$

where

$$G(\nu) = \int_{-\infty}^{\infty} u(t) \exp\{-\mathrm{i}2\pi\nu t\}\mathrm{d}t = \mathscr{F}\{u(t)\} \tag{8.50}$$

We see that $|G(\nu)|^2$ is the spectral distribution of the light and that

$$P(\nu) = \frac{|G(\nu)|^2}{I_0} \tag{8.51}$$

is the same normalized spectral distribution as in Equation (3.63).

8.6 SPECKLE-SHEARING INTERFEROMETRY

Laser speckle methods can be utilized in many different ways. One method, the speckle-shearing interferometer, is particularly interesting because it enables direct measurements of displacement derivatives which are related to the strains (Hung and Taylor 1973; Leendertz and Butters 1973).

The principle of speckle-shearing interferometry, also termed shearography, is to bring the rays scattered from one point of the object into interference with those from a neighbouring point. This can be obtained in a speckle-shearing interferometric camera as depicted in Figure 8.14. The set-up is the same as that used in ordinary speckle photography, except that one half of the camera lens is covered by a thin glass wedge. In that way, the two images focused by each half of the lens are laterally sheared with respect to each other. If the wedge is so oriented that the shearing is in the x-direction, the rays from a point P(x, y) on the object will interfere in the image plane with those from a neighbouring point P($x + \delta x, y$). The shearing δx is proportional to the wedge angle.

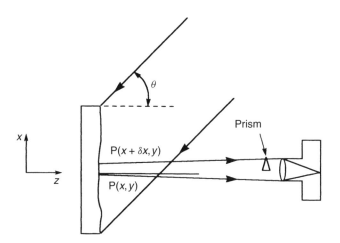

Figure 8.14   Speckle-shearing interferometric camera

When the object is deformed there is a relative displacement between the two points that produces a relative optical phase change $\Delta\phi$ given by

$$\Delta\phi = k\{(1 + \cos\theta)[w(x + \delta x, y) - w(x, y)]$$
$$+ \sin\theta[u(x + \delta x, y) - u(x, y)]\} \tag{8.52}$$

where θ is the angle of incidence and u and w are the displacement components in the x-and z-directions, respectively. If the shear δx is small, the relative displacements may be approximated by the displacement derivatives and thus Equation (8.52) becomes

$$\Delta\phi = k\left[(1 + \cos\theta)\frac{\partial w}{\partial x} + \sin\theta\frac{\partial u}{\partial x}\right]\delta x \tag{8.53}$$

By rotating the camera $90°$ about the z-axis, u in Equation (8.53) is replaced by v, the displacement component in the y-direction.

By double exposure of the object before and after deformation, a speckle fringe pattern depicting $\Delta\phi$ of Equation (8.53) will be generated. This pattern will be clearly visible by placing the negative in an optical filtering set-up and blocking out the zero order component. Dark fringes occur when

$$\Delta\phi = (2n + 1)\pi, \quad \text{for} \quad n = 0, 1, 2\ldots \tag{8.54}$$

Equation (8.53) shows that to measure $\partial w/\partial x$ only, one may employ normal illumination ($\theta = 0°$), which gives

$$\Delta\phi = \frac{4\pi}{\lambda}\frac{\partial w}{\partial x}\delta x \tag{8.55}$$

but it is not possible to isolate $\partial u/\partial x$. This can be obtained by recording two fringe patterns using two different illumination angles. Then $\partial u/\partial x$ can be pointwise separated from $\partial w/\partial x$.

Compared to speckle interferometry, shearography has a lot of advantages. Good quality fringes are easily obtained and the vibration isolation and the coherence length requirements are greatly reduced. Like holography, shearography has been adapted to electronic cameras combined with digital image processing. A convenient means for obtaining the shear is to place a Michelson interferometer set-up with a tilt of one of the mirrors in front of the image sensor. By an ingenious set-up constructed by the group at Luleå University of Technology simultaneous observation of the out-of-plane displacement and slope in real time is made possible. The set up is shown in Figure 8.15 where an electronic holography configuration is combined with a shearography configuration so that the displacement and slope interferograms are displayed in each half of the monitor. The result of an experiment (Mohan *et al.* 1993) conducted on a rectangular aluminium plate of dimensions 65 mm × 100 mm × 0.6 mm is shown in Figure 8.16.

Figure 8.17 shows another advantage of shearography. Figure 8.17(a) shows the phase map of a ceramic object using shearography, while Figure 8.17(b) is a phase map of the same object using holography (Saldner 1994). A crack in the centre of the object becomes visible in the upper image due to the fact that shearography is highly sensitive to local variations in the deformation field. In the holographic phase map of Figure 8.17(b), however, it is difficult or impossible to see the same crack.

Finally we mention that speckle-shearing interferometry can be obtained by intentional misfocusing of the object. Then it is possible for rays from neighbouring points on the object to interfere in the image plane.

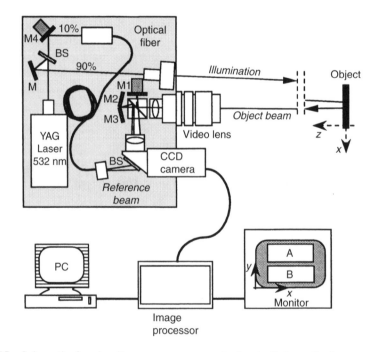

Figure 8.15 Schematic for simultaneous measurement of out-of-plane displacement and slope. (M, mirror; BS, beam splitter). (From Saldner, H.O. (1994) Electronic holography and Shearography in experimental mechanics. Licentiate thesis, Luleå University of Technology.)

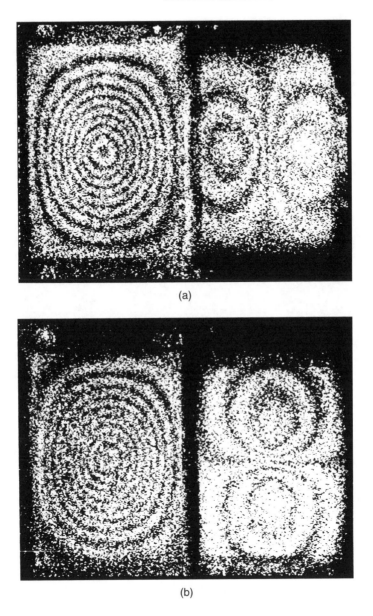

(a)

(b)

Figure 8.16   (a) Out-of-plane displacement fringes (w) and slope fringes ($\delta w/\delta x$) as seen from the monitor for a rectangular aluminium plate clamped along its edges and loaded at the centre. The shear Δx is 6 mm, and the displacement $w = 2.5$ μm and (b) Out-of-plane displacement fringe pattern (w) and slope pattern ($\delta w/\delta y$) for the same object as in (a). The shear Δy is 7 mm. (From Mohan, N.K., Saldner, H. and Molin, N.E. (1993) Electronic speckle pattern interferometry for simultaneous measurements of out-of-plane displacement and slope, Opt letters, **18**, 1861–3. Reproduced by courtesy of The Optical Society of America and of H. Saldner and N.E. Molin

(a)

(b)

Figure 8.17   Electronic shearography (ES) used for non-destructive testing of a ceramic material. (a) A vertical crack is clearly visualized by ES as a fringe in the centre of the sample and (b) The crack is not detected using TV holography. ((From Saldner, H.O. (1994) Electronic holography and Shearography in experimental mechanics. Licentiate thesis, Luleå University of Technology.)

8.7 WHITE-LIGHT SPECKLE PHOTOGRAPHY

The term 'white-light speckle photography' has been introduced because this technique utilizes the same principles as laser speckle photography. But here a white light source is used instead of a laser. The 'speckles' in this method are some type of structure attached to the object surface. A common technique is to cover the object surface with retroreflective paint. This paint consists of small glass spheres embedded in the emulsion and produces reflected points of light which are very bright when photographed from a position close to the illumination direction. These spheres represent a random distribution of points containing all spatial frequencies up to the reciprocal of the smallest sphere separation. This highest spatial frequency is therefore limited by the size of the glass spheres. However, if the spheres are small enough, or photographed from a sufficient distance, the highest spatial frequency will be limited by the impulse response of the imaging lens, i.e. the diameter of the Airy disc (see Section 4.6) in the same way as in laser speckle photography. There may be several ways of creating a random pattern on the object surface. One method is to spray the object with black paint on a white background or vice versa.

The working principle of white light speckle photography is the same as for laser speckle photography. Both the Fourier fringe and the Young fringe methods can be applied using the same equations in evaluating the patterns. The main difference is that the detailed structure of the white light speckles is fixed on the surface, not in space as with laser speckles. Consequently the requirement for high-quality recording optics is more severe. The random pattern on the surface must be resolved by the camera which requires a large aperture lens with minimum aberration. On the other hand, if the surface is curved, a good depth of focus is required to resolve the pattern over the whole object area. The depth of focus, however, varies inversely as the square of the lens aperture (see Section 4.6). These conflicting requirements are partly overcome by an ingenious modification of the camera made by Burch and Forno (1975). They have mounted a mask with symmetrically disposed parallel slots inside the camera lens. In that way they tune the camera response to a particular spatial frequency determined by the slot separation. Thereby both the resolution and the depth of focus is increased for this tuned frequency.

White light speckle photography is easily applied in industrial environments. Figure 8.18 shows a white light speckle camera assembly. It consists of an ordinary photographic camera with a standard 50 mm lens mounted on a plate together with an electronic flash and a beamsplitter. This is so arranged to make the retroreflected light from the painted object surface fall inside the lens aperture. Figure 8.19 shows an example of the application of this camera. It is a 3 m high concrete tube of 0.75 m diameter and 50 mm tube wall thickness. Between the two exposures the tube has been horizontally loaded by 5000 kg with a jack at the upper right end. The picture shows Young fringe readings at some selected points on the negative. The displacement per fringe is 0.22 mm and results have been obtained over an object depth of about 30 cm.

In the same way as for laser speckles, we can talk of white-light subjective and objective speckles. To obtain the latter, we have to place the film in contact with the object surface. Figure 8.20 shows the results of such an experiment. Figure 8.20(a) shows the object, a notched plate made of Plexiglas loaded in tension in a miniature rig. A film plate was clamped to one side of the Plexiglas® plate which was rubbed to get speckles. The plate was illuminated from the other side with an ordinary lamp and loaded between the two exposures. Figure 8.20(b, c) shows the Fourier fringe pattern displaying the vertical and horizontal displacements obtained from this experiment.

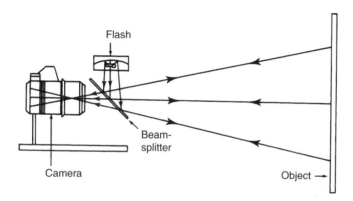

Figure 8.18   White light speckle camera

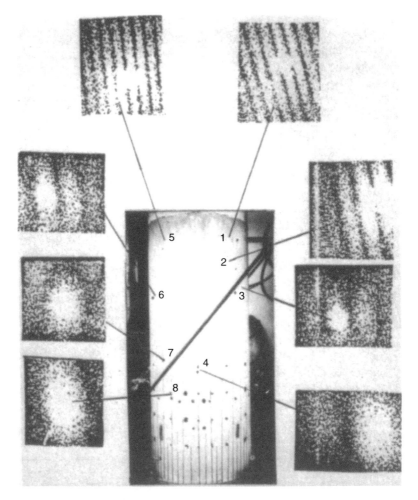

Figure 8.19 Young fringes obtained by white light speckle photography on a 3-metre high concrete tube loaded at the upper right end

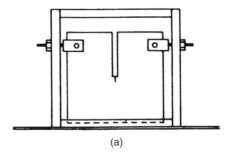

(a)

Figure 8.20 (a) Notched Plexiglas® plate under tension in miniature rig. The resulting Fourier fringe patterns displaying the resulting; (b) vertical; and (c) horizontal displacements

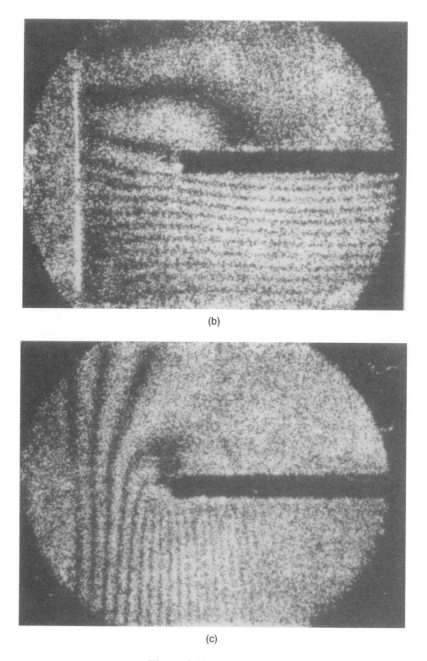

(b)

(c)

Figure 8.20   (*continued*)

PROBLEMS

8.1 In speckle photography we image a surface onto a film using a camera with aperture number F and a magnification of m. Assume the recorded intensity distribution (using a He−Ne laser) in the first recording is given by $I(x, y)$. Before the second recording the surface is displaced in the x-direction a distance s. After development the film is placed in the optical filtering set-up (using an He−Ne laser) shown in Figure 4.14 a distance z from the filter plane.

(a) Find (to an arbitrary multiplicative constant) the intensity distribution in the filter plane.

(b) The fringe period in the filter plane is d. Find s.

(c) For $d = 2$ mm, $m = 0.1$ and $z = 100$ mm, calculate s.

(d) What is the minimum measurable displacement s when $F = 4$, and when $F = 22$.

8.2 Suppose we are applying the white-light speckle method on an object with a transversal magnification m of our imaging optics. The sensitivity limit for the in-plane deformation is given by the Rayleigh criterion

$$s_{min} = 1.22\lambda(1 + m)F$$

where F is the aperture number. The depth of focus for the lens is given by Equation (4.60), and the longitudinal magnification $m_L = m^2$. We assume that we obtain the sensitivity limit s_{min} also for object points which lies within the depth of focus given by Equation (4.60). (a) What is the sensitivity limit when we wish to measure on an object with depth variations up to Δa?

8.3 When imaging a speckle pattern a mask consisting of a double slit as sketched in Figure P4.4 is placed in the plane of the exit pupil.
    Calculate the resulting power spectral density $W(f_x, 0)$.

9

Photoelasticity and Polarized Light

9.1 INTRODUCTION

Up to now we have treated the light field as a scalar quantity. The electromagnetic field, however, is a vector quantity which is perpendicular to the direction of propagation and with a defined orientation in space. This property is known as the polarization of light. In our treatments of interferometry and holography it is silently understood that the interfering waves have the same polarization. In practice, however, this condition is fulfilled to a greater or lesser degree. Unequal polarization of the interfering waves results in a bias intensity which reduces the contrast of the fringes and, in the limit of opposite polarization, we get no fringes at all. In analysing finer diffraction effects, the vector property of the light must be taken into account, for example in the reconstruction of a hologram (Gåsvik 1976). Special techniques have been used to reconstruct an arbitrary state of polarization (Lohmann, Kurtz, Gåsvik). Although it is possible to get along quite well with optical metrology without knowing anything about polarization, it is in many cases very important to understand the vectorial properties of light.

In this chapter we learn how to describe polarization and we develop a useful formalism based on Jones vectors and matrices. We also learn how to control and measure the state of polarization by means of polarization filters. The points in a light path where the change in polarization is most difficult to predict are at reflecting and refracting interfaces. Therefore this subject is treated in Section 9.5. Finally, a specific optical measurement technique is based on the light's polarization. This technique, called photoelasticity, is described at the end of this chapter.

9.2 POLARIZED LIGHT

Consider a plane wave propagating in the z-direction. The field amplitude is a vector denoted by

$$\mathbf{u} = \mathbf{U} \, e^{ikz} \tag{9.1}$$

of length U at an arbitrary angle ϕ to the x-axis, see Figure 9.1. The components along the x- and y-axes therefore are

$$U_x = U \cos \phi \tag{9.2a}$$

$$U_y = U \sin \phi \tag{9.2b}$$

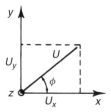

Figure 9.1

We then have

$$\mathbf{u} = [U_x e^{i\delta_x} \mathbf{e}_x + U_y e^{i\delta_y} \mathbf{e}_y] e^{ikz} \tag{9.3}$$

where $\mathbf{e}_x$ and $\mathbf{e}_y$ are the unit vectors along the x- and y-axis respectively. By including the phase constants δ_x and δ_y we have taken care of the fact that the two components may not have the same phase. The factor e^{ikz} we omit, since it merely gives the orientation of the z-axis. Equation (9.3) then can be written as

$$\mathbf{u} = e^{i\beta/2}[U_x e^{-i\delta/2} \mathbf{e}_x + U_y e^{i\delta/2} \mathbf{e}_y]$$
$$= e^{i\beta/2}[u_x \mathbf{e}_x + u_y \mathbf{e}_y] \tag{9.4}$$

where

$$\beta = \delta_x + \delta_y \tag{9.5a}$$

$$\delta = \delta_y - \delta_x \tag{9.5b}$$

$e^{i\beta/2}$ is a common phase factor which can be omitted since it does not affect the orientation of $\mathbf{u}$. This orientation, termed the direction of polarization or the state of polarization, is therefore completely fixed by two independent quantities, i.e.

$$\tan \phi = U_y/U_x$$
$$\text{and } \delta \tag{9.6}$$

The intensity of this wave is given as

$$I = |u|^2 = U_x^2 + U_y^2 \tag{9.7}$$

It is evident and easy to prove that the intensity always becomes equal to the sum of the squares of the field components in a Cartesian coordinate system independent of the orientation of the coordinate axes.

When such a polarized wave passes a fictional plane perpendicular to the z-axis, the tip of the U-vector will in the general case describe an ellipse in that plane. A general state of polarization is therefore termed 'elliptically polarized light'. When $\delta = \pm\pi/2$ and $U_x = U_y$, the ellipse degenerates to a circle and when $\delta = 0$ or π to a straight

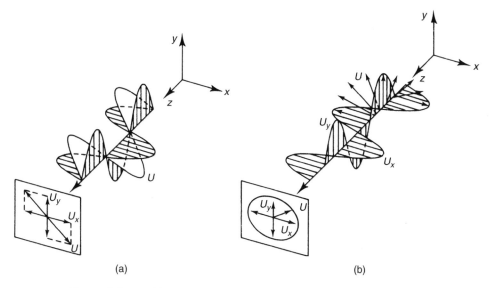

Figure 9.2   (a) Linear polarized light and (b) Circularly polarized light

line. These two special cases therefore represent so-called circular and plane (or linear) polarized light respectively (see Figure 9.2). More specifically, $\delta = +\pi/2$ represents right circularly polarized light, whilst $\delta = -\pi/2$ represents left circularly polarized light. This has to do with the direction of rotation of the tip of the field vector.

9.3 POLARIZING FILTERS

To alter and to analyse the state of polarization of the light, one places in the beam different types of polarization filters. Below, we consider the most important types.

9.3.1 The Linear Polarizer

A linear or plane polarizer (often called simply a polarizer) has the property of transmitting light which field vector is parallel to the transmission direction of the polarizer only. The field u_t transmitted through the polarizer therefore becomes equal to the component of the incident field onto this direction. If the incident wave U is plane-polarized at an angle α to the transmission direction of the polarizer, we therefore have

$$u_t = U \cos \alpha \tag{9.8}$$

and the intensity

$$I_t = |u_t|^2 = U^2 \cos^2 \alpha \tag{9.9}$$

Equation (9.9) is called Malus' law.

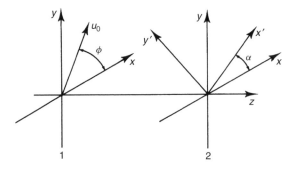

Figure 9.3

This is the case of an ideal polarizer with amplitude transmittances $t_1 = 1$ and $t_2 = 0$ parallel to an normal to the transmission direction respectively. In a real polarizer, $t_1 \leq 1$ and $t_2 \geq 0$, but $t_1 \gg t_2$. Consider two such identical linear polarizers placed in tandem in a light beam as shown in Figure 9.3. Here polarizer 1 has its transmission direction along the x-axis while the transmission axis of polarizer 2 is inclined an angle α to the x-axis. Assume that the wave field u_o incident on polarizer 1 is linearly polarized at an angle ϕ to the x-axis with amplitude equal to 1.

Then

$$\mathbf{u}_o = \cos \phi \mathbf{e}_x + \sin \phi \mathbf{e}_y \tag{9.10}$$

After passing polarizer 1, the field amplitude becomes

$$\mathbf{u}_1 = t_1 \cos \phi \mathbf{e}_x + t_2 \sin \phi \mathbf{e}_y \tag{9.11}$$

To calculate the effect of polarizer 2, we first decompose $\mathbf{u}_1$ into the coordinate system of unit vectors $\mathbf{e}_{x'}$, $\mathbf{e}_{y'}$, along and normal to the transmission axis of polarizer 2 respectively:

$$\mathbf{u}_1 = (t_1 \cos \phi \cos \alpha + t_2 \sin \phi \sin \alpha)\mathbf{e}_{x'} + (-t_1 \cos \phi \sin \alpha + t_2 \sin \phi \cos \alpha)\mathbf{e}_{y'} \tag{9.12}$$

After passing polarizer 2, the field amplitude becomes

$$\mathbf{u}_2 = t_1(t_1 \cos \phi \cos \alpha + t_2 \sin \phi \sin \alpha)\mathbf{e}_{x'} + t_2(-t_1 \cos \phi \sin \alpha + t_2 \sin \phi \cos \alpha)\mathbf{e}_{y'} \tag{9.13}$$

and the intensity

$$I = |u_2|^2 = (t_1^4 \cos^2 \phi + t_2^4 \sin^2 \phi) \cos^2 \phi$$
$$+ 2t_1 t_2 (t_1^2 - t_2^2) \sin \phi \cos \phi \sin \alpha \cos \alpha + t_1^2 t_2^2 \sin^2 \alpha \tag{9.14}$$

If the incident light is unpolarized (see Section 9.4), all polarization angles ϕ from 0 to π will be equally represented. To find the transmitted intensity I_{up} in this case, we therefore have to average Equation (9.14) over this range of ϕ, i.e.

$$I_{up} = \frac{1}{\pi} \int_0^\pi I \, d\phi = (1/2)(T_1^2 + T_2^2) \cos^2 \alpha + T_1 T_2 \sin^2 \alpha \tag{9.15}$$

where we have substituted the intensity transmittances $T_1 = t_1^2$ and $T_2 = t_2^2$ for the corresponding amplitude transmittances of the polarizer. Equation (9.15) can be regarded as the generalized Malus law and can be used to determine T_1 and T_2 from measuring the intensity at, for example $\alpha = 0°$ and $90°$. Normally, T_1 and T_2 are wavelength dependent and the ratio T_2/T_1 is a measure of the quality of a linear polarizer. Values of T_2/T_1 as low as $10^{-5} - 10^{-6}$ can be reached with high-quality crystal polarizers.

A linear polarizer can be pictured as a grid of finely spaced parallel conducting metal threads. When the field of the incident wave oscillates parallel to the threads, it will induce currents in the threads. The energy of the field is therefore converted into electric current, which is converted into heat and the incident wave is absorbed. Because of the non-conducting spacings of the grid, currents cannot flow perpendicularly to the threads. Fields oscillating in the latter direction will therefore not produce any current and the light is transmitted. Although easy to understand, this type of polarizer is difficult to fabricate.

The most widely used linear polarizer is the polaroid sheet-type polarizer invented by E. H. Land. It can be regarded as the chemical version of the metal thread grating. Instead of long, thin threads it consists of long, thin molecules, i.e. long chains of polymeric molecules that contain many iodine atoms. These long molecules are oriented almost completely parallel to each other and because of the conductivity of the iodine atoms, the electric field oscillating parallel to the molecules will be strongly absorbed.

Another type of linear polarizer is made from double-refracting crystals. Double-refraction is a phenomenon where the incident light is split into two orthogonally linearly polarized components. By proper cutting and cementing of such crystals, usually calcite, one of the components is isolated and the other is transmitted, thereby giving a linear polarizer. Other phenomena utilized for the construction of linear polarizers are reflection and scattering.

9.3.2 Retarders

A retarder (or phase plate) is a polarization element with two orthogonal, characteristic directions termed the principal axes. When light passes a retarder, the field components parallel to the principal axes will acquire different phase contributions resulting in a relative phase difference δ. This is due to the phenomenon of double-refraction or birefringence which means that light fields oscillating parallel to the two principal axes 'experience' different indexes of refraction n_1 and n_2. A retarder plate of thickness t therefore produces a phase difference (retardance) equal to

$$\delta = \delta_2 - \delta_1 = k(n_2 - n_1)t \tag{9.16}$$

Figure 9.4 sketches a retarder with the principal axes h_1 and h_2 parallel to the x- and y-axis respectively. Consider a wave field U linearly polarized at an angle α to the x-axis normally incident on the retarder. The transmitted wave u_t then becomes

$$\mathbf{u}_t = U(\cos \alpha e^{i\delta_1} \mathbf{e}_x + \sin \alpha e^{i\delta_2} \mathbf{e}_y)$$
$$= U e^{i(\delta_1 + \delta_2)/2}(\cos \alpha e^{-i\delta/2} \mathbf{e}_x + \sin \alpha e^{i\delta/2} \mathbf{e}_y) \tag{9.17}$$

Note that the intensity $|u_t|^2 = U^2$ is unchanged by passing the retarder. In the case of

$$\delta = \pm \pi/2 \tag{9.18}$$

and $\cos \alpha = \sin \alpha$; i.e. $\alpha = 45°$, u_t becomes circularly polarized.

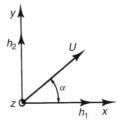

Figure 9.4

A retarder with $\delta = \pm\pi/2$ is called a quarterwave plate because it corresponds to an optical path length difference equal to $\lambda/4$ (cf. Equation (9.16)). The combination of a linear polarizer followed by a quarterwave plate with their axes inclined at 45° therefore is termed a circular polarizer.

By putting $\delta = \pi$, Equation (9.17) becomes

$$\mathbf{u}_t = -iU e^{i(\delta_1+\delta_2)/2}(\cos\alpha\,\mathbf{e}_x - \sin\alpha\,\mathbf{e}_y) \tag{9.19}$$

We see that the outcoming wave is linearly polarized and that the field vector is the mirror image of the incoming field vector about the x-axis, i.e. the polarization angle has changed by 2α. A retarder with $\delta = \pm\pi$ is called a halfwave plate because it corresponds to an optical path-length difference equal to $\lambda/2$. A halfwave plate therefore offers a convenient means for rotating the polarization angle of a linearly polarized light wave by turning the axes of the halfwave plate by the desired amount.

A retarder is usually made from double-refracting crystals such as quartz or mica. In contrast to the construction of linear polarizers, where one of the doubly refracting components is isolated, both components are transmitted collinearly by proper cutting and orientation of the crystal. A retarder can also be made from stretched sheets of polyvinyl alcohol (PVA) in the same way as polaroids. In fact, cellophane sheets can be used as a retarder.

In the retarders mentioned so far, the retardance δ is fixed by the plate thickness. It would be highly desirable, however, to have a retarder in which the retardance could be continuously varied. This can be done by a device called a Babinet–Soleil compensator shown in Figure 9.5. Here two retarders of the same crystal of thicknesses t_1 and t_2 with their axes inclined at 90° are mounted together. The total retardance of the unit therefore becomes proportional to the thickness difference $t_1 - t_2$. The upper plate consists of two wedges, one of which can be moved relative to the other, thereby varying the effective thickness t_1. This movement is controlled by a micrometer screw. The result is a retarder with variable retardance which is uniform over the whole field of the compensator.

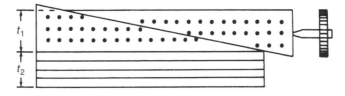

Figure 9.5 Babinet-soleil compensator

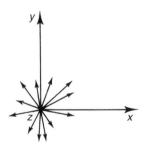

Figure 9.6   Unpolarized light

9.4 UNPOLARIZED LIGHT

Light from ordinary light sources is said to be unpolarized. This is something of a mis-nomer, but it means that the instantaneous direction of polarization will vary rapidly and randomly in time between 0 and 2π. A common way of illustrating unpolarized light propagating in the z-direction is as illustrated in Figure 9.6. This picture could be mis-leading; a better illustration would be a collection of ellipses of random orientations and eccentricities.

Mathematically, unpolarized light can be represented in terms of two arbitrary, orthog-onal, linearly polarized waves of equal amplitudes for which the relative phase difference varies rapidly and randomly. This aspect is strongly related to the coherence properties of the light (see Section 3.3). By analogy with the degree of coherence, one also speaks about the degree of polarization.

We shall not here go into any further details of this phenomenon. It should, however, be easy to realize that unpolarized light will (1) be unaffected with regard to intensity when transmitted through a retarder, and (2) become linearly polarized by transmitting a linear polarizer, but with an intensity independent of the transmission direction of the polarizer.

9.5 REFLECTION AND REFRACTION AT AN INTERFACE

When light is incident at an interface between two media of different refractive indices, both the reflected and the transmitted light will in general change its state of polarization relative to the state of polarization of the incident light.

Consider Figure 9.7 where the light is incident from a medium of refractive index n_1, on a medium of refractive index n_2. The incident light field is decomposed into the components u_{ip} and u_{in} parallel and normal to the plane of incidence respectively. (The plane of incidence is defined as the plane spanned by the surface normal at the point of incidence and the incident light ray.) The corresponding quantities of the reflected light are denoted u_{rp} and u_{rn} and their amplitudes U are related by

$$U_{rp} = r_p U_{ip} \tag{9.20a}$$

$$U_{rn} = r_n U_{in} \tag{9.20b}$$

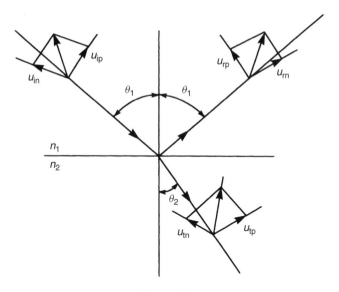

Figure 9.7

where r_p and r_n are the amplitude reflection coefficients. These so-called Fresnel reflection coefficients are given as

$$r_p = \frac{\tan(\theta_1 - \theta_2)}{\tan(\theta_1 + \theta_2)} \tag{9.21a}$$

$$r_n = \frac{-\sin(\theta_1 - \theta_2)}{\sin(\theta_1 + \theta_2)} \tag{9.21b}$$

where $\theta_1 = $ the angle of incidence and $\theta_2 = $ the angle of refraction. By using Snell's law of refraction (cf. Equation (1.16)) $n_1 \sin \theta_1 = n_2 \sin \theta_2$, we get the following alternative expressions for the reflection coefficients

$$r_p = \frac{n^2 \cos \theta - \sqrt{n^2 - \sin^2 \theta}}{n^2 \cos \theta + \sqrt{n^2 - \sin^2 \theta}} \tag{9.22a}$$

$$r_n = \frac{\cos \theta - \sqrt{n^2 - \sin^2 \theta}}{\cos \theta + \sqrt{n^2 - \sin^2 \theta}} \tag{9.22b}$$

where $n = n_2/n_1$, is the relative refractive index and where we also have dropped the subscript 1 on the angle of incidence.

The corresponding amplitude transmission coefficients are given as

$$t_p = \frac{2 \sin \theta_2 \cos \theta_1}{\sin(\theta_1 + \theta_2) \cos(\theta_1 - \theta_2)} \tag{9.23a}$$

$$t_n = \frac{2 \sin \theta_2 \cos \theta_1}{\sin(\theta_1 + \theta_2)} \tag{9.23b}$$

with the alternative expressions

$$t_p = \frac{2n \cos \theta}{n^2 \cos \theta + \sqrt{n^2 - \sin^2 \theta}} \tag{9.24a}$$

$$t_n = \frac{2 \cos \theta}{\cos \theta + \sqrt{n^2 - \sin^2 \theta}} \tag{9.24b}$$

Figure 9.8 shows plots of r_p and r_n versus the angle of incidence for $n = 1.52$ (typically for an air – glass interface). We see that $r_p = 0$ at an angle of incidence equal to θ_B. From Equation 9.21(a) we see that this occurs when

$$\theta_1 + \theta_2 = \pi/2 \tag{9.25}$$

which, when inserted into Snell's law gives

$$\tan \theta_B = \frac{n_2}{n_1} \tag{9.26}$$

θ_B is called the polarization angle or the Brewster angle.

This is our reason for wearing polarization sunglasses. When the sunlight, which is unpolarized, strikes a dielectric surface (like the sea) near the Brewster angle, the reflected light becomes linearly polarized. When the transmission axis of the polarization sunglasses is properly oriented, this specularly polarized reflected component will be blocked out.

As mentioned in Section 1.9, when light is incident from a medium of higher onto a medium of lower refractive index, we get total internal reflection at the critical angle

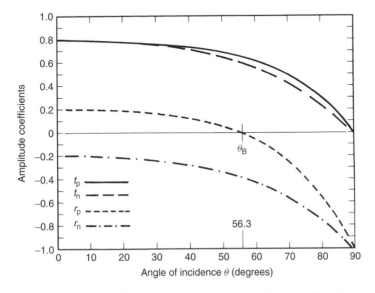

Figure 9.8 The amplitude coefficients r_p, r_n, t_p and t_n as a function of incident angle θ at a dielectric interface (air-glass, $n_2/n_1 = 1.5$)

given by

$$\sin \theta_c = \frac{n_2}{n_1} \tag{9.27}$$

The field amplitude of a plane wave transmitted into the second medium can be described by (see Equation (1.9))

$$u_t = U_t e^{ik_2(x \sin \theta_2 + z \cos \theta_2)} \tag{9.28}$$

When using Snell's law we get

$$\sin \theta_2 = \frac{n_1}{n_2} \sin \theta_1 = \frac{\sin \theta_1}{n} \tag{9.29a}$$

$$\cos \theta_2 = \pm \sqrt{1 - \left(\frac{\sin \theta_1}{n} \right)^2} \tag{9.29b}$$

or, since we are concerned with the case where $\sin \theta_1 > n$

$$\cos \theta_2 = \pm i \sqrt{\left(\frac{\sin \theta_1}{n} \right)^2 - 1} \equiv \pm i\beta \tag{9.30}$$

Hence

$$u_t = U_t e^{ik_2 x \sin \theta_1 / n} e^{\pm k_2 \beta z} \tag{9.31}$$

Neglecting the positive exponential which is physically untenable we have a wave whose amplitude drops off exponentially as it penetrates the less dense medium. The field advances in the x-direction as a so-called surface or evanescent wave. Notice that the wavefronts or surfaces of constant phase (parallel to the yz-plane) are perpendicular to the surfaces of constant amplitude (parallel to the xy-plane) and therefore the wave is said to be inhomogeneous. Its amplitude decays rapidly in the z-direction, becoming negligible at a distance into the second medium of only a few wavelengths.

Precautions should be taken when calculating the transmittance of the interface. Transmittance is the ratio of the transmitted over the incident flux and is given by

$$T = \frac{I_t \cos \theta_2}{I_i \cos \theta_1} \tag{9.32}$$

where the projected areas of the incident beams are taken into account since they are unequal. Since u_i and u_t are propagating in media of different refractive indices we must also take into account the correct proportionality factor between the field amplitude squared and the irradiance, see Section 1.8, Equation (1.15). Then we get

$$\frac{I_t}{I_i} = \frac{n_2 |U_t|^2}{n_1 |U_i|^2} \tag{9.33}$$

and the transmittance

$$T = \frac{n_2 \cos \theta_2 |U_t|^2}{n_1 \cos \theta_1 |U_i|^2} = \frac{n_2 \cos \theta_2}{n_1 \cos \theta_1} t^2 \tag{9.34}$$

Since both the incident and reflected beams are propagating in the same medium and their projected areas are equal, we get for the reflectance

$$R = r^2 \qquad (9.35)$$

or in component forms

$$R_p = r_p^2 \qquad (9.36a)$$

$$R_n = r_n^2 \qquad (9.36b)$$

$$T_p = \frac{n_2 \cos \theta_2}{n_1 \cos \theta_1} t_p^2 = \frac{\sqrt{n^2 - \sin^2 \theta}}{\cos \theta} t_p^2 \qquad (9.36c)$$

$$T_n = \frac{n_2 \cos \theta_2}{n_1 \cos \theta_1} t_n^2 = \frac{\sqrt{n^2 - \sin^2 \theta}}{\cos \theta} t_n^2 \qquad (9.36d)$$

It can be shown that

$$R_p + T_p = 1 \qquad (9.37a)$$

$$R_n + T_n = 1 \qquad (9.37b)$$

as it should be.

9.6 THE JONES MATRIX FORMALISM OF POLARIZED LIGHT

Equation (9.4) can be written in a more compact form as

$$|U\rangle = e^{i\beta/2} \begin{pmatrix} u_x \\ u_y \end{pmatrix} \qquad (9.38)$$

where

$$u_x = U_x e^{-i\delta/2} \qquad (9.39a)$$

$$u_y = U_y e^{i\delta/2} \qquad (9.39b)$$

$|U\rangle$ is called a Jones vector or state vector, representing the state of polarization of the wave. Evidently, the state of polarization, and thereby the Jones vector, remains unchanged by a multiplicative constant. Therefore

$$|U'\rangle = c|U\rangle \qquad (9.40)$$

and $|U\rangle$ are equal state vectors representing the same state of polarization. Equation (9.38) can be decomposed to read

$$|U\rangle = U_x e^{-i\delta/2}|P_x\rangle + U_y e^{i\delta/2}|P_y\rangle \qquad (9.41)$$

where

$$|P_x\rangle = \begin{pmatrix} 1 \\ 0 \end{pmatrix} \tag{9.42a}$$

$$|P_y\rangle = \begin{pmatrix} 0 \\ 1 \end{pmatrix} \tag{9.42b}$$

are base vectors representing waves, linearly polarized in the x- and y-directions respectively. Here we have omitted the common phase factor $e^{i\beta/2}$ for reasons given after Equation (9.5). Equation (9.41) is a general expression for an arbitrary state of polarization decomposed into the orthogonal basis $|P_x\rangle$, $|P_y\rangle$.

As discussed in Section 9.2, Equation (9.41) represents a wave, linearly polarized at an angle ϕ to the x-axis, and is termed a P-state, when $\delta = 0$ and $U_x = \cos\phi$, $U_y = \sin\phi$, which yields

$$|P\rangle = \cos\phi|P_x\rangle + \sin\phi|P_y\rangle \tag{9.43}$$

For circularly polarized states, we further have $\delta = \pm\pi/2$ and $U_x = U_y = 1/\sqrt{2}$ (normalized intensity). Specifically, a right circularly polarized wave (an R-state) is given by

$$|R\rangle = \frac{1}{\sqrt{2}}(e^{-i\pi/4}|P_x\rangle + e^{i\pi/4}|P_y\rangle) \tag{9.44a}$$

and a left circularly polarized wave (an L-state) by

$$|L\rangle = \frac{1}{\sqrt{2}}(e^{i\pi/4}|P_x\rangle + e^{-i\pi/4}|P_y\rangle) \tag{9.44b}$$

A state vector may equally well be represented by a set of orthonormal base vectors other than $|P_x\rangle$, $|P_y\rangle$. For example $|R\rangle$ and $|L\rangle$ will form such a set. That two vectors

$$|A\rangle = \begin{pmatrix} a_1 \\ a_2 \end{pmatrix} \tag{9.45a}$$

$$|B\rangle = \begin{pmatrix} b_1 \\ b_2 \end{pmatrix} \tag{9.45b}$$

are orthonormal means that the scalar products

$$\langle A|A\rangle = \langle B|B\rangle = 1 \tag{9.46a}$$

$$\langle A|B\rangle = \langle B|A\rangle = 0 \tag{9.46b}$$

Here $\langle A|$ and $\langle B|$ means the row matrices

$$\langle A| = (a^*a_2^*) \tag{9.47a}$$

$$\langle B| = (b_1^*b_2^*) \tag{9.47b}$$

and the scalar product is given by ordinary matrix multiplication, e.g.

$$\langle A|B\rangle = (a_1^*a_2^*)\begin{pmatrix} b_1 \\ b_2 \end{pmatrix} = a_1^*b_1 + a_2^*b_2 \tag{9.48}$$

where the asterisks denote complex conjugation, see Appendix A.

We have seen that, for example, for a P_x-state, there exists a P_x-filter, i.e. a linear polarizer with its transmission direction along the x-axis, for an R-state there exists an R-filter, i.e. a right circular polarizer, and so on. Generally, it can be shown that to elliptically polarized light, an E-state, it is possible to construct a corresponding E-filter. The physical interpretation of the scalar product $\langle A|B \rangle$ is therefore given as the probability that an A-filter will be transmitted by a B-state and the transmitted intensity is given as the absolute square of the probability, i.e.

$$I = |\langle A|B \rangle|^2 \tag{9.49}$$

As an example, we calculate the intensity when a P_x-filter is transmitted by a P-state given by Equation 9.43

$$I = |\langle P_x|P \rangle|^2 = |\cos\phi \langle P_x|P_x \rangle + \sin\phi \langle P_x|P_y \rangle|^2 = \cos^2\phi \tag{9.50}$$

since

$$\langle P_x|P_x \rangle = 1 \quad \text{and} \quad \langle P_x|P_y \rangle = 0$$

which again gives Malus' law.

When the state of polarization of a light wave is changed by a polarization filter (or any other optical phenomenon) this change can be described as an input-output relation by a 2×2 Jones matrix, viz.

$$|E_u \rangle = M|E_i \rangle \tag{9.51}$$

where $|E_i \rangle$ and $|E_u \rangle$ are the state vectors of the incoming and outcoming light respectively, and M is the matrix representing the polarization change. It is the power of Jones matrix theory that when light passes through several polarizing filters in succession, represented by, for example, the matrices $M_1, M_2, \ldots, M_n$, the total polarizing effect is described by the product of the matrices of the individual filters, i.e.

$$|E_u \rangle = M_n \ldots M_2 M_1 |E_i \rangle \tag{9.52}$$

Note the order of matrices of the product.

It is easily verified that the Jones matrix for a linear polarizer with its transmission direction along the x-axis represented in the $|P_x \rangle$, $|P_y \rangle$ basis, must be

$$P_x = \begin{pmatrix} 1 & 0 \\ 0 & 0 \end{pmatrix} \tag{9.53}$$

and that of a retarder with retardance δ with its fast axis along the x-axis must be

$$M_x = \begin{pmatrix} e^{-i\delta/2} & 0 \\ 0 & e^{i\delta/2} \end{pmatrix} \tag{9.54}$$

To find the general expressions of these matrices when their axes are oriented at an arbitrary angle ϕ to the x-axis, we proceed as follows. First rotate the coordinate system an angle ϕ such that the x'-axis coincides with the filter axis, then apply the matrix of the filter and finally rotate the coordinate system back to its original position. This is expressed mathematically as:

For the polarizer

$$P = R^{-1} P_x R \tag{9.55}$$

For the retarder

$$M = R^{-1} M_x R \tag{9.56}$$

where

$$R = \begin{pmatrix} \cos\phi & \sin\phi \\ -\sin\phi & \cos\phi \end{pmatrix} \tag{9.57}$$

is the rotation matrix and

$$R^{-1} = \begin{pmatrix} \cos\phi & -\sin\phi \\ \sin\phi & \cos\phi \end{pmatrix} \tag{9.58}$$

is its inverse. This gives

$$P = \begin{pmatrix} a & b \\ b & c \end{pmatrix} \tag{9.59}$$

where

$$a = \cos^2\phi$$
$$b = \sin\phi\cos\phi$$
$$c = \sin^2\phi \tag{9.60}$$

and

$$M = \begin{pmatrix} p & q \\ q & r \end{pmatrix} \tag{9.61}$$

where

$$p = e^{-i\delta/2}\cos^2\phi + e^{i\delta/2}\sin^2\phi$$
$$q = -2i\sin(\delta/2)\sin\phi\cos\phi \tag{9.62}$$
$$r = e^{-i\delta/2}\sin^2\phi + e^{i\delta/2}\cos^2\phi$$

As examples, by putting $\delta = \pi/2$ and $\delta = \pi$ we find the matrices for the $\lambda/4$ and $\lambda/2$ plates respectively.

9.7 PHOTOELASTICITY

9.7.1 Introduction

Photoelasticity is a method for stress analysis of specimens subject to load. The standard methods rely on the technique of making a model which is a copy (often on a reduced scale) of the specimen under investigation.

This model, usually made from plastics such as epoxy or polyester resins, becomes birefringent when subject to loads. Thus, a two-dimensional photoelastic model exerted by forces in its own plane will behave as a general retarder, the retardance and the direction of the retarder axes being continuous variables across the model plane. The retardance δ is given as

$$\delta = kC(\sigma_1 - \sigma_2)t \tag{9.63}$$

where σ_1 and σ_2 are the principal stresses, C is the stress-optic coefficient characteristic of the model material and t is the thickness of the model. The retarder axes h_1, h_2 coincide with the axes of the principal stresses σ_1, σ_2.

When this loaded model is placed in a polariscope which consists of a light source and properly arranged polarizing elements, a system of fringes is observed. In the next subsections we shall find the relation between this fringe system and the stresses.

In a related technique termed the photoelastic coating method, a birefringent coating is attached to the real specimen. The optical principles here are essentially the same as for the standard methods except that the fringes have to be observed in reflection instead of transmission.

9.7.2 The Plane Polariscope

In Figure 9.9 the photoelastic model is placed in plane 2 between two linear polarizers in planes 1 and 3. This configuration is termed a plane polariscope. When the two polarizers are crossed (i.e. their transmission axes are inclined at 90°) as in the figure, we have a dark-field plane polariscope, a term which refers to the field without the model. When the two polarizers are parallel, we have a light-field plane polariscope.

Let us for the time being assume the model to be a uniform retarder with its principal axis h_1 making an angle α to the x-axis and try to find the intensity of the light behind plane 3.

The transmission direction of P_1 is along the x-axis. Hence the field behind P_1 is

$$\mathbf{u}_1 = \mathbf{e}_x \tag{9.64}$$

where we have put the amplitude equal to 1. The components of u_1 along h_1 and h_2 are $\cos\alpha$ and $-\sin\alpha$ respectively. The field behind the retarder therefore becomes

$$\mathbf{u}_2 = \cos\alpha e^{-i\delta/2}\mathbf{e}_1 - \sin\alpha e^{i\delta/2}\mathbf{e}_2 \tag{9.65}$$

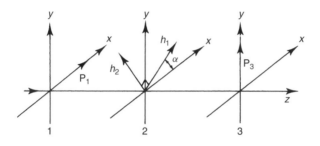

Figure 9.9 Plane polariscope. Schematic

where $\mathbf{e}_1$ and $\mathbf{e}_2$ are the unit vectors along h_1 and h_2. The transmission direction of P_3 is along the y-axis. To find the field behind plane 3, we therefore have to calculate the component of $\mathbf{u}_2$ along the y-axis which gives

$$\mathbf{u}_3 = \sin\alpha\cos\alpha(e^{-i\delta/2} - e^{i\delta/2})\mathbf{e}_y \tag{9.66}$$

and hence the intensity

$$
\begin{aligned}
I_3 &= |\mathbf{u}_3|^2 = 4\sin^2\alpha\cos^2\alpha\sin^2(\delta/2)\\
&= \sin^2 2\alpha \sin^2(\delta/2)
\end{aligned}
\tag{9.67}
$$

It is easily verified that if the transmission direction of P_3 is along the x-axis, the intensity becomes

$$I_3' = (1 - \sin^2 2\alpha \sin^2(\delta/2)) \tag{9.68}$$

These calculations can be written in a very compact and elegant form using the state vectors $|P_x\rangle$ and $|P_y\rangle$ and the matrix M of a retarder (Equation (9.61); cf. Equation (9.49))

$$I_3 = |\langle P_y|M|P_x\rangle|^2 = |q|^2 = \sin^2(2\alpha)\sin^2(\delta/2) \tag{9.69}$$

$$I_3' = |\langle P_x|M|P_x\rangle|^2 = |p|^2 = 1 - \sin^2(2\alpha)\sin^2(\delta/2) \tag{9.70}$$

For a uniform retarder the intensity across the field of view is uniform and determined by Equations (9.69) and (9.70). For a photoelastic model, however, both α and δ vary over the xy-plane resulting in a system of fringes across the field of view. From Equation (9.69) and (9.70) we see that this fringe system consists of two sets: one depending on α, the other on δ. In a dark-field polariscope we see from Equation (9.69) that $I_3 = 0$ when $\alpha = 0$, i.e. we get a dark fringe. These fringes depending on α are called isoclines, which means 'equal inclinations', and are loci of points where the principal stress axes coincide with the axes of the polarizers. They show up as broad dark bands on the model. By synchronous rotation of the two polarizers, we vary α and the isoclinics move, thereby determining the direction of the principal stresses over the whole model. From Equation (9.69) and (9.70) we see that $I_3 + I_3' = 1$. A dark isocline in a dark-field polariscope therefore turns into a bright isocline in a light-field polariscope. The other fringe system depending on δ does not change its position by varying α. We will examine these fringes more closely in the next subsection.

9.7.3 The Circular Polariscope

Figure 9.10 shows the same set-up as in Figure 9.9 except that two quarterwave plates with their axes oriented $45°$ to the transmission directions of the polarizers are placed between the polarizers and the retarder. We thus have a circular polarizer on each side of the retarder (the model). This configuration is called a circular polariscope, and is termed a dark-field polariscope when the two circular polarizers have opposite handedness and a light-field polariscope when the two circular polarizers have the same handedness.

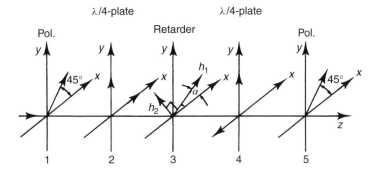

Figure 9.10 Circular polariscope. Schematic

To calculate the intensity behind plane 5, we now may derive the field amplitude behind each element in steps, in the same manner as above. This, however, is a tedious task. A much simpler way is to apply Jones matrix algebra, which gives:

For circular polarizers of opposite handedness

$$I_5 = |\langle L|M|R\rangle|^2 = |p - r - 2iq|^2 = \sin^2(\delta/2) = \tfrac{1}{2} - \tfrac{1}{2}\cos\delta \qquad (9.71)$$

For circular polarizers of the same handedness

$$I'_5 = |\langle R|M|R\rangle|^2 = |p + r|^2 = \cos^2(\delta/2) = \tfrac{1}{2} + \tfrac{1}{2}\cos\delta \qquad (9.72)$$

From these equations we see that in a circular polariscope the fringe pattern is only dependent on δ, i.e. the isoclinics vanish. From Equation (9.63) we further see that δ and thereby the intensity is dependent on the wavelength λ. Therefore, by using a white light source in a circular polariscope, the resulting pattern will show up as fringes of different colours. These fringes, therefore are termed isochromatics which means equal colours. As can be seen from Equation (9.63), these isochromatics are loci of equal principal stress difference $\sigma_1 - \sigma_2$.

From Equation (9.71) we see that in a dark-field circular polariscope, we get a dark isochromatics when

$$\delta = \frac{2\pi}{\lambda}C(\sigma_1 - \sigma_2)t = 2\pi \cdot n, \quad n = 0, 1, 2, \ldots \qquad (9.73)$$

i.e. when

$$\sigma_1 - \sigma_2 = n \cdot S \qquad (9.74)$$

and bright isochromatics when

$$\frac{2\pi}{\lambda}C(\sigma_1 - \sigma_2)t = 2\pi(n + \tfrac{1}{2}), \quad n = 0, 1, 2, \ldots \qquad (9.75)$$

i.e. when

$$\sigma_1 - \sigma_2 = (n + \tfrac{1}{2})S \qquad (9.76)$$

where n is the order of the isochromatics and where we have put

$$S = \frac{\lambda}{Ct} \tag{9.77}$$

When using the same material of the same thickness and a monochromatic light source of the same wavelength, S is a system constant. It is usually calibrated by measuring the fringe order in the centre of a diametrally loaded circular disc and inserting its value into the formula

$$S = \frac{8F}{n\pi D} \tag{9.78}$$

where F is the applied force, D is the diameter of the disc, and n is the measured fringe order.

The principal stress difference in an arbitrary model is then found by just counting the fringe order n using Equation (9.74)–(9.76). To do this, of course, one needs to know where to start the counting, i.e. which is the zero-order fringe. From Equation (9.73) we see that for $n = 0$, δ becomes independent of the wavelength. Therefore, by using a white light source, fringes of order $n > 0$ become coloured while the zero-order fringe is easily recognized as a dark colourless fringe. The higher-order fringes are most accurately determined by using monochromatic light, often a sodium lamp. A standard polariscope is therefore usually equipped with two light sources.

Figure 9.11 shows a typical pattern of isochromatics in a dark-field polariscope. The model is a strip with a hole under uniaxial tension.

Finally, we mention that on a load-free boundary of the model, the principal stress normal to the boundary is zero, and the value of the other principal stress is found directly.

9.7.4 Detection of Isochromatics of Fractional Order. Compensation

We have shown that in a circular polariscope, an isochromatic fringe pattern giving the stress distribution is observed. It may happen, however, that an interesting measuring point lies between two isochromatic fringes. To determine the fringe order with sufficient accuracy at such a point there exist several methods. These are as follows.

(1) Varying the load (whose value must be known) until an isochromatic fringe intersects the point of interest. This is not always possible.

(2) Colour matching, i.e. knowing the change in colour with increasing order using a white light source or a spectral lamp. This method relies on human judgement of colours and is possible only for low orders.

(3) Photometric methods, i.e. accurately measuring the intensity at the point of interest when rotating the polarizing elements with known amounts. These methods have some inherent sources of error.

(4) Compensation. This can be done by first orienting, for example, a Babinet–Soleil compensator parallel to the direction of the principal stresses, i.e. the isoclinics, and then adjust the compensator's retardance until the measurement point becomes dark. Non-expensive compensators can be made from loaded strips of photoelastic material.

Figure 9.11   Dark-field isochromatic pattern from a strip with a hole under uniaxial tension

The most common method of compensation, however, is by means of quarterwave plates, known as the Tardy of Senarmont methods. Here we shall describe the latter in more detail.

In Figure 9.12(a) the point of interest P lies between two isochromatics of order n and $n + 1$. To determine the order ($= n + p$ where $0 < p < 1$) at point P we proceed as follows:

(1) Change to a dark-field plane polariscope configuration (Figure 9.9) and rotate the polarizers P_1 and P_3 in synchronization until a bright isoclinic intersects P (see Figure 9.12(b)). We then have $\alpha = 45°$ and $I_3 = \sin^2(\delta/2)$ (cf. Equation 9.69)).

(2) Place a quarterwave plate with its axes parallel to the transmission directions of P_1 and P_3 between the model and P_3, i.e. in a plane 2b between planes 2 and 3. To find the field amplitude $\mathbf{u}_{2b}$ behind the quarterwave plate, we consider Equation 9.65

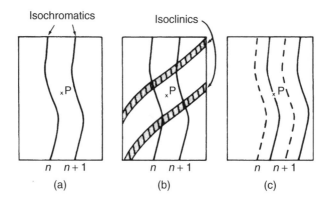

Isochromatics   Isoclinics

n $n+1$ n $n+1$ n $n+1$

(a)   (b)   (c)

Figure 9.12

which gives the field incident on the quarterwave plate. We find (for $\alpha = 45°$)

$$\mathbf{u}_{2b} = e^{i\pi/4}\cos(\delta/2)\mathbf{e}_x - ie^{-i\pi/4}\sin(\delta/2)\mathbf{e}_y \qquad (9.79)$$

(3) Rotate P_3 (alone) on angle χ. The field amplitude u_3 behind P_3 then becomes

$$\mathbf{u}_3 = (e^{i\pi/4}\sin\chi\cos(\delta/2) - ie^{-i\pi/4}\cos\chi\sin(\delta/2))\mathbf{e}_x$$
$$= e^{i\pi/4}\sin(\chi - \delta/2)\mathbf{e}_x \qquad (9.80)$$

where $\mathbf{e}_x$ is the unit vector along the transmission direction of P_3. The intensity now becomes equal to

$$I_3 = |\mathbf{u}_3|^2 = \sin^2(\chi - \delta/2) \qquad (9.81)$$

For the nth-order isochromatics, $\delta = n \cdot 2\pi$ and

$$I_3 = \sin^2\chi \qquad (9.82)$$

which for $\chi = 0$ gives $I_3 = 0$ as it should. By rotating P_3 an angle $\chi = \pi$, we again have $I_3 = 0$. We will observe that during this rotation, the nth-order isochromatics would have moved to the initial position of the $(n + 1)$th-order isochromatics.

For an intermediate value χ_p of the rotation angle, the nth-order isochromatics will intersect our point P (see Figure 9.12(c)). A rotation angle $\chi = \pi$ therefore corresponds to increasing the order number by unity, while a rotation angle χ_p corresponds to an increase of the order number equal to

$$p = \chi_p/\pi \qquad (9.83)$$

where χ_p is measured in radians. In conclusion, we have found the isochromatic order in point P to be $n + \chi_p/\pi$.

In practice, this procedure can be executed as follows. The polarizer P_1, the model, the quarterwave plate Q and the analyser P_3 are placed in that order as described above.

P_1 and P_2 are crossed (dark-field polariscope) and the axes of Q are parallel to the transmission directions of P_1 and P_3. In this position, the quarterwave plate does not affect the intensity (see Equation (9.17)). The scales of P_1, Q and P_3 are divided to read 1 for a 180° rotation.

Initially, all scales are set to zero. The filters are coupled together, e.g. by means of toothwheel transmissions. Firstly, all three filters are rotated until a dark isoclinic intersects P and the resulting scale reading is a. This is taken as the first step because it is easier to determine the intersection of a dark isoclinic rather than a bright one. Then the filters are rotated 45°, i.e. to a scale reading equal to $b = a + 0.25$. We then have reached point (1) in the description above, i.e. a bright isoclinic intersects P. Finally, the analyser P_3 is released and rotated alone (point 3) until the nth-order isochromatic intersects P and the new scale reading is c. The isochromatic order at P is then given as

$$n + |c - b| \qquad (9.84)$$

9.8 HOLOGRAPHIC PHOTOELASTICITY

As shown in the previous chapter, the directions and the difference $\sigma_1 - \sigma_2$ of the principal stresses can be determined by using a standard polariscope. For a complete solution of the stress distribution inside a two-dimensional model, however, the absolute values of σ_1 and σ_2 must be known.

Up to now we have omitted a common phase factor $e^{i\beta/2}$ imposed on the light when passing through the stressed model. This phase is unaffected by the birefringence of the model, and is given by

$$\beta = (2\pi/\lambda)[K(\sigma_1 + \sigma_2) + 2n_0]t \qquad (9.85)$$

where K is a constant of the model material and n_0 is the refractive index of the unstressed model. If β could be determined, the principal stress sum $\sigma_1 + \sigma_2$ is given, which together with the values of $\sigma_1 - \sigma_2$ gives a complete solution of the stress-distribution problem. This phase can be found by interferometric methods. Here the method of holographic interferometry will be considered in more detail.

For this purpose, a set-up like that shown in Figure 9.13 can be applied. Here the light passing through the model placed in a light-field circular polariscope constitutes the object wave. The method consists of making two exposures of the model, first in its unloaded and then in its loaded condition. To get a registration of the unloaded model, it is therefore essential to have a light-field polariscope.

The matrix representing the unloaded model will be

$$M_0 = e^{i\phi_0} I \qquad (9.86)$$

where $\phi_0 = kn_0 t$ is the uniform phase due to the unloaded model and I is the identity matrix

$$I = \begin{pmatrix} 1 & 0 \\ 0 & 1 \end{pmatrix} \qquad (9.87)$$

The complex amplitude recorded in the first hologram exposure therefore can be written as the state vector

$$|S_0\rangle = SM_0|R\rangle = e^{i\phi_0}|P_x\rangle \qquad (9.88)$$

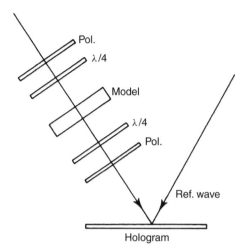

Figure 9.13   Holographic photoelasticity set up. $\lambda/4 =$ quarterwave plates

where we have assumed the incident light to be in a right circular state and where S represents the quarterwave plate-polarizer combination behind the model.

The field amplitude when the model is loaded in the second exposure then becomes

$$|S\rangle = SM|R\rangle = e^{i\beta/2} \cos(\delta/2)|P_x\rangle \qquad (9.89)$$

With a reference wave polarized in a P_x-state, these two exposures will be recorded with maximum modulation. The second polarizer in Figure 9.13 can be removed since the reference wave will act as an analyser when interfering with the object wave. By preparing the reference wave in a right circular polarized state, the second quarterwave plate can also be removed. But then the recorded modulation will be reduced and unwanted polarization effects will occur.

By reconstruction of this doubly exposed hologram, the reconstructed field will be proportional to the sum of the partial exposures and the intensity becomes

$$
\begin{aligned}
I &= |e^{i\phi_0} + e^{i\beta/2} \cos(\delta/2)|^2 \\
&= 1 + 2\cos(\delta/2)\cos(\phi_0 - \beta/2) + \cos^2(\delta/2) \\
&= 1 + 2\cos\left[k\frac{C}{2}(\sigma_1 - \sigma_2)t\right]\cos\left[k\frac{K}{2}(\sigma_1 + \sigma_2)t)\right] + \cos^2\left[k\frac{C}{2}(\sigma_1 - \sigma_2)t\right] (9.90)
\end{aligned}
$$

The resulting fringe pattern therefore consists of two types: the isochromatics depending on $\sigma_1 - \sigma_2$ and the so-called isopachics depending on $\sigma_1 + \sigma_2$.

From the second term of Equation (9.90), we see that the isopachics are modulated by the isochromatics and the analysis of this combined pattern might in some cases become quite complicated.

A dark isochromatic occurs whenever

$$\delta = (2n + 1)\pi, \quad \text{for} \quad n = 0, 1, 2, \ldots \qquad (9.91)$$

with a resulting intensity

$$I = 1 \tag{9.92}$$

The dark isochromatics therefore possess non-zero intensity and show up as half-tone fringes in the resulting pattern. A bright isochromatic occurs whenever

$$\delta = n \cdot 2\pi, \quad \text{for} \quad n = 0, 1, 2, \ldots \tag{9.93}$$

with a resulting intensity

$$I = 2[1 + (-1)^n \cos(\phi_0 - \beta/2)] \tag{9.94}$$

Assume for a moment n to be an even number k. The intensity is then

$$I_k = 2[1 + \cos(\phi_0 - \beta/2)] \tag{9.95}$$

To simultaneously have a dark isopachic we must demand

$$\phi_0 - \beta/2 = (m + \tfrac{1}{2})2\pi, \quad \text{for} \quad m = 0, 1, 2, \ldots \tag{9.96}$$

which gives $I_{k,m} = 0$. When tracing this dark isopachic from the kth bright isochromatic into the next $(k + 1)$th bright isochromatic, the intensity becomes $I_{k+1,m} = 4$ which implies that the isopachic has changed from a dark to a bright fringe in crossing the isochromatic. Moreover, the condition of a dark isopachic now becomes

$$\phi_0 - \beta/2 = m \cdot 2\pi \tag{9.97}$$

which shows that the isopachic has changed by one half-order in crossing the dark isochromatic.

These characteristics of the combined pattern are shown in Figure 9.14 for the case where the isochromatics and the isopachics cross each other nearly perpendicularly.

Examples of this phase shift are also seen on the photograph in Figure 9.15. In the case of perpendicular crossing of the two patterns, their analysis becomes fairly easy, but when the two patterns are nearly parallel, the quantitative interpretation can give erroneous results. Attempts have therefore been made to separate the two patterns. One solution is to let the object wave pass twice through the model and a Faraday rotator as shown Figure 9.16. By passing the rotator the orthogonal polarizations are reversed and the birefringence effect is cancelled in the second pass of the model resulting in a isopachics pattern only.

9.9 THREE-DIMENSIONAL PHOTOELASTICITY

9.9.1 Introduction

Up to now we have considered two-dimensional states of stress, i.e. the stress in the z-direction is constant. When the stress also varies in this direction, we have a three-dimensional stress state. This cannot generally be investigated by the two-dimensional procedure of observing the isochromatics and isoclinics in a polariscope. This is because

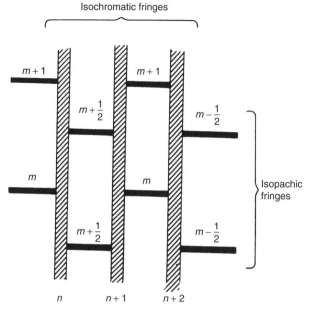

Figure 9.14

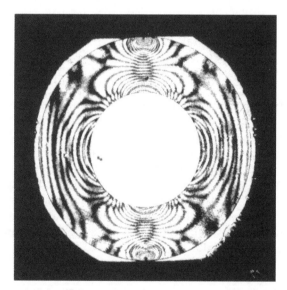

Figure 9.15   Combined isochromatics-isopachics pattern obtained with double-exposure holography

the birefringence effect integrated over the optical path is generally so complex that it is impossible to relate it to the stresses which produce it.

Several methods are available for the investigation of three-dimensional problems. Below we briefly discuss two of them, the frozen stress method and the scattered light method.

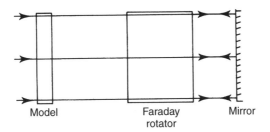

Figure 9.16

9.9.2 The Frozen Stress Method

This is the most widely used three-dimensional method; it is restricted, however, to static loading by external forces or constant body forces such as gravitational or centrifugal forces. In this method, advantage is taken of the multiphase (diphase) nature of plastics used as model materials to conserve strain and birefringence in the model after the load has been removed. Then, slices can be cut from the model and analysed in a polariscope as a two-dimensional model without disturbing the frozen-in stress pattern.

The procedure of strain freezing consists of heating the model to a temperature slightly above the softening point or critical temperature followed by slow cooling under load to room temperature. In practice, it is most convenient to load the model before placing it in the oven. While the rate of heating is immaterial, the cooling must be carried out slowly, typically at less than $2\,°C$ per hour.

In the slicing process, sharp tools should be used to avoid overheating, which may disturb the frozen stress distribution.

Consider a slice cut from a three-dimensional model in such a way that the directions of two of the three principal stresses, e.g. σ_1 and σ_2, lie in the plane of the slice (see Figure 9.17). This pre-supposes that the direction of at least one of the principal stresses, e.g. σ_3, is known, which is frequently the case, for example on a section of symmetry or a free surface of the model. By placing this slice in a polariscope, the isochromatics will determine $\sigma_1 - \sigma_2$ while the isoclinics will indicate their directions. To determine the principal stress difference in the other two principal planes, a subslice may be cut from the original slice parallel to one of the principal stresses, say σ_1 (see Figure 9.17). If this subslice is observed in a polariscope in the direction of σ_2, the resulting isochromatics will

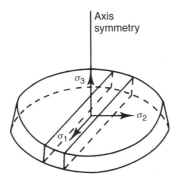

Figure 9.17

give the value of $\sigma_1 - \sigma_3$. The remaining value $\sigma_2 - \sigma_3$ may then be found by subtracting the second value from the first.

When none of the directions of the three principal stresses is *a priori* known, the situation is more complicated. For the detailed description of the various methods for solving this problem, we direct the reader to the specialized literature given in the Bibliography.

9.9.3 The Scattered Light Method

If a beam of light passes through a transparent isotropic medium, it will be scattered by small particles in suspension or by the molecules of the medium. The field amplitude of the light scattered from a certain point, will be proportional to the component of the amplitude of the incident light normal to the scattering direction.

Consider Figure 9.18 where a thin pencil of light propagates in the z-direction through a three-dimensional photoelastic model. At each point along this primary beam we observe the scattered light in a direction normal to z, say the x-axis. The field amplitude of the scattered light then becomes equal to the y-component of the field of the primary beam at the scattering point. Now assume that the principal stresses, say σ_1 and σ_2, are normal to the z-axis, and that their directions remain constant along the primary beam and make an angle ϕ to the x-axis. Further assume that the incident light is linearly polarized at an angle α to the x-axis. The phase difference between the field components parallel to σ_1 and σ_2 accumulated from the point of entry to the scattering point we denote by Δ. The calculation of the scattered intensity now becomes equivalent to the problem of calculating the intensity of the light transmitting a retarder given by Equation (9.61) placed between a polarizer

$$|P\rangle = \begin{pmatrix} \cos\alpha \\ \sin\alpha \end{pmatrix} \tag{9.98}$$

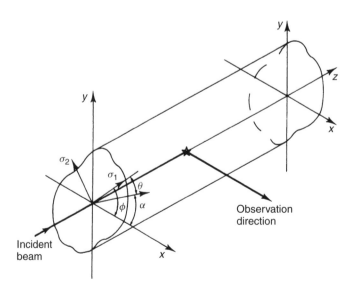

Figure 9.18

and an analyser $|P_y\rangle = \begin{pmatrix} 0 \\ 1 \end{pmatrix}$ (see Figure 9.18). This gives

$$I = |\langle P_y|M|P\rangle|^2 = |q\cos\alpha + r\sin\alpha|^2$$
$$= \sin^2\alpha + \sin^2(\Delta/2)\sin 2\phi \sin 2(\phi - \alpha) \qquad (9.99)$$

which, by introducing the angle $\theta = \phi - \alpha$ between the retarder axis and the polarization direction, becomes

$$I = \tfrac{1}{2}(1 - \cos 2\phi \cos 2\theta - \sin 2\phi \sin 2\theta \cos\Delta) \qquad (9.100)$$

Therefore, by observing the scattered light in a direction perpendicular to the primary beam along its path, the intensity will vary from

$$I_{\min} = \tfrac{1}{2}[1 - \cos 2(\phi - \theta)] = \sin^2\alpha \qquad (9.101)$$

at points where

$$\Delta = n \cdot 2\pi \quad \text{for} \quad n = 0, 1, 2, \ldots$$

to

$$I_{\max} = \tfrac{1}{2}[1 - \cos 2(\phi + \theta)] \qquad (9.102)$$

at points where

$$\Delta = (2n + 1)\pi$$

This gives for the visibility

$$V = \frac{I_{\max} - I_{\min}}{I_{\max} + I_{\min}} = \frac{\sin 2\phi \sin 2\theta}{1 - \cos 2\phi \cos 2\theta} \qquad (9.103)$$

From this we see that the fringe visibility is maximum (i.e. unity) when $\theta = \phi$, i.e. when the direction of observation is parallel to the polarization direction or when the axis of σ_1 bisects the angle between the polarization and observation directions. In the first case, the intensity becomes

$$I = \tfrac{1}{2}\sin^2 2\phi(1 - \cos\Delta) \qquad (9.104)$$

and in the second case $(\theta = -\phi)$

$$I = \tfrac{1}{2}\sin^2 2\phi(1 + \cos\Delta) \qquad (9.105)$$

Moreover, the maximum variation of I occurs for $\sin 2\phi = 1$, i.e. when the principal stress axis is inclined $45°$ to the direction of observation.

From Equation (9.100) we see that the intensity is constant and independent of Δ when either the polarization or observation direction is parallel to one of the principal stresses. No interference effects will then be observed along the path of the primary beam. Moreover, if $\theta = \phi = 0°$ or $90°$, the intensity becomes a minimum. The directions of the principal stresses can therefore be determined by making the polarization and observation directions parallel and then rotating the model about the axis of the primary beam until a uniform minimum intensity is observed. With an elementary path length dz along the primary beam the principal stress difference $\sigma_1 - \sigma_2$ can be regarded as constant. The stress optic law, Equation (9.63) then gives

$$d\Delta = kC(\sigma_1 - \sigma_2)\, dz \tag{9.106}$$

or

$$\sigma_1 - \sigma_2 = \frac{1}{kC}\frac{d\Delta}{dz}$$

Expressed in terms of the fringe order, this equation can be written as

$$\sigma_1 - \sigma_2 = f(dn/dz) \tag{9.107}$$

The principal stress difference at any point along the primary beam is therefore proportional to the gradient or inversely proportional to the spacing of the fringes in the scattered light. The possibility of errors in fringe counting owing to a change of sign of the principal stress difference at any point may be avoided by introducing an additional phase difference by means of a compensator. The fringes on opposite sides of such a point will then move in opposite directions. Alternatively, a change of sign may be indicated by reversal of the colour sequence of the fringes when using white light.

Another variant of the scattered light method is to apply an unpolarized primary beam propagating in, for example, the z-direction as in Figure 9.19. The light scattered in the

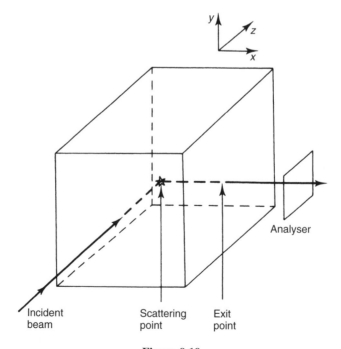

Figure 9.19

x-direction from any point will then be linearly polarized along the *y*-axis. Each point thus acts as a secondary source of linearly polarized light, and if the model is stressed, this light will be influenced by the stresses between the scattering point and the point of exit from the model. If this scattered light emerging from the model is viewed through an analyser, a fringe pattern is seen which has exactly the same meaning as that which would be produced in a plane polariscope. The problem is that the total birefringence will be accumulated from the scattering point to the point of exit from the model.

To apply the method, the primary beam is moved parallel to itself in small steps along the line of observation from the surface towards the interior of the model. For each step, the phase difference of the emergent light components is measured. If the results are plotted against the position of the scattering point, the slope of the resulting curve multiplied by the stress-optic coefficient gives the principal stress difference at that point.

The basic optical requirement in a scattered light polariscope is a fine pencil or thin ribbon of highly collimated light. Since the efficiency of the scattering process is low, a source of high intensity is required. These requirements are met by mercury arc lamps while lasers are particularly suitable. The model is usually placed in an immersion tank to avoid refraction and polarization changes at the model surface.

Provision for rotational and translational movement of the model in the tank is needed. This makes the method rather complicated in practice.

9.10 ELLIPSOMETRY

Ellipsometry (Passaglia 1964; Neal and Fane 1973) is a method for the measurement of optical constants (refractive index) of surfaces and for measuring the thickness of thin layers. It is based on the fact that light generally changes its state of polarization upon reflection; see Section 9.5.

Since the two reflection coefficients are not equal, the light reflected from a surface will generally become elliptically polarized, hence the name ellipsometry. A fraction of the light, determined by the transmittance coefficients, will propagate into the second medium. When the surface consists of a thin layer, by taking account of the reflection and transmittance coefficients and of multiple reflections in the layer, the state of polarization of the reflected light will depend both on the optical constants and the thickness of the layer.

Figure 9.20 shows an ellipsometer with a polarizer P in the incident and an analyser A in the reflected beam. P and A can be combinations of linear polarizers and retarders. The measurement principle is analogous to the one applied in photoelasticity (see Section 9.7). We shall not discuss ellipsometry in more detail here, but mention that layer thickness down to the angstrom (10^{-10} m) range can be measured.

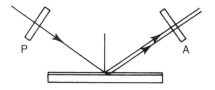

Figure 9.20   Principle of an ellipsometer. P = polarizer A = analyser

PROBLEMS

9.1 Determine the emergent state in each of the following cases:

(a) A P-state incident on a quarter-wave plate with the line of the P-state midway between the principal axes of the plate.

(b) A P-state incident on a half-wave plate with the line of the P-state midway between the principal axes of the plate.

(c) R- and L-states incident on a quarter-wave plate.

(d) R- and L-states incident on a half-wave plate.

9.2 Given that a general polarization state is given by

$$|E\rangle = A_x e^{-i\phi/2}|P_x\rangle + A_y e^{i\phi/2}|P_y\rangle$$

show that the orthogonal state $|E'\rangle$, such that $\langle E'|E\rangle = 0$ is given by

$$|E'\rangle = A_y e^{-i\phi/2}|P_x\rangle - A_x e^{i\phi/2}|P_y\rangle$$

In the case $|E\rangle = |R\rangle$, $|E'\rangle = |L\rangle$ show that the above forms are consistent with those given by Equations (9.44a,b).

9.3 If T is the matrix of a polarization element in the $|P_x\rangle$, $|P_y\rangle$-basis, the matrix T' of the same element in the $|R\rangle$, $|L\rangle$-basis is given by

$$T' = UTU^+$$

where

$$U = \begin{pmatrix} \langle R|P_x\rangle & \langle R|P_y\rangle \\ \langle L|P_x\rangle & \langle L|P_y\rangle \end{pmatrix}$$

(U^+, the adjoint of U means the transposed and complex conjugate of U)

(a) Calculate U.

(b) Find the expression for the matrix M_x of a retarder given in Equation (9.54) in the $|R\rangle$, $|L\rangle$-basis.

9.4 Just as (linear) retarders resolve beams into P-states and transmit them at different speeds there exist circular retarders which resolve beams into R- and L-states and transmit them at different speeds. Such materials are said to be optically active. The matrix N of a circular retarder with retardance β is therefore given in the $|R\rangle$, $|L\rangle$-basis as

$$N = \begin{pmatrix} e^{-i\beta/2} & 0 \\ 0 & e^{i\beta/2} \end{pmatrix}$$

(a) What is the matrix N' of the same circular retarder in the $|R\rangle$, $|L\rangle$-basis? *Hint*: Make an 'obvious' change in the matrix U.

(b) A P_x-state is incident on an optical active material with $\beta = 180°$. In what state is the output beam?

9.5 Equations (9.71) and (9.72) are derived in the $|P_x\rangle$, $|P_y\rangle$-basis. Do the same calculations in the $|R\rangle$, $|L\rangle$-basis.

9.6 The indices of refraction for calcite and quartz for the sodium yellow line are

$$\text{calcite} : n_0 = 1.658, \quad n_e = 1.486$$

$$\text{quartz} : n_0 = 1.544 \quad n_e = 1.553$$

Calculate the thickness of a quarterwave plate made from these materials.

9.7 Consider a quarterwave plate designed for wavelength λ_0. Discuss its behaviour for wavelengths λ near λ_0, that is for small $\varepsilon = (\lambda_0/\lambda - 1)$. Imagine the use of this plate as the 'photoelastic model' in the dark-field plane polariscope in Figure 9.9. When rotating the quarter-wave plate the output intensity will vary. Compute the amplitude of this variation for $\varepsilon = 0$ and $\varepsilon = 0.1$.

9.8 Show that the Brewster angles for internal and external reflection at a given interface are complementary, i.e. $\theta_{B1} + \theta_{B2} = 90°$.

9.9 Prove that for a plane parallel slab if Brewster's condition is satisfied at the top interface it will be satisfied also for internal incidence onto the second interface.

9.10 Show that when $\theta_1 > \theta_c$ at a dielectric interface, r_p and r_n are complex and $r_n r_n^* = r_p r_p^* = 1$.

9.11 Find an expression for the relative phase shift between u_{rp} and u_{rn} when $n_1 > n_2$. At which angle of incidence is the phase shift maximum for $\theta_1 > \theta_c$?

9.12 The distance travelled by an evanescent wave at total internal reflection to a depth where the intensity is $1/e$ of its initial value is called the 'skin depth' δ.

(a) Find an expression for δ.

(b) How many wavelengths is the skin depth for $\theta_1 = 45°$ when $n_1 = 1.5$, $n_2 = 1$.

10

Digital Image Processing

10.1 INTRODUCTION

The electronic camera/frame grabber/computer combination has made it easy to digitize, store and manipulate images. These possibilities have made a great impact on optical metrology in recent years. Digital image processing has evolved as a specific scientific branch for many years. Many of the methods and techniques developed here are directly applicable to problems in optical metrology. In this chapter we go through some of the standard methods such as edge detection, contrast stretching, noise suppression, etc. Besides, algorithms for solving problems specific for optical metrology are needed. Such methods will be treated in Chapter 11.

10.2 THE FRAME GRABBER

A continuous analogue representation (the video signal) of an image cannot be conveniently interpreted by a computer and an alternative representation, the digital image, must be used. It is generated by an analogue-to digital (A/D) converter, often referred to as a 'digitizer', a 'frame-store' or a 'frame-grabber'. With a frame grabber the digital image can be stored into the frame memory giving the possibility of data processing and display. The block diagram of a frame grabber module is shown in Figure 10.1. These blocks can be divided into four main sections:

(1) the video source interface;

(2) multiplexer and input feedback LUT;

(3) frame memory;

(4) display interface.

The video source interface

The video source interface performs three main operations: (1) signal conditioning, (2) synchronization/timing and (3) digitalization. In the signal condition circuitry the signal is low-pass filtered with a cut-off frequency of 3–5 MHz to avoid aliasing, see Section 5.8.3. Some frame grabbers also have programmable offset and gain.

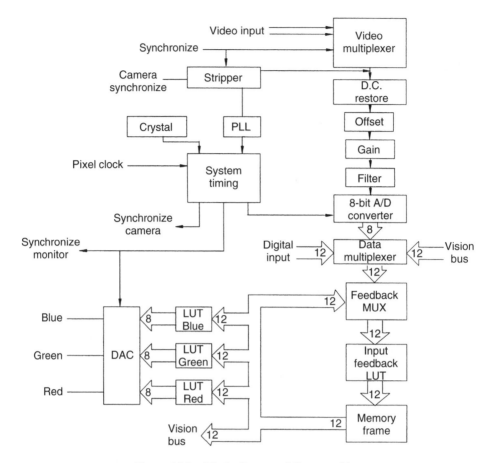

Figure 10.1 Block diagram of frame grabber

Before A/D conversion, the video signal must be stripped from its horizontal and vertical sync signals. The pixel clock in the frame grabber defines the sampling interval of the A/D converter and generates an internal Hsync signal. A phase-locked loop (PLL) tries to match this Hsync with the Hsync signal from the camera by varying the pixel clock frequency. This matching is iterative so it takes some time before the Hsyncs fit. And even then this process keeps on oscillating and produce a phenomenon called line jitter. Line jitter is therefore a mismatch and a wrong sampling of the analogue signal and has its largest effect in the first TV lines (upper part of the picture). The error can be as high as one pixel and therefore may ruin measurements which aims at subpixel accuracy.

Most frame grabbers have an 8-bit flash converter but 12- and 16-bit converters exists. The converter samples the analogue video signal at discrete time intervals and converts each sample to a digital value called a pixel element or pixel. The incoming signal is an analogue signal ranging from 0 to 714 mV at a frequency range from 7 to 20 MHz (with no prefiltering). The 8-bit converter produces samples with intensity levels between 0 and 255, i.e. 256 different grey levels.

Multiplexer and input feedback LUT

The frame memory of an 8-bit converter has a depth of 12 bits. The 4-bit spare allows the processor to use look-up table (LUT) operations. The LUT transforms image data before it is stored in the frame memory. The LUTs are mostly used for colouring (false colours) of images and can also be used to draw graphics on the screen without changing the underlying image. This can be done by protecting the lower 8 bits (the image) and draw only in the upper 4 bits. It is therefore possible to grab a new image without destroying the graphics. The LUT operations can be done in real time and its therefore possible to correct images radiometrically before storing them. LUTs can not be used geometrically because their memory is pixel organized, not space oriented. For geometrical transformations one therefore has to make special programs. The multiplexer (MUX) in combination with the feedback/input LUT allows some feedback operations like real-time image differencing, low pass filtering, etc.

Frame memory

The frame memory is organized as a two-dimensional array of pixels. Depending on the size of the memory it can store one or more frames of video information. When the memory is a 12-bit memory, 8 bits are used for the image and 4 bit planes for generating graphic overlay or LUT operations. In normal mode the memory acquires and displays an image using read/modify/write cycles. The memory is XY addressed to have an easy and fast access to single pixels.

Display interface

The frame memory transports the contents to the display interface every memory cycle. The display interface transforms the digital 12-bit signal from the frame memory into an analogue signal with colour information. This signal is passed to the RED, GREEN and BLUE ports and from there to the monitor.

10.3 DIGITAL IMAGE REPRESENTATION

By means of an electronic camera and a frame grabber, an image will be represented as a two-dimensional array of pixels, each pixel having a value $g(x, y)$ between 0 and 255 representing the grey tone of the image in the pixel position. Most current commercial frame grabbers have an array size of 512×512 pixels. Due to the way the image is scanned, the customary XY coordinate axis convention is as indicated in Figure 10.2.

10.4 CAMERA CALIBRATION

The calibration of the camera/lens combination is the process of determining the correct relationships between the object and image coordinates (Tsai 1987; Lenz and Tsai 1988). Since the elements of such a system are not ideal, this transformation includes

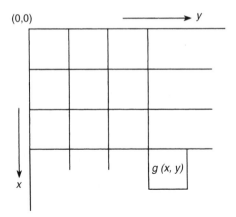

Figure 10.2   Digital image representation

parameters that must be calibrated experimentally. Because we are mainly concerned with relative measurements, we confine our discussion to three parameters that will affect our type of measurements. That is lens distortion, image centre coordinates and perspective transformations.

10.4.1 Lens Distortion

For an ideal lens, the transformation from object (x_o, y_o) to image (x_i, y_i) coordinates is simply

$$x_i = mx_o \tag{10.1a}$$

$$y_i = my_o \tag{10.1b}$$

where m is the transversal magnification. It is well known, however, that real lenses possesses distortion to a smaller or larger extent (Faig 1975; Shih *et al.* 1993). The transfer from object to image coordinates for a distorting lens is (see Figure 10.3)

$$r_i = mr_o + d_1 r_o^3 \tag{10.2a}$$

where

$$r_i = \sqrt{x_i^2 + y_i^2}, \quad r_o = \sqrt{x_o^2 + y_o^2} \tag{10.2b}$$

Higher odd order terms of r_o may be added, but normally they will be negligible. By multiplying Equation (10.2a) by $\cos\phi$ and $\sin\phi$, we get (since $x = r\cos\phi$, $y = r\sin\phi$)

$$x_i = mx_o + d_1 x_o (x_o^2 + y_o^2) \tag{10.3a}$$

$$y_i = my_o + d_1 y_o (x_o^2 + y_o^2) \tag{10.3b}$$

This results in the well-known barrel (positive d_1) and pin-cushion (negative d_1) distortion.

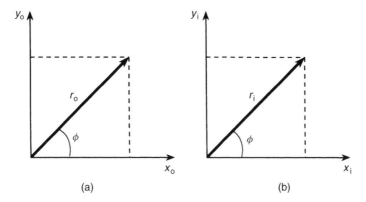

Figure 10.3   (a) Object and (b) image coordinates

In a digital image-processing system we want to transform the distorted coordinates (x_d, y_d) to undistorted coordinates (x_u, y_u). This transformation becomes

$$x_u = x_d + dx_d(x_d^2 + \varepsilon^2 y_d^2) \tag{10.4a}$$

$$y_u = y_d + dy_d(x_d^2 + \varepsilon^2 y_d^2) \tag{10.4b}$$

where ε is the aspect ratio between the horizontal and vertical dimensions of the pixels.

The origin of the xy-coordinates is at the optical axis. When transforming to the frame-store coordinate system XY (see Section 10.3) by the transformation

$$x = X - X_s \tag{10.5a}$$

$$y = Y - Y_s \tag{10.5b}$$

Equation (10.4) becomes

$$X_u = X_d + d(X_d - X_s)[(X_d - X_s)^2 + \varepsilon^2(Y_d - Y_s)^2] \tag{10.6a}$$

$$Y_u = Y_d + d(Y_d - Y_s)[(X_d - X_s)^2 + \varepsilon^2(Y_d - Y_s)^2] \tag{10.6b}$$

where (X_s, Y_s) are the centre coordinates. The distortion factor d has to be calibrated by e.g. recording a scene with known, fixed points or straight lines. The magnitude of d is of the order of $10^{-6} - 10^{-8}$ pixels per mm^3.

It has been common practice in the computer vision area to choose the center of the image frame buffer as the image origin. For a 512×512 frame buffer that means $X_s = Y_s = 255$. With a CCIR video format, the center coordinates would rather be (236, 255) since only the first 576 of the 625 lines are true video signals, see Table 5.4. A mispositioning of the sensor chip in the camera could add further to these values. The problem is then to find the coordinates of the image center. Many methods have been proposed, one which uses the reflection of a laser beam from the frontside of the lens (Tsai 1987). When correcting for camera lens distortion, correct image center coordinates are quite important.

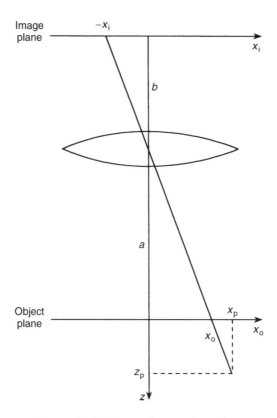

Figure 10.4 Perspective transformation

10.4.2 Perspective Transformations

Figure 10.4 shows a lens with the conjugate object- and image planes and with object and image distances a and b respectively. A point with coordinates (x_p, z_p) will be imaged (slightly out of focus) with image coordinate $-x_i$, the same as for the object point $(x_o, 0)$. From similar triangles we find that

$$x_p = \frac{-x_i(z_p + a)}{b} \tag{10.7a}$$

$$y_p = \frac{-y_i(z_p + a)}{b} \tag{10.7b}$$

Equation (10.7) is the perspective transformation and must be taken into account when e.g. comparing a real object with an object generated in the computer.

10.5 IMAGE PROCESSING

Broadly speaking, digital image processing can be divided into three distinct classes of operations: point operations, neighbourhood operations and geometric operations. A point operation is an operation in which the grey level of each pixel in the output image is a function of the grey level of the corresponding pixel in the input image, and only of that pixel. Typical point operations are photometric decalibration, contrast stretching and thresholding. A neighbourhood operation generates an output pixel on the basis of the grey level of the corresponding pixel in the input image and its neighbouring pixels. Geometric operations change the spatial relationships between points in an image, i.e. the relative distances between points a, b and c will typically be different after a geometric operation or 'warping'. Correcting lens distortion is an example of geometric operations. Digital image processing is a wide and growing topic with an extensive literature (Vernon 1991; Baxes 1994; Niblack 1988; Gonzales and Woods 2002; Pratt 1991; Rosenfeld and Kak 1982). Here we'll treat only a small piece of this large subject and specifically consider operation, which can be very useful for enhancing interferograms, suppress image noise, etc.

10.5.1 Contrast Stretching

In a digitized image we may take the number of pixels having the same grey level and make a plot of this number of pixels as a function of grey level. Such a plot is called a grey-level histogram. For an 8-bit (256 grey levels) image it may look such as that in Figure 10.5(a). In this example the complete range of grey levels is not used and the contrast of this image will be quite poor. We wish to enhance the contrast so that all levels of the grey scale are utilized. If the highest and lowest grey value of the image are denoted g_H and g_L respectively, this is achieved by the following operation:

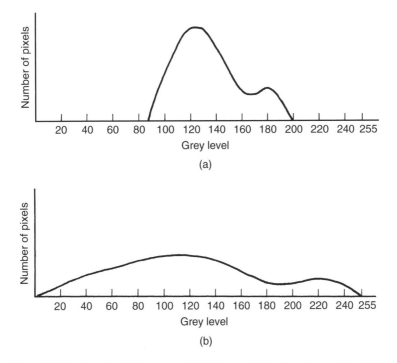

Figure 10.5 Grey-level histogram (a) before and (b) after contrast stretching

$$g_N = (g_o - g_L) \frac{255}{(g_H - g_L)} \tag{10.8}$$

where g_o is the original grey value and g_N is the new grey value. This is called contrast stretching and the resulting histogram when applied to the image of Figure 10.5(a) is given in Figure 10.5(b).

Grey-level histograms can be utilized in many ways. Histogram equalization is a technique which computes the histogram of an image and reassigns grey levels to pixels in an effort to generate a histogram where there are equally many pixels at every grey level, thereby producing an image with a flat or uniform grey-level histogram.

10.5.2 Neighbourhood Operations. Convolution

Neighbourhood processing is formulated in the context of so-called mask operations (the terms template, window or filter are also often used to denote a mask). The idea behind mask operations is to let the value assigned to a pixel be a function of itself and its neighbours. The size of the neighbourhood may vary, but techniques using 3×3 or 5×5 neighbourhoods centred at the input pixel are most common. The neighbourhood operations are often referred to as filtering operations. This is particularly true if they involve the convolution of an image with a filter kernel or mask. Other neighbouring operations are concerned with modifying the image, not by filtering in the strict sense, but by applying some logical test based on e.g. the presence or absence of object pixels in the local neighbourhood surrounding the pixel in question. Object thinning or skeletonizing is a typical example of this type of operation, as are the related operations of erosion and dilatation, which respectively seek to contract and enlarge an object in an orderly manner.

Recall the two-dimensional convolution integral

$$g(x, y) = f(x, y) \otimes h(x, y) = \int_{-\infty}^{\infty} \int_{-\infty}^{\infty} f(\xi, \eta) h(x - \xi, y - \eta) \, d\xi \, d\eta \tag{10.9}$$

When the variables x, y are not continuous but merely discrete values m, n as the pixel numbers in the x- and y-direction as of a digitized image, the double integral has to be replaced by a double sum:

$$g(i, j) = f \otimes h = \sum_m \sum_n f(m, n) h(i - m, j - n) \tag{10.10}$$

As mentioned in Appendix B, the geometrical interpretation of the convolution $f \otimes h$ is the area of overlap between the functions f and h as a function of the position of h as h is translated from $-\infty$ to ∞. Therefore the summation is taken only over the area where f and h overlap. This multiplication and summation is illustrated graphically in Figure 10.6(a). Here the filter kernel h is a 3×3 pixel mask with the origin $h(0, 0)$ at the centre representing a mask of nine distinct weights, $h(-1, -1), \ldots, h(+1, +1)$, see Figure 10.6(b). Note that the convolution formula requires that the mask h be first rotated $180°$, but this can be omitted when the mask is symmetric.

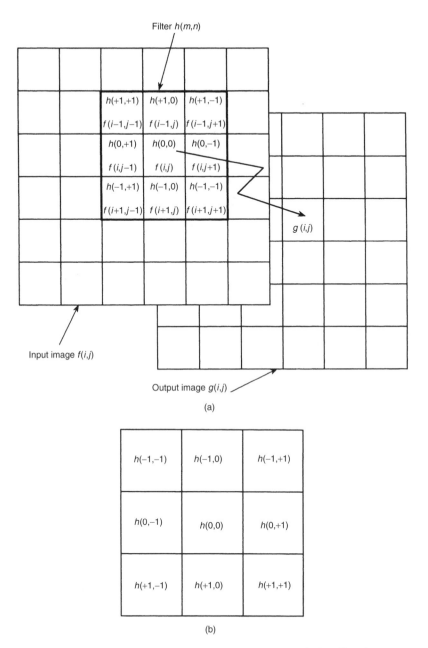

Figure 10.6   (a) convolution and (b) 3×3 convolution filter h

10.5.3 Noise Suppression

High spatial frequencies in an image manifests itself as large variations in grey tone values from one pixel to the next. This may be due to unwanted noise in the image. Such variations can be smoothed out by a convolution operation using an appropriate mask, the

1 / 9	1 / 9	1 / 9
1 / 9	1 / 9	1 / 9
1 / 9	1 / 9	1 / 9

(a)

120 × 1 / 9	105 × 1 / 9	100 × 1 / 9				
109 × 1 / 9	142 × 1 / 9	120 × 1 / 9				
130 × 1 / 9	128 × 1 / 9	126 × 1 / 9				

(b)

Figure 10.7   (a) 3 × 3 average mask and (b) image smoothing using local average mask

mask values constitute the weighting factors which will be applied to the corresponding image point when the convolution is being performed. For example, each of the mask values might be equally weighted, in which case the operation is simply the evaluation of the local mean of the image in the vicinity of the mask. Such a mask therefore acts as a low-pass filter. Figure 10.7(a) shows this local neighbourhood average mask and Figure 10.7(b) illustrates the application of the mask to part of an image. Referring to Figure 10.7(b) we find the value of the output pixel which replace the input pixel corresponding to the centre position of the mask to be

$$1/9[102 + 105 + 100 + 109 + 142 + 120 + 130 + 128 + 126] = 118$$

Thus the central point becomes 118 instead of 142 and the image will appear much smoother.

It may be more useful to apply this smoothing subject to some condition, e.g. the centre pixel is only assigned if the difference between the average value and the original pixel value is greater than a previously set threshold. This remove some of the noise without smoothing out too much of the detail in the original image.

Of the other noise-suppression methods we mention median filtering. This is a noise-reduction technique whereby a pixel is assigned the value of the median in some local neighbourhood. This is, however, a computationally time consuming procedure so neighbourhoods in excess of 3×3 or 5×5 may be impractical. In general, the median filter is superior to the mean filter in that image blurring is minimized.

10.5.4 Edge Detection

Detection of edges of objects in an image is a fundamental problem, not only for dimensional measurements, but in object recognition, military surveillance, etc. In the latter applications the main concern is if an edge exists or not, while in optical metrology one is also interested in the exact location of the edge. A lot of different edge detection techniques are described in the literature (Roberts 1965; Prewitt 1970; Fram and Deutsch 1975; Canny 1986). The different approaches can be divided into:

(1) gradient- and difference-based operators;

(2) template matching;

(3) edge fitting;

(4) statistical edge detection.

Here we will not treat all of these methods, but confine ourselves to describe in some detail the 'classical' method based on gradient- and difference-based operators.

An image of an edge will result in an intensity distribution consisting of a step-like function, see Figure 10.8(a). The actual position of the edge might be questioned but it is commonly adopted that the position is where the gradient has its maximum absolute value, see Figure 10.8(b) (if the edge transition is from bright to dark, the derivative becomes negative). So how do we calculate the gradient in a digitized image? From the definition of the derivative we have

$$f'(x) = \lim_{\Delta x \to 0} \frac{f(x + \Delta x) - f(x)}{\Delta x} \tag{10.11}$$

In an image of distinct pixel values the closest we can come to the limit $\Delta x \to 0$ is unity. The best approximation of the (partial) derivative in the x-direction we can give is therefore simply

$$S_x = f(x + 1, y) - f(x, y) \tag{10.12a}$$

and in the y-direction

$$S_y = f(x, y + 1) - f(x, y) \tag{10.12b}$$

The gradient S is a vector

$$S = S_x e_x + S_y e_y \tag{10.13}$$

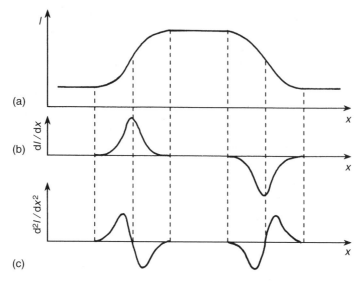

Figure 10.8 (a) intensity distribution I across two edges; (b) the derivative of I; and (c) the second derivative of I

of magnitude

$$S = \sqrt{S_x^2 + S_y^2} \qquad (10.14a)$$

and a direction given by

$$\tan \theta = \frac{S_y}{S_x} \qquad (10.14b)$$

The gradient magnitude and direction can be found from the two components along any two orthogonal directions.

The essential difference between all gradient-based edge detectors are the directions which the operators use, the manners in which they approximate the one-dimensional derivatives along these directions, and the manner in which they combine these approximations to form the gradient magnitude.

An operator due to Roberts estimates the derivatives diagonally over a 2 × 2 neighbourhood and the gradient is approximated by

$$S(x, y) = \sqrt{[f(x, y) - f(x + 1, y + 1)]^2 + [f(x, y + 1) - f(x + 1, y)]^2} \qquad (10.15)$$

One of the main problems with the Roberts operator is its susceptibility to noise because of the manner in which it estimates the directional derivatives. This has prompted alternative estimations of the gradient by combining the differencing process with local averaging. For example, the Sobel operator estimates the partial derivative in the x- and y-direction over a 3 × 3 region centred at $f(x, y)$ by:

$$S_x = [f(x + 1, y - 1) + 2f(x + 1, y) + f(x + 1, y + 1)]$$
$$- [f(x - 1, y - 1) + 2f(x - 1, y) + f(x - 1, y + 1)] \qquad (10.16a)$$

$$S_y = [f(x - 1, y + 1) + 2f(x, y + 1) + f(x + 1, y + 1)]$$
$$- [f(x - 1, y - 1) + 2f(x, y - 1) + f(x + 1, y - 1)] \qquad (10.16b)$$

The gradient may then be estimated as before by either calculating the RMS

$$S = \sqrt{S_x^2 + S_y^2} \qquad (10.17)$$

or by taking the absolute values

$$S = |S_x| + |S_y| \qquad (10.18)$$

From Figure 10.9 we see that Equation (10.16a) is obtained by applying the convolution mask illustrated in Figure 10.9(b) (left) and similarly Equation (10.16b) is obtained by applying the mask in Figure 10.9(b) (right). The same figure also gives the convolution masks for the Roberts and Prewitt edge operators.

After applying the convolution mask we get an intensity distribution as sketched in Figure 10.8(b). To determine the position of the edge with subpixel accuracy, one of the techniques described in Section 11.2.4 can be applied. A simple but reliable solution is to use the method of fitting a quadratic curve to the three points $f(i - 1)$, $f(i)$ and $f(i + 1)$ where pixel i is the pixel of the highest intensity $f(i)$. Then the location of the maximum

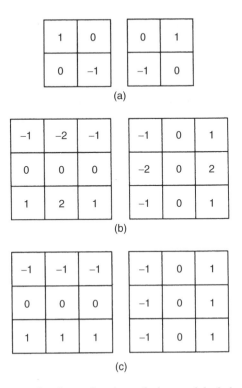

Figure 10.9 Convolution masks for estimation of the partial derivatives with (a) Roberts; (b) Sobel; and (c) Prewitt edge detection operators

of the quadratic curve is given by

$$i + \frac{f(i+1) - f(i-1)}{2[(f(i) - f(i-1)) - (f(i+1) - f(i))]} \tag{10.19}$$

To avoid detecting false edges, the magnitude of $f(i)$ can be compared to a pre-determined threshold value and edges excluded with a maximum derivative below this threshold. Obviously the choice of threshold is important and in noisy images, threshold selection involves a trade-off between missing valid edges and including noise-induced false edges.

Finally we mention that edges can be detected using the second derivative by using the Laplacian:

$$\nabla^2 = \frac{\partial^2}{\partial x^2} + \frac{\partial^2}{\partial y^2} \tag{10.20}$$

i.e. the sum of second-order, unmixed partial derivatives. The standard approximation is given by

$$L(x, y) = f(x, y) - \tfrac{1}{4}[f(x, y+1) + f(x, y-1) + f(x+1, y) + f(x-1, y)] \tag{10.21}$$

From Figure 10.8(c) we see that the second derivative crosses zero at the position of the edge. Therefore the edge is detected by finding the points where the intensity goes from positive to negative. The Laplacian has, however, one significant disadvantage: it responds very strongly to noise. To reduce noise, the image can be smoothed. This was proposed by Marr and Hildreth (1980) by first smoothing the image by convolving it with a two-dimensional Gaussian function. This is however a rather computationally expensive operation.

Figure 10.10 shows the result of an experiment for detecting a straight edge using an interlaced frame transfer CCD video camera. Here the edge is detected by use of the Sobel edge operator and sub-pixel accuracy is obtained by fitting three measured points around the intensity maximum to a quadratic curve. Figure 10.10(a, b) shows the result from detecting a vertical and a horizontal edge respectively. The two curves in each figure show the result before and after the data being corrected for camera lens distortion. In the left part of Figure 10.10(a) we see the effect of line jitter. It takes some time before the phase-locked loop (PLL) matches the internal Hsync signal. The left part of the corrected curve therefore will be overcompensated for lens distortion and gives an erroneous result. When the edge is horizontal as in Figure 10.10(b) a better result is obtained although the resolution in the vertical direction is poorer.

10.6 THE DISCRETE FOURIER TRANSFORM (DFT) AND THE FFT

In Section 5.8.1 we considered a continuous function $g(x)$ which we sample at regular intervals $x_s = np$ where n is an integer and p is a constant called the sampling period. The sampled function then is given as

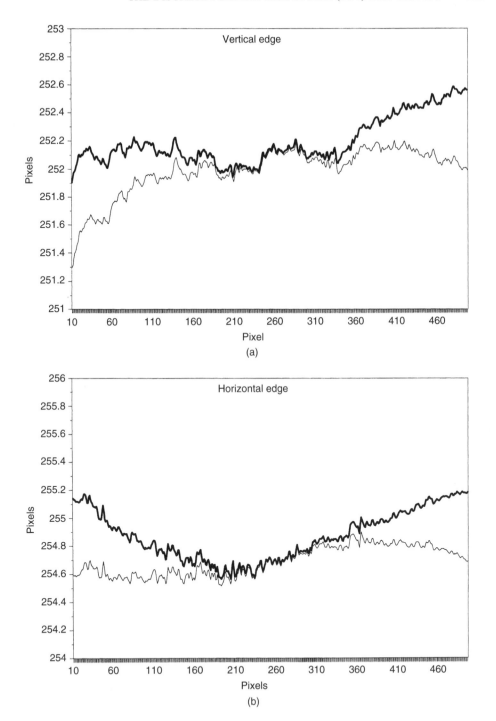

Figure 10.10   Detection of edge using the Sobel operator with sub-pixel accuracy. Thick curve before and thin curve after correcting for lens distortion: (a) vertical edge and (b) horizontal edge

$$g_s(x) = g(x) \sum_{n=-\infty}^{\infty} \delta(x - np) = \sum_{n=-\infty}^{\infty} g(np)\delta(x - np) = \sum_{n=-\infty}^{\infty} g_s[n]\delta(x - np) \quad (10.22)$$

where $g_s[n]$ is called the discrete space signal and is just the sequence of sample values $\{g(np)\}$, e.g. $g_s[0], g_s[1], g_s[2], \ldots$.

In Section 5.8.1 we found that the Fourier transform (the spectrum) $G_s(u)$ of $g_s(x)$ is periodic with period $1/p$. It can therefore be written as a Fourier series (see Appendix C):

$$G_s(u) = \sum_{n=-\infty}^{\infty} g_s[n] \exp\{-i2\pi npu\} \quad (10.23)$$

with

$$g_s[n] = p \int_0^{1/p} G_s(u) \exp\{i2\pi npu\} \, du \quad (10.24)$$

In practice we can have only a finite number N of samples and therefore Equation (10.23) becomes

$$G_s(u) = \sum_{n=0}^{N-1} g_s[n] \exp\{-i2\pi npu\} \quad (10.25)$$

Next, we sample along the frequency axis for N samples over one period. The spectral spacing u_p then equals $1/N$, $Nu_p = p$ and $u = ku_p$ with $k = 0, 1, 2, \ldots, N - 1$. This inserted into Equation (10.25) gives the sampled frequency domain representation:

$$G[k] = \sum_{n=0}^{N-1} g_s[n] \exp\{-i2\pi nk/N\} \quad (10.26)$$

This describes the discrete Fourier transform (DFT) of $g_s[n]$. Note that the DFT actually describes a set of N equations. Since $\exp(i2\pi/N)$ is periodic with period N, so is the DFT $G[k]$. To develop the inverse relation, we start with Equation (10.24). Choosing N samples of $G_s(u)$ over one period again gives the spectral spacing $u_p = 1/N$. With $du = u_p$, $u = ku_p$, the integral may be approximated by the summation

$$g_s[n] = \frac{1}{N} \sum_{k=0}^{N-1} G[k] \exp\{i2\pi nk/N\} \quad (10.27)$$

This is the inverse discrete Fourier transform (IDFT). We note that: (1) the IDFT relation also describes a set of N equations; (2) the IDFT is periodic with period N. This means there is implied periodicity in both the DFT and the IDFT.

Clearly, the DFT is only an approximation to the actual spectrum of the underlying analogue signal. As shown in Section 5.8.1, the spectrum of a sampled function $g(x)$ is given as

$$G_s(u) = \sum_{n=-\infty}^{\infty} G\left(u - \frac{n}{p}\right) \quad (10.28)$$

where $G(u)$ is the true Fourier transform of $g(x)$ and p is the sampling period. Equation (10.28) shows that the spectrum of the sampled function g_s is periodic with period $1/p$. If the spectrum $G(u)$ of $g(x)$ vanishes outside some interval $[-W, W]$, i.e. $g(x)$ is band-limited, we see from Figure 5.28(c) that if

$$\frac{1}{p} \geq 2W \tag{10.29}$$

the spectra $G(u - n/p)$ do not overlap. Equation (10.29) can also be stated as

$$f_p \geq f_n \tag{10.30}$$

where $f_p = 1/p$ is the sampling frequency and $f_n = 2W$ is the Nyquist frequency. The Nyquist theorem states that to recover a band-limited signal, the sampling frequency must at least be twice the bandwidth and is given by Equation (10.30). If this condition is not fulfilled, the repeated spectra will overlap and we get the problem of aliasing as discussed in Section 5.8.3, which gives a false recovered signal.

Now, a space-limited signal can never be band-limited. Since the DFT is computed by taking a finite number of samples, the 'true' function $g(x)$ is in effect multiplied by a rect-function (see Section 11.4.4). A method for reducing this problem is to multiply $g(x)$ by a smoothing or tapered space-domain window: see Section 11.4.4 and Ambardar (1995), page 529. It is important to remember that for N sample points, $G[k]$ consists of N frequencies. For example, when dividing the image into subimages of size 8×8 pixels, for example, the DFT consists of eight discrete frequencies (in each direction). When recording an image of a granular pattern (speckle), for example, one can hardly expect to recover the original pattern by the DFT.

For more details of the DFT, see the book by Ambradar (1995). Here, however, we shall point out a problem related to the periodic extension of signals. Optical metrology is greatly concerned with fringes. For simplicity, let us consider a pure sinusoidal signal. In Figure 10.11(a) this signal is sampled with $N = 8$, but only over one half-period, which gives a sampling frequency well above the Nyquist frequency. But due to the implied periodicity, the DTF 'sees' this as a full rectified sine with a doubled fundamental frequency and computes a spectrum as given to the right in the figure. This phenomenon is called leakage. Figure 10.11(b) shows the same signal with $N = 16$ sampled over one period. This gives a periodic extension whose DTF result is identical to the true Fourier transform. Figure 10.11(c) shows the same signal with $N = 24$ sampled over 1.5 periods and the resulting spectrum. Finally, Figure 10.11(d) shows the same signal with $N = 82$ sampled over 10.25 periods. Still there is leakage, but the estimate of the original signal has improved. In conclusion, to recover a sinusoidal signal exactly, it must be sampled over an integral number of periods. If we do not know the period, the best solution is to sample over as many periods as possible.

Since both the DFT and the IDFT consist of N equations, the DFT relation is suitably expressed in matrix form. This is utilized in the ingenious algorithm known as the fast Fourier transform (FFT), which is a highly efficient implementation of the DFT on computers. We shall not go into details of the FFT here, just mention that it reduces the number of computations (compared to the DFT) considerably. For details of the FFT we refer to Ambardar (1995) and Morrison (1994).

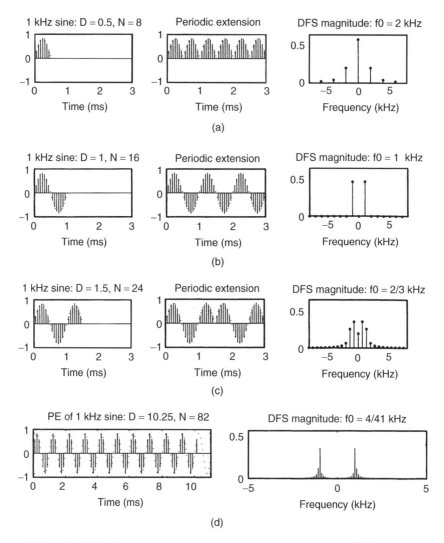

Figure 10.11 The DFT of sampled periodic signals. From Ambardar, A. (1995) *Analog and Digital Signal Processing*. Reproduced by permission of PWS Publishing Co., Boston, MA

PROBLEMS

10.1 A common measure for the transmission of digital data is the baud rate, defined as the number of bits transmitted per second. Usually each byte (8 bits) is accompanied by a start bit and a stop bit. What then, is the transmit time for

(a) a 512 × 512 image with 256 grey levels at 300 baud?

(b) The same at 14 400 baud?

(c) Repeat (a) and (b) for a 1024 × 1024 image with 256 grey levels.

10.2 How does contrast stretching influence edge detection by a gradient-based operator?

10.3 Obtain the Fourier transforms of (a) $df(x)/dx$, (b) $[\partial f(x, y)/\partial x + \partial f(x, y)/\partial y]$, and (c) $[\partial^2 f(x, y)/\partial x^2 + \partial^2 f(x, y)/\partial y^2]$. Assume that x and y are continuous variables.

10.4 The basic approach used to compute the digital gradient involves taking differences of the form $f(x, y) - f(x + 1, y)$.

    (a) Obtain the filter transfer function, $H(f_x, f_y)$, for performing the equivalent process in the frequency domain.

    (b) Show that this is a high-pass filter.

10.5 The result obtained by a single pass through an image of some two-dimensional masks can also be achieved by two passes using one-dimensional masks.

    (a) Show that the result obtained by a single pass of the mask given in Figure 10.7(a) also can be obtained by first passing the mask [1 1 1] through the image followed by a vertical counterpart.

    (b) Show that the Sobel operator can be implemented by one pass of a differencing mask of the form [−1 0 1] (or its vertical counterpart) followed by a smoothing mask of the form

$$\begin{pmatrix} 1 \\ 2 \\ 1 \end{pmatrix}$$

    (or its horizontal counterpart).

    (c) If the mask is represented by $h(x, y)$, what is the condition that it can be implemented with one horizontal one-dimensional mask followed by one vertical one-dimensional mask.

10.6 Write down the 3×3 mask representing the Laplacian operator given in Equation (10.21).

10.7 The two-dimensional Gaussian function is given by

$$G(x, y) = \frac{1}{2\pi\sigma^2} \exp[-(x^2 + y^2)/2\sigma^2]$$

An approach to edge detection (Marr and Hildreth) is by first smoothing the image by convolving it with a two-dimensional Gaussian function and subsequently isolating the zero-crossings of the Laplacian of this image:

$$\nabla^2\{I(x, y) \otimes G(x, y)\}$$

    (a) Show that the two operations commute such that we can derive a single filter: the Laplacian of Gaussian.

    (b) Show that this operation can be implemented by first an operation in the x-direction followed by an operation in the y-direction.

11
Fringe Analysis

11.1 INTRODUCTION

In Chapters 3 and 6–9 we have given a description of classical interferometry, holographic interferometry, moiré, speckle and photoelasticity. The outcome of all these techniques is a set of fringes called interferograms. For many years, the analysis of these interferograms has been a matter of manually locating the positions and numbering of the fringes. With the development and decreasing cost of digital image processing equipment, a lot of effort has been made into what is termed digital fringe pattern measurement techniques. It is three main reasons for this effort: (1) to obtain better accuracy; (2) to increase the speed; (3) to automate the process.

In this chapter, some of the basic principles of digital fringe pattern analysis will be described. In Section 11.2 we describe techniques which intend to be a direct replacement of the human brain – eye combination by detecting the positions of the fringes. In Sections 11.3, 11.4 and 11.5, techniques for continuous determination of the phase of the fringe function are described. For more detailed discussions of digital fringe pattern measurement techniques, the book edited by Robinson and Reid (1993) is recommended. A comparison of the different techniques is performed by Kujawinska 1993b and Perry and McKelvie (1993).

11.2 INTENSITY-BASED ANALYSIS METHODS

11.2.1 Introduction

Before the development of the phase-measurement techniques described in Sections 11.3 and 11.4, intensity-based techniques were the only image-processing tools available for the automatic analysis of interferograms. They still are important methods in fringe analysis and are sometimes the only viable technique for interferograms which have been retrieved from photographic records or which have been obtained from interferometers in which it is impossible or impractical to introduce phase-measuring techniques. They may also be appropriate simply because quantitative results are not even needed.

For intensity-based methods it is very important to minimize the influence of noise, including speckle noise. Therefore preprocessing of the interferograms is highly recommended. Techniques most commonly used and described in Section 10.5.3 are low-pass filtering and median filtering. Specially designed for fringe analysis are the so-called spin-filters (Yu *et al.* 1994). Another method (if possible) is to combine two interferograms of opposite phase. When an interferogram is shifted in phase by π radians, the resulting dark fringes occupy the previous location of bright fringes and vice versa. If the noise is stationary, the noise distribution is unaffected by this shift. Then if the two interferograms are subtracted, the noise will subtract to give zero, while the fringes will combine to give a higher contrast than in the original. When the noise has approximately the same spatial frequency as the fringes, this method might be the only workable one for noise reduction.

11.2.2 Prior Knowledge

When interferometric techniques are used for repetitive calibration, non-destructive testing or inspection, it is sometimes possible to design a comparatively simple fringe analysis system. After identifying the characteristics of the fringe pattern which are peculiar to the application, the capabilities of the analysis system can then be confined to those required for the measurement in question.

A very simple analysis method can be applied to moiré technique using projected fringes where the shape of a manufactured component can be compared with that of a master component. For a manufactured component identical to the master, the resulting image becomes uniformly dark. As the manufactured component deviates from the master, fringes appear in the image and hence the total intensity of the image increases. Der Hovanesian and Hung (1982) used this technique to control component shape by limiting the total image intensity to a threshold value.

Many specialised fringe analysis procedures can be developed by utilizing prior knowledge of the fringe pattern. The Young fringe method (see Section 8.4.2) resulting in a set of parallel fringes is a typical example which has undergone a lot of investigations (Halliwell and Pickering (1993)). Another example is holographic interferometry applied to testing of honeycomb panels (Robinson 1983). Debrazing of the honeycomb produces groups of nearly circular fringes of a particular size and fringe density. As shown in Figure 11.1 the procedure starts with counting the number of fringes along a number of horizontal lines through the image. When a comparatively large number of fringes appear on a given line, a flaw is presumed to exist on that line. Short vertical scans along the line are then used to search for the location of the flaw by looking for a comparatively large number of fringes in the vertical direction. Having identified the probable existence and location of a flaw, the system carries out a further check by counting fringes along each of four short vectors angularly spaced at $45°$, centred on the probable flaw site. If the same number of fringes appear on each of the four vectors, then the existence of a flaw is confirmed. As shown in Figure 11.2 the flaw sites are marked with small crosses.

11.2.3 Fringe Tracking and Thinning

Boundary (contour) tracking and object thinning (skeletoning) are standard operations in digital image processing. In fringe analysis we talk about fringe tracking and thinning, which have a slightly different meaning. A number of fringe tracking and thinning procedures have been proposed.

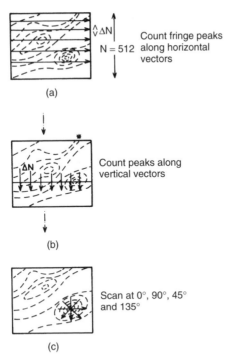

Count fringe peaks along horizontal vectors (a)

Count peaks along vertical vectors (b)

Scan at 0°, 90°, 45° and 135° (c)

Figure 11.1 Procedure for the analysis of Figure 11.2 (From Robinson, D. W. (1983) Automatic fringe analysis with a computer image processing system, Applied Optics, **22**, 2169–76. Reproduced by permission of The Optical Society of America and by Courtesy of D. W. Robinson)

Figure 11.2 Result of analysing a holographic interferogram by the method shown in Figure 11.1 (From Robinson, D. W. (1983) Automatic fringe analysis with a computer image processing system, Applied Optics, **22**, 2169–76. Reproduced by permission of the Optical Society of America and by Courtesy of D. W. Robinson)

Fringe tracking involves a search for the locus of the fringe maxima (or minima) by examining the pixel values in all directions from the starting point (often determined manually) and moving the pixel locus in the direction along which the sum of the intensity is maximized (or minimized) or alternatively the gradient is a minimum. (As opposed to boundary tracking where one searches for the maximum gradient value.) In this way only a limited set of the whole image array is examined (Button *et al.* 1985), see Figure 11.3.

Thinning and skeletonizing techniques use similar approaches to detect fringe peaks (or minima) but instead of following the peaks with a roving pixel locus, the whole image is subjected to a peak detection matrix. A procedure due to Yatagai *et al.* (1982) uses two-dimensional peak detection, locally performed within a 5×5 pixel matrix as shown in Figure 11.4(a). With respect to the four directions shown in Figure 11.4(b), the peak conditions are defined as

$$P_{00} + P_{0-1} + P_{01} > P_{-21} + P_{-20} + P_{-2-1}$$
$$P_{00} + P_{0-1} + P_{01} > P_{21} + P_{20} + P_{2-1}$$

$$(11.1)$$

for the x-direction, with similar expressions for the y-direction and the xy- and yx- directions. When the peak conditions are satisfied for any two or more directions, the object point is recognized to be a point on a fringe skeleton.

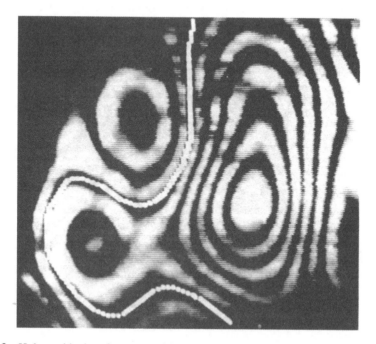

Figure 11.3 Holographic interferogram with a computer tracked fringe. (From Button, B. L., Cutts, J., Dobbins, B. N., Moxan, G. J. and Wykes, C. (1985) The identification of fringe positions in speckle patterns, Opt. and Lasers Technology, 17, 189–92. Reproduced by permission of Elsevier Science Ltd)

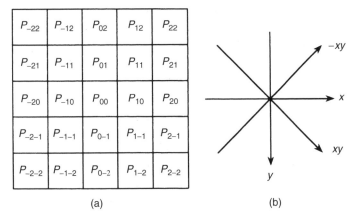

Figure 11.4   (a) 5 × 5 pixel matrix and (b) directions for fringe peak detection

11.2.4 Fringe Location by Sub-Pixel Accuracy

By the methods described above, fringes are located with an accuracy of one pixel. We now describe two methods by which fringe positions can be determined with an accuracy of a fraction of a pixel. The first method consists off fitting a curve around the fringe maximum (or minimum)

Curve fitting

Assume that the intensity distribution of the fringe pattern locally is given by

$$I(x) = a + b \cos \frac{2\pi}{p}(x - x_0) \tag{11.2}$$

with a maximum at $x = x_0$, i.e. $x_0 = np$ where n is an integer. Near the maximum we can approximate the intensity by a Taylor expansion of the intensity around x_0 up to the second order:

$$I(x) = a + b\left[1 - \frac{1}{2}\left(\frac{2\pi}{p}\right)^2 x^2\right] \tag{11.3}$$

In Appendix D it is shown how such a quadratic curve can be fitted to N observation points by using the least squares solution. By using three observation points $I(i-1)$, $I(i)$ and $I(i+1)$ where i is the pixel number with the highest intensity $I(i)$, the position of the fringe maximum is given by, cf. Equation (D.19):

$$x_{\text{max}} = i + \frac{I(i-1) - I(i+1)}{2I(i-1) - 4I(i) + 2I(i+1)} \tag{11.4}$$

Thereby determining the position with sub-pixel accuracy.

The accuracy of this method is, among other things, dependent on the fringe period p. By using three measuring points, the length of p should at least be six pixels. If p is much longer than six pixels, it will be a good idea to fit the curve to four or more measurement points. It could also be advantageous to include four and higher-order terms in the expression for the intensity.

Zero-crossing

Another method for sub-pixel location of fringe maxima (or minima) is to find the points where the intensity crosses the mean intensity value (Gåsvik *et al.* 1989). The principle is shown in Figure 11.5. On the left side of the maximum, the last pixel x_{lu} with intensity $I(x_{lu})$ below the mean intensity I_m and the first pixel x_{lo} with intensity $I(x_{lo})$ over I_m is found. Then a straight line

$$I = \frac{I(x_{lo}) - I(x_{lu})}{x_{lo} - x_{lu}} x + \frac{I(x_{lu})x_{lo} - I(x_{lo})x_{lu}}{x_{lo} - x_{lu}}$$
$$= [I(x_{lo}) - I(x_{lu})]x + [I(x_{lu}) - I(x_{lo})]x_{lu} + I(x_{lu}) \qquad (11.5)$$

connecting the two points is calculated. (The last equality follows since $x_{lo} - x_{lu} = 1$.) The intersection between this line and the mean intensity I_m is given by

$$x_l = \frac{I_m - I(x_{lu})}{I(x_{lo}) - I(x_{lu})} + x_{lu} \qquad (11.6)$$

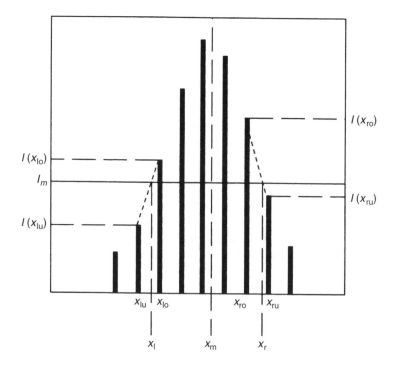

Figure 11.5   Illustration of the detection of the crossover points x_l and x_r with the mean intensity I_m

In the same way, the crossover point x_r on the right side of the maximum is found to be

$$x_r = \frac{I_m - I(x_{ru})}{I(x_{ru}) - I(x_{ro})} + x_{ru} \tag{11.7}$$

where x_{ro} is the last pixel with intensity $I(x_{ro})$ over I_m and x_{ru} is the first pixel with intensity $I(x_{ru})$ below I_m. The position x_m of the fringe maximum is then taken to be the midpoint

$$x_m = \frac{x_l + x_r}{2} = \frac{1}{2}\left[\frac{I_m - I(x_{lu})}{I(x_{lo}) - I(x_{lu})} - \frac{I_m - I(x_{ru})}{I(x_{ro}) - I(x_{ru})}\right] + \frac{x_{lu} + x_{ru}}{2} \tag{11.8}$$

The mean intensity can be found in different ways. From Figure 11.6 we see that the area under the intensity curve and the area under the straight line

$$I_m = ax + b \tag{11.9}$$

representing the mean intensity should be approximately equal. If the interval from x_1 to x_3 is divided into two equal subintervals at x_2, we have

$$a = \frac{4(A_2 - A_1)}{N^2} \tag{11.10a}$$

$$b = \frac{3A_1 - A_2}{N} - \frac{4(A_2 - A_1)}{N^2}x_1 \tag{11.10b}$$

where $N = x_3 - x_1$, A_1 is the area of the first interval from x_1 to x_2 and A_2 is the area of the second interval from x_2 to x_3. These areas are found by simply taking the sum of the intensities for each pixel

$$A_1 = \sum_{i=x_1}^{x_2} I_i \tag{11.11a}$$

$$A_2 = \sum_{i=x_2}^{x_3} I_i \tag{11.11b}$$

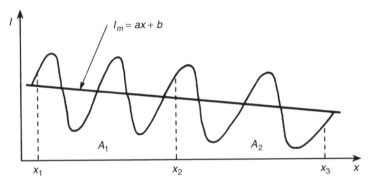

Figure 11.6 Intensity distribution and mean intensity I_m along a TV line

To get the best possible representation of the mean intensity, each scan should be divided into as many intervals as possible, giving a sequence of straight lines representing I_m. Each subinterval must however not be shorter than one fringe period.

Another way of calculating the mean intensity is by the so-called bucket and bin method (Choudry 1981). It consists of finding the value of I_{min} and I_{max} (to the nearest integer pixel) for successive minima and maxima and then taking $I_m = (I_{max} + I_{min})/2$. For fringes with a high level of noise it could however be difficult to discriminate against spurious maxima and minima.

The accuracy of the zero-crossing method has been analysed with respect to the error in I_m and the fringe period (Gåsvik and Robbersmyr 1994). It shows that the absolute accuracy is highest for a fringe period of six pixels, while the relative accuracy is relatively constant for fringe periods of six pixels and higher. This analysis is done assuming noise-free fringes.

For more details about intensity-based analysis methods, see Yatagai (1993).

11.3 PHASE-MEASUREMENT INTERFEROMETRY

11.3.1 Introduction

By means of a digital image-processing system, we have the possibility of storing an image of the interferogram into the computer memory and do manipulations on the individual pixels. When looking at the general expression for an interferogram, Equation (3.7):

$$I = I_1 + I_2 + 2\sqrt{I_1 I_2}|\gamma| \cos \Delta\phi \qquad (11.12)$$

it would be tempting to solve this expression with respect to $\Delta\phi$ and let the computer calculate:

$$\Delta\phi = \cos^{-1} \frac{I - (I_1 + I_2)}{2\sqrt{I_1 I_2}|\gamma|} \qquad (11.13)$$

thereby calculating the phase at each pixel, and by knowledge of the geometrical and optical configuration of the interferometer, calculating the parameter sought in each pixel of the whole image. To do this, we see from Equation (11.13) that we must know the intensities I_1 and I_2 and the degree of mutual coherence $|\gamma|$ of the interfering waves, and, moreover, we must know these quantities for each pixel. This assumption is unrealistic in most cases, and even if we knew these quantities of the ideal interferogram, the intensity distribution from a complex imaging system will always be accompanied by uncontrollable noise. In this section and in Sections 11.4 and 11.5 we will treat methods that are dependent only on the recorded intensity at each pixel and which are less sensitive to noise.

11.3.2 Principles of TPMI

A general expression for the recorded intensity in an interferogram can be written:

$$I(x, y) = a(x, y) + b(x, y) \cos \phi(x, y) \qquad (11.14)$$

where both I, a, b and ϕ are functions of the spatial coordinates. Here $a(x, y)$ is the mean intensity, $V = a(x, y)/b(x, y)$ is equal to the visibility (see Section 3.4) and ϕ is the phase difference between the interfering waves.

To recover the phase we now describe a group of methods called phase-measurement interferometry (PMI). PMI is the most widely used technique today for the measurement of wavefront phase in interferometers and it has also been successfully applied in holographic interferometry and moiré (Rosvold 1990; Takeda and Mutoh 1983). PMI techniques can be divided into two main categories: those which take the phase data sequentially, and those which take the phase data simultaneously. Methods of the first type are known as temporal PMI or TPMI, and those of the second type are known as spatial PMI and will be treated in Section 11.4.

The starting point for all PMI techniques is the expression for the interferogram intensity:

$$I = a + b\cos(\phi + \alpha) \tag{11.15}$$

where we have introduced an additional phase term α. The essential feature of all PMI techniques is that α is a modulating phase which is introduced and controlled experimentally.

Techniques for determining the phase can be split into two basic categories: electronic and analytic. For analytical techniques, intensity data are recorded while the phase is temporally modulated, sent to a computer and then used to compute the relative intensity measurements. Electronic techniques are also known as heterodyne interferometry, see Section 3.6.4. An example of this technique is described in Section 3.6.3 about the dual-frequency Michelson interferometer. This method is used extensively in distance-measuring interferometers where the phase at a single point with a fast update is required. But the technique can also be used to determine the phase over an area, see Section 6.8.3. The detector then has to be scanned or there must be multiple detectors with all the necessary circuitry.

The analytic methods can be subdivided into two techniques, one that integrates the intensity while the phase is increased linearly, and a second where the phase is altered in steps between intensity measurements. The first method is referred to as integrating bucket phase-shifting, while the second is termed phase-stepping. The phase-step method has clearly become the most popular in recent years and below we give a brief description of this technique.

Equation (11.15) contains three unknowns, a, b and ϕ, requiring a minimum of three intensity measurements to determine the phase. The phase shift between adjacent measurements can be anything between 0 and π degrees. By arbitrary phase shifts α_1, α_2 and α_3 we get

$$I_1 = a + b\cos(\phi + \alpha_1)$$
$$I_2 = a + b\cos(\phi + \alpha_2) \tag{11.16}$$
$$I_3 = a + b\cos(\phi + \alpha_3)$$

from which we find

$$\phi = \tan^{-1} \frac{(I_2 - I_3)\cos\alpha_1 - (I_1 - I_3)\cos\alpha_2 + (I_1 - I_2)\cos\alpha_3}{(I_2 - I_3)\sin\alpha_1 - (I_1 - I_3)\sin\alpha_2 + (I_1 - I_2)\sin\alpha_3} \tag{11.17}$$

With $\alpha_1 = \pi/4$, $\alpha_2 = 3\pi/4$, $\alpha_3 = 5\pi/4$, i.e. a phase shift of $\pi/2$ per exposure, we reach a particularly simple expression

$$\phi = \tan^{-1} \frac{I_2 - I_3}{I_2 - I_1} \tag{11.18}$$

Three intensity measurements to solve for ϕ gives an exactly determined system. As mentioned in Appendix D, this often gives numerically unstable solutions and it is often wise to overdetermine the system by providing more measurement points.

In general, for the ith stepped phase, the resulting intensity can be written as (Greivenkamp 1984; Morgan 1982)

$$I_i = a + b\cos(\phi + \alpha_i) = a_0 + a_1 \cos \alpha_i + a_2 \sin \alpha_i \tag{11.19a}$$

where

$$\begin{aligned} a_0 &= a = I_0 \\ a_1 &= b \cos \phi \\ a_2 &= -b \sin \phi \end{aligned} \tag{11.19b}$$

By making N phase steps ($i = 1, 2, \ldots, N$), Equation (11.19a) can be written in matrix form as

$$\begin{pmatrix} I_1 \\ I_2 \\ \vdots \\ I_N \end{pmatrix} = \begin{pmatrix} 1 & \cos \alpha_1 & \sin \alpha_1 \\ 1 & \cos \alpha_2 & \sin \alpha_2 \\ \vdots & \vdots & \vdots \\ 1 & \cos \alpha_N & \sin \alpha_N \end{pmatrix} \begin{pmatrix} a_0 \\ a_1 \\ a_2 \end{pmatrix} \tag{11.20}$$

The coefficients a_0, a_1 and a_2 can be found using the least squares solution to Equation (11.20), see Appendix D

$$\begin{pmatrix} a_0 \\ a_1 \\ a_2 \end{pmatrix} = A^{-1} B \tag{11.21}$$

where

$$A = \begin{pmatrix} N & \Sigma \cos \alpha_i & \Sigma \sin \alpha_i \\ \Sigma \cos \alpha_i & \Sigma \cos^2 \alpha_i & \Sigma \cos \alpha_i \sin \alpha_i \\ \Sigma \sin \alpha_i & \Sigma \cos \alpha_i \sin \alpha_i & \Sigma \sin^2 \alpha_i \end{pmatrix} \tag{11.22}$$

and

$$B = \begin{pmatrix} \Sigma I_i \\ \Sigma I_i \cos \alpha_i \\ \Sigma I_i \sin \alpha_i \end{pmatrix} \tag{11.23}$$

From Equation (11.19b) we find that

$$\phi = \tan^{-1}\left(\frac{-a_2}{a_1}\right) \tag{11.24}$$

$$V = \frac{b}{a} = \frac{\sqrt{a_1^2 + a_2^2}}{a_0} \tag{11.25}$$

We include the expression for the visibility because pixels with too low visibility can give invalid phase data.

11.3.3 Means of Phase Modulation

A phase shift or modulation in an interferometer can be induced by moving a mirror, tilting a glass plate, moving a grating, rotating a half-wave plate or analyzer (see Figure 11.7) using an acousto-optic or electro-optic modulator, or using a Zeeman laser. Phase shifters such as moving mirrors, gratings, tilted glass plates, or polarization components (Jin *et al.* 1994) can produce continuous as well as discrete phase shifts between the object and reference beams. Phase shifters may either be placed in one arm of the interferometer or positioned so that they shift the phase of one of two orthogonally polarized beams.

11.3.4 Different Techniques

From Equations (11.20)–(11.25) we can derive the expressions for different techniques, e.g.

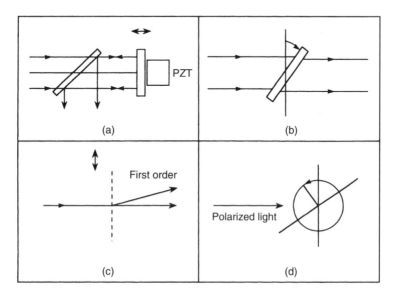

(a)

(b)

(c)

(d)

Figure 11.7   Means of modulating or shifting the phase of the light in an interferometer (a) moving mirror, (b) tilted glass plate; (c) moving diffraction grating; (d) rotating waveplate

Three-frame technique

(1) $\alpha_i = \pi/4,\ 3\pi/4,\ 5\pi/4$ (Wyant *et al.* 1984)

$$\phi = \tan^{-1}\left(\frac{I_3 - I_2}{I_1 - I_2}\right) \tag{11.26a}$$

$$V = \frac{\sqrt{(I_3 - I_2)^2 + (I_1 - I_2)^2}}{\sqrt{2}I_0} \tag{11.26b}$$

(2) $\alpha_i = -2\pi/3,\ 0,\ 2\pi/3$

$$\phi = \tan^{-1}\left[\sqrt{3}\frac{I_1 - I_3}{2I_1 - I_2 - I_3}\right] \tag{11.27a}$$

$$V = \frac{\sqrt{3(I_1 - I_3)^2 + (2I_2 - I_1 - I_3)^2}}{3I_0} \tag{11.27b}$$

Four-frame technique (Wyant 1982; Schwider et al. 1993)

$\alpha_i = 0,\ \pi/2,\ \pi,\ 3\pi/2$

$$\phi = \tan^{-1}\left(\frac{I_4 - I_2}{I_1 - I_3}\right) \tag{11.28a}$$

$$V = \frac{\sqrt{(I_4 - I_2)^2 + (I_1 - I_3)^2}}{2I_0} \tag{11.28b}$$

Five-frame technique (Hariharan et al. 1987; Schwider et al. 1983)

$\alpha_i = -\pi,\ -\pi/2,\ 0,\ \pi/2,\ \pi$

$$\phi = \tan^{-1}\left(\frac{7(I_2 - I_4)}{-4I_1 + I_2 + 6I_3 + I_4 - 4I_6}\right) \tag{11.29a}$$

$$V = \frac{\sqrt{(-4I_2 + I_2 + 6I_3 + I_4 - 4I_5)^2 + 49(I_2 - I_4)^2}}{14I_0} \tag{11.29b}$$

Hariharan uses slightly different formulas, not obtained by the least squares method.

Carré technique

In a technique due to Carré (1966) the amount of phase shift need not be known, but must be constant in the four steps, e.g. $\alpha_i = -3\alpha/2,\ -\alpha/2,\ \alpha/2,\ 3\alpha/2$. This gives

$$\alpha = 2\tan^{-1}\left(\sqrt{\frac{3(I_2 - I_3) - (I_1 - I_4)}{(I_2 - I_3) + (I_1 - I_4)}}\right) \tag{11.30a}$$

$$\phi = \tan^{-1}\left[\tan\left(\frac{\alpha}{2}\right)\frac{(I_1 - I_4) + (I_2 - I_3)}{(I_2 + I_3) - (I_1 - I_4)}\right] \tag{11.30b}$$

To calculate the phase modulo π, the above two equations are combined to yield

$$\phi = \tan^{-1} \frac{\sqrt{[(I_1 - I_4) + (I_2 - I_3)][3(I_2 - I_3) - (I_1 - I_4)]}}{(I_2 + I_3) - (I_1 + I_4)} \tag{11.31}$$

An obvious advantage of the Carré technique is that the phase does not need to be calibrated. It also has the advantage of working when a linear phase shift is introduced in a converging or diverging beam where the amount of phase shift varies across the beam. Unknown phase shift techniques with different number of steps are introduced by Vikram *et al.* (1993) and Lassahn *et al.* (1994).

Ideally, two frames with a relative phase shift of 2π should be identical. This is the case for the five-frame technique where the phase difference between the first and the last frame is 2π. This 'extra' frame therefore can be used to correct for phase shifter miscalibration and detector non-linearities (Larkin and Oreb 1992). Techniques with a 2π phase difference between the first and the last frame are called $(N + 1)$-frame techniques (Surrel 1993).

A technique due to Vikhagen (1990) requires the collection of many frames of intensity data with random phase shifts. From a number of recordings, the maximum and minimum intensity value at each detector point are determined. When the number of frames becomes large, these values will approach I_{max} and I_{min} which in turn are used to determine the mean intensity I_0 and the visibility V (see Equation 3.29). Once a large number of frames (about 20) of random phases has been recorded, the phase at each pixel of intensity I_i can be calculated according to

$$\phi = \cos^{-1} \left(\frac{I_i - I_0}{V I_0} \right) \tag{11.32}$$

11.3.5 Errors in TPMI Measurements

The most common source of error in TPMI is vibration and air turbulence (Magee and Welsh 1994). In order to get good measurements, the optical system must be isolated from vibrations and shielded from air turbulence. Other factors that influence the measurement accuracy are phase-shifter errors, non-linearities due to the detector, quantization of the detector signal and TV-line jitter, see Section 10.2.

It has been shown that the errors in phase due to a calibration error or non-linearity in the phase shifter will decrease as the number of measurements increases (Schwider *et al.* 1983; Larkin and Oreb 1992; Joenathan 1994). For a consistent phase-shift error, such as miscalibration, a periodic error is seen in the calculated phase which has a spatial frequency of twice the fringe frequency. Figure 11.8 shows the peak-to-valley (P–V) phase error versus amount of linear phase-shifter error for a number of different algorithms (Creath 1993). As can be seen, the Carré and five-frame algorithms are the least sensitive to miscalibration.

Non-linear phase-shift errors are not as easy to detect or remove (Creath 1986), but also here the five-frame algorithm is the least sensitive to miscalibration.

With regard to detector non-linearities, the algorithms with four frames and more are quite insensitive to such errors. The three-frame algorithm, however, can cause large phase errors because of detector non-linearities.

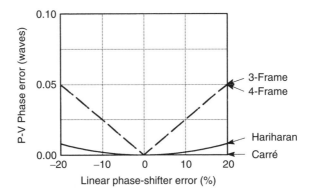

Figure 11.8 P–V phase error versus percent linear phase-shifter error. (From Creath 1993 in Robinson, D. W. and Reid, G. T. (eds) (1993) Interferogram Analysis. Digital Fringe Pattern Measurement Techniques, Institute of Physics Publishing. Reproduced by permission of Institute of Physics Publishing and by courtesy of K. Creath)

For an intensity signal digitized to eight bits or more, the quantization error has an insignificant influence on the phase error.

For more details about TPMI techniques, see Creath (1993).

11.4 SPATIAL PHASE-MEASUREMENT METHODS

11.4.1 Multichannel Interferometer

The temporal phase-stepping methods described in Section 11.3 has its spatial counterpart.

Consider Figure 11.9, which is a polarization (Michelson) interferometer. In the object and reference paths are placed two quarterwave plates (QWP) with their axes inclined at 45° to each other. The polarization matrix for a QWP with its fast axis inclined an angle ψ to the x-axis is (cf. Equation 9.61)

$$M_\psi = \begin{pmatrix} 1 - i\cos 2\psi & -i\sin 2\psi \\ -i\sin 2\psi & 1 + i\cos 2\psi \end{pmatrix} \tag{11.33}$$

The resultant matrix after two passes (first through the QWP and back after being reflected from the mirror or object) therefore becomes

$$W_\psi = M_\psi M_\psi = \begin{pmatrix} -\cos 2\psi & -\sin 2\psi \\ -\sin 2\psi & \cos 2\psi \end{pmatrix} \tag{11.34}$$

which is the same as for a halfwave plate. The resultant matrix for a QWP with its axis inclined at 45° to that of Equation (11.34) becomes

$$W_{\psi+45} = \begin{pmatrix} \sin 2\psi & -\cos 2\psi \\ -\cos 2\psi & -\sin 2\psi \end{pmatrix} \tag{11.35}$$

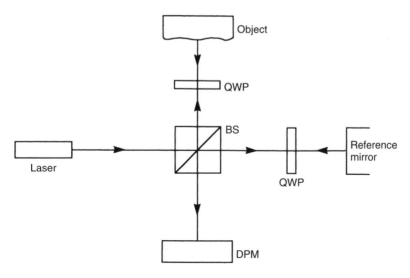

Figure 11.9 Polarization interferometer for direct phase measurement. BS = beamsplitter, QWP = quarter-wave plate, DPM = direct phase measurement module

Now assume that the state of polarization of the laser beam incident on the interferometer is represented by

$$|U\rangle = \begin{pmatrix} u_x \\ u_y \end{pmatrix} \tag{11.36}$$

If M_ψ is placed in the reference path and $M_{\psi+45}$ is placed in the object path, the state of polarization of the reference and object wave becomes

$$|U_r\rangle = W_\psi |U\rangle = \begin{pmatrix} u_{rx} \\ u_{ry} \end{pmatrix} = \begin{pmatrix} -u_x \cos 2\psi - u_y \sin 2\psi \\ -u_x \sin 2\psi + u_y \cos 2\psi \end{pmatrix} \tag{11.37a}$$

$$|U_o\rangle = W_{\psi+45} |U\rangle = \begin{pmatrix} u_{ox} \\ u_{oy} \end{pmatrix} = \begin{pmatrix} u_x \sin 2\psi - u_y \cos 2\psi \\ -u_x \cos 2\psi - u_y \sin 2\psi \end{pmatrix} \tag{11.37b}$$

The condition that these two states are orthogonal is, cf. Equations (9.46) and (9.48)

$$\langle U_r | U_o \rangle = \langle U_o | U_r \rangle = 0 \tag{11.38}$$

which becomes

$$u_x^* u_y - u_x u_y^* = 0 \tag{11.39}$$

This condition is fulfilled only when the incident laser beam is linearly polarized (u_x and u_y are real). Therefore, when placing two QWPs in the object- and reference paths with their axes inclined at 45° to each other and the incident beam is linearly polarized, the state of polarization of the resultant object- and reference waves becomes orthogonal (and linear). In this analysis we have neglected changes in the state of polarization due to reflections from the beamsplitter and possible depolarization by scattering from the object.

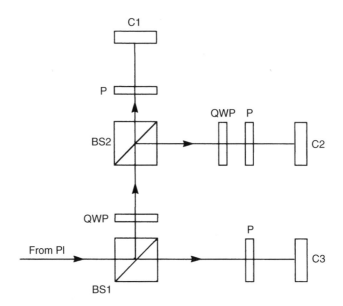

Figure 11.10   Direct phase measurement module DPM with three interference channels in parallel. PI = polarization interferometer, C1–C3 = cameras

The light from the interferometer is directed to a direct phase measurement module (DPM). A three channel DPM (Bareket 1985) is illustrated in Figure 11.10. A camera imaging the interferogram is placed in each channel. In front of the camera is a linear polarizer with its transmission axis at 45° to the polarization direction of the object and reference waves, thereby mixing the two components. QWPs oriented parallel to the polarization direction shift the relative phase by $\pi/2$ and π in channel 1 and 2 respectively. The detected intensities in the three channels therefore become

$$I_1 = a + b\cos(\phi + \pi/2) = a - b\sin\phi$$
$$I_2 = a + b\cos(\phi + \pi) = a - b\cos\phi \tag{11.40}$$
$$I_3 = a + b\cos\phi$$

from which we solve

$$\phi = \tan^{-1}\left(\frac{2I_1 - I_2 - I_3}{I_2 - I_3}\right) \tag{11.41}$$

The DPM in Figure 11.10 can be extended further to four channels (Smythe and Moore 1984) or more and retarders with phase shifts other than $\pi/2$ can be inserted instead of the QWPs to get different types of spatial phase-stepping methods in analogy with the different types of temporal phase-stepping methods described in Section 11.3.4. But why introduce this technique with the rather complicated set-up described in Figures 11.9 and 11.10? The answer is that we are in this way able to record many phase-shifted interferograms simultaneously, thereby allowing analysis of dynamic events and pulsed wavefront sensing. It enables analysis in one frame time and therefore the measurements may be performed in adverse conditions.

Spatial phase-stepping can also be obtained by means of gratings (Kwon and Shough 1985). The transmission function of a square grating can be written as

$$t(x, y) = \sum_{n} a_n e^{i2\pi n f x} \tag{11.42}$$

where a_n are the coefficients of the Fourier series of the grating, f is the frequency, $f = 1/d$, and d is the grating period. A lateral translation of the grating by x_o results in an additional phase term described by

$$t(x - x_o, y) = \sum_{n} a_n e^{i2\pi n f (x - x_o)} \tag{11.43}$$

If a wavefront $U_o e^{i\phi}$ is incident on the grating, the field behind the grating is

$$u = U_o e^{i\phi} \sum_{n} a_n e^{i2\pi n f (x - x_o)} \tag{11.44}$$

In this way a series of phase-shifted wavefronts propagate behind the grating. The magnitude of the phase shift is given by

$$\delta_n = 2\pi n f x_o \tag{11.45}$$

Therefore, a three-channel interferometer can be obtained from the $n = +1, 0, -1$ orders of the diffracted beams. For $90°$ phase shift, the amount of lateral translation of the grating equals $d/4$. Grating phase-stepping has been successfully applied in two basic configurations. That is in (1) three-channel interferometers: point-diffraction (PDI) (Kwon and Shough 1985), radial-shear (RSI) (Kwon $et\ al.$, 1987), holographic (Kujawinska and Robinson 1988) and speckle (Kujawinska $et\ al.$ 1989) interferometers, and (2) four-channel grating lateral shear interferometers (Kwon and Shough 1985).

Apart from polarization and grating methods, also a colour technique in projection moiré has been demonstrated as a means for spatial phase-stepping (Harding $et\ al.$ 1988).

11.4.2 Errors in Multichannel Interferometers

Multichannel interferometers are susceptible to errors due to conditions that are not common to all channels, such as are associated with using separate detectors. Errors may result from variations of the optical transmission among channels, detector responsivity variations and amplifier gain and bias errors. These can, however, be adjusted either optically, electronically or in software. Another source of errors is connected with pixel mismatch between the fringe patterns due to incorrect relative spatial adjustment of the cameras. The overall accuracy of the spatial phase-stepped techniques is usually expected to be lower than for the temporal methods.

11.4.3 Spatial-Carrier Phase-Measurement Method

Spatial-carrier phase measurement methods are based on the idea of superposing a carrier fringe pattern onto the interferogram fringes. This can be done in e.g. holographic interferometry by tilting the reference wave in the second exposure. The fringe pattern is then given by

$$g(x, y) = a(x, y) + b(x, y) \cos[\phi(x, y) + 2\pi f_0 x] \qquad (11.46)$$

where f_0 is the carrier frequency in the x-direction. The two basic approaches to the spatial carrier technique include the Fourier transform method (FTM), in which the processing is performed in the frequency domain, and methods which are equivalent but the processing is performed in spatial coordinates.

The Fourier transform method was originally conceived and demonstrated by Takeda *et al.* (1982) who employed the fast fourier transform (FFT) method. Following Takeda's method we write Equation (11.46) in the form

$$g(x, y) = a(x, y) + c(x, y)e^{i2\pi f_0 x} + c^*(x, y)e^{-2\pi f_0 x} \qquad (11.47a)$$

where

$$c(x, y) = \tfrac{1}{2}b(x, y)e^{i\phi(x,y)} \qquad (11.47b)$$

The fringe pattern is Fourier transformed with respect to x, which gives

$$G(f_x, y) = A(f_x, y) + C(f_x - f_0, y) + C^*(f_x + f_0, y) \qquad (11.48)$$

where the upper-case letters denote Fourier spectra and f_x is the spatial frequency in the x-direction. To have the method work, the spatial variations of $\phi(x, y)$ must be slow compared to f_0 and the Fourier spectra will be separated as shown schematically in Figure 11.11(a). By use of a filter function $H(f_x - f_0, y)$ in the frequency plane the function $C(f_x - f_0, y)$ can be isolated and translated by f_0 towards the origin to remove the carrier and obtain $C(f_x, y)$ as shown in Figure 11.11(b). Next the inverse Fourier transform of this function is computed and as a result the complex function $c(x, y)$ from Equation (11.47b) is obtained. The phase may then be determined by two equivalent

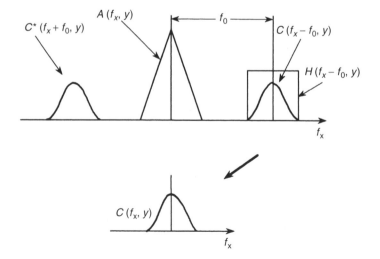

Figure 11.11   Separated Fourier spectra of a tilted fringe pattern

operations. In the first one a complex logarithm of $c(x, y)$ is calculated

$$\log[c(x, y)] = \log[1/2b(x, y)] + i\phi(x, y) \tag{11.49}$$

The phase in the imaginary part is completely separated from the amplitude variation $b(x, y)$ in the real part. In the second method (more commonly used) the phase is obtained by

$$\phi(x, y) = \tan^{-1} \frac{\text{Im}[c(x, y)]}{\text{Re}[c(x, y)]} \tag{11.50}$$

where Re and Im represent the real and imaginary parts of $c(x, y)$.

Figure 11.12 (Takeda *et al.* 1982) shows the various stages in the Fourier transform process. Before computing the FFT the data are multiplied by a Hamming window (Figure 11.12b) to eliminate the influence of discontinuities in the data at both ends.

The Fourier transform method has also been extended using a full two-dimensional spatial filtering (Nugent 1985; Bone *et al.* 1986).

The Fourier transform method has become quite popular and has also been applied to moiré grating techniques (Morimoto and Fujisawa 1994; Morimoto *et al.* 1994; Simova and Stoev 1993; Yoshizawa and Tomisawa 1993) and photoelasticity (Quan *et al.* 1993).

11.4.4 Errors in the Fourier Transform Method

Many of the error sources in the FTM are common to Fourier transform methods in general and have been studied extensively by many authors. The most serious source of error is energy leakage. A simplified explanation of this phenomenon goes as follows.

The FFT has to be computed over a sample window of finite width D, i.e. the function $g(x)$ of Equation (11.46) is in effect multiplied by a rect-function:

$$g_s(x) = g(x) \text{ rect} \left(\frac{x}{D}\right) \tag{11.51}$$

and the Fourier transform gives

$$\mathscr{F}\{g_s\} = G(f_x, y) \otimes D \text{ sinc} (Df_x) \tag{11.52}$$

From this expression we see that the spectrum $G(f_x, y)$ is convolved with a sinc-function. Due to the side-lobes of the sinc-function, parts of the spectrum of $G(f_x, y)$ will 'leak out' into neighbouring frequency components because of this convolution operation, resulting in a false spectrum. When multiplying the data with a Hamming window as proposed by Takeda *et al.*, the steep discontinuities at the boundaries of the fringe data are avoided, see Figure 11.12(b). The Hamming window is given by

$$w(x) = \left(\frac{1}{2} + \frac{1}{2} \cos \frac{2\pi}{D} x\right) \text{ rect} \left(\frac{x}{D}\right) \tag{11.53}$$

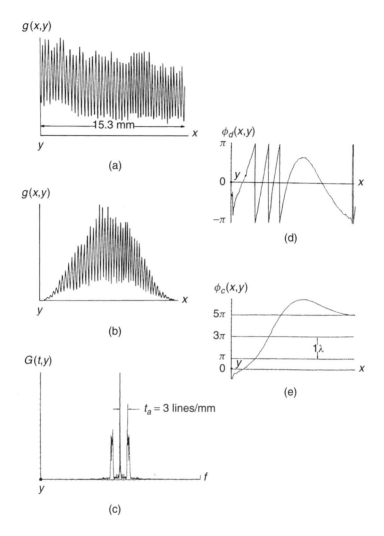

Figure 11.12 Various stage in the Fourier transform method: (a) 1-D intensity distribution; (b) the intensity weighted by Hamming window; (c) the modulus of its Fourier transform; (d) the wrapped phase function; (e) the reconstructed phase function after the unwrapping procedure. (From Takeda, M., Ina, H. and Kobayashi, S. (1982) Fourier-transform method of fringe pattern analysis for computer-based topography and interferometry, J. Opt. Soc. Am., 72(1), 156–160. Reproduced by permission of the Optical Society of America and by courtesy of M. Takeda)

and its Fourier transform

$$W(f_x) = \tfrac{1}{2}\operatorname{sinc}(Df_x) + \tfrac{1}{4}\operatorname{sinc}(Df_x - 1) + \tfrac{1}{4}\operatorname{sinc}(Df_x + 1) \tag{11.54}$$

This window function has lower side-lobes and therefore when convolving it with $G(f_x, y)$ we get less leakage. The Hamming window is most commonly used, however Hann, bell and $\cos^4$ windows also give good results (Malcolm *et al.* 1989; Frankowski *et al.* 1989).

However, by using such windows, another source of error comes up. Because the fringe amplitude approaches zero at the ends of the window (see Figure 11.12(b)), the determination of the phase becomes quite sensitive to noise in these regions.

11.4.5 Space Domain

A process equivalent to the Fourier transform method can be performed in the space domain. From Equation (11.48) the spectrum passed by the filter function $H(f_x - f_0, y)$ is written as

$$C(f_x - f_0, y) = H(f_x - f_0, y)G(f_x, y) \tag{11.55}$$

By taking the inverse Fourier transform of Equation (11.55) we get

$$c(x, y)e^{i2\pi f_0 x} = g(x, y) \otimes h(x, y)e^{i2\pi f_0 x} \tag{11.56}$$

Many different approaches to the solution of Equation (11.56) have been proposed (Womack 1984; Singh and Sirkis 1994). A solution by means of a dedicated hardware system has been proposed and (fully) demonstrated in the pioneering work by Ichioka and Inuiya (1972). The intensity given by Equation (11.46) is transformed into the following time varying electrical signal by the video tube:

$$g(t) = a(t) + b(t) \cos[\omega t - \phi(t)] \tag{11.57}$$

The procedure is then exactly the same as for synchronous demodulation of a modulated carrier signal, see Section 3.6.4, where the signal is divided into two signals and multiplied by $\cos \omega t$ and $\sin \omega t$:

$$g(t) \cos \omega t = \frac{b}{2} \cos \phi + a \cos \omega t + \frac{b}{2} \cos(2\omega t - \phi) \tag{11.58a}$$

$$g(t) \sin \omega t = \frac{b}{2} \sin \phi + a \sin \omega t + \frac{b}{2} \sin(2\omega t + \phi) \tag{11.58b}$$

By low-pass filtering of these two signals, one is left with

$$C(t) = \frac{b}{2} \cos \phi \tag{11.59a}$$

$$S(t) = \frac{b}{2} \sin \phi \tag{11.59b}$$

from which the phase can be computed electronically:

$$\phi(t) = \tan^{-1}\left(\frac{S(t)}{C(t)}\right) \tag{11.60}$$

For more details about spatial PMI techniques, see Kujawinska (1993a).

11.5 PHASE UNWRAPPING

11.5.1 Introduction

As mentioned in Section 11.3.2 a general expression for the intensity in an interferogram can be written

$$I = a + b\cos\phi \tag{11.61}$$

In Sections 11.3 and 11.4 we have described techniques for solving Equation (11.61) by collecting extra information from the interferometer to solve the sign ambiguity and reduce the influence of stationary noise. All these methods result in an equation of the form

$$\phi = \tan^{-1}\left(\frac{C}{D}\right) \tag{11.62}$$

where C and D are functions of the recorded intensity from a set of interferograms at the image point where the phase is being measured. Because of the multivalued arctan function, the solution for ϕ is a sawtooth function (see Figure 11.13(a)) and discontinuities occur every time ϕ changes by 2π. If ϕ is increasing, the slope of the function is positive and vice versa for decreasing phase. The term 'phase unwrapping' arises because the final step in the fringe pattern measurement process is to unwrap or integrate the phase along a line (or path) counting the 2π discontinuities and adding 2π each time the phase angle

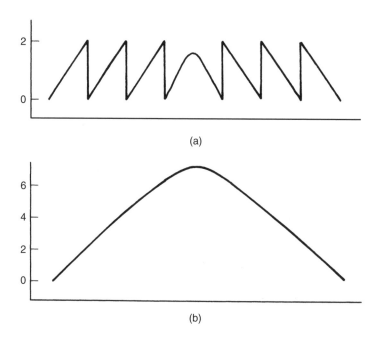

(a)

(b)

Figure 11.13 (a) Characteristic 'saw-tooth' wrapped phase function. (b) Continuous phase function obtained by 'unwrapping' the data in (a)

jumps from 2π to zero or subtracting 2π if the change is from zero to 2π. Figure 11.13(b) shows the data in Figure 11.13(a) after unwrapping.

The key to reliable phase-unwrapping algorithms is the ability to accurately detect the 2π phase jumps. In the case of noise-free wrapped phase data and where this data is adequately sampled (i.e. the phase gradients is significantly less than 2π), then a simple approach to phase unwrapping will be adequate and all that is required is a sequential scan through the data (line by line) to integrate the phase by adding or subtracting 2π at the phase jumps.

In many measurement problems, however, noise in the sampled data is a major contributing factor in the false identification of phase jumps. Figure 11.14 shows the effect of the addition of noise to unwrapped data. It is clear that as the amplitude of the noise approaches 2π, the real phase jumps become obscured. In the case of one-dimensional data, the only solution to this problem (other than averaging more data sets to integrate out the noise over time) is to smooth the raw sinusoidal fringe data with a low-pass filter. However, this is not always satisfactory and information is always lost in filtering operations.

For simple unwrapping methods to work, the data must be continuous across the whole image array (no holes in the data) and extend to the boundaries of the sample window. A phase discontinuity might be caused by a rapid change in the measurement parameter, such as a large height step in a component under test. Such defects might appear as a sudden change in the fringe spacing or as a point where the fringe stops (referred to sometimes as a fringe break). In these circumstances, errors in the phase unwrapping are propagated from the defect or hole in the data through the rest of the data array. This is particularly serious if the data is being scanned line by line in one direction. The problems presented by defects or holes in the data becomes even more complex when the shape

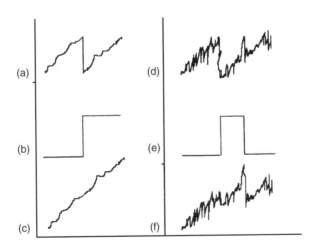

Figure 11.14 Effect of additive noise in the wrapped phase data: (a) added noise with a maximum amplitude less than π-the 2π phase discontinuity is correctly located (b) and the data unwrapped; (c) high amplitude noise added to the same data (d) can result in false detection of a 2π phase jump. (e) The resulting unwrapped data (f) retains an erroneous 2π phase jump. (From Robinson, D. W. and Reid, G. T. (eds) (1993) Interferogram Analysis. Digital Fringe Pattern Measurement Techniques, Institute of Physics Publishing. Reproduced by permission of Institute of Physics Publishing and by courtesy of D. W. Robinson)

of the hole becomes complicated. It is to find methods for automatically detecting and compensating for these problems that sophisticated phase-unwrapping algorithms have been developed.

When the terms C and D in Equation (11.62) are both below a certain threshold (i.e. tend to zero) the result of the arctangent calculation becomes indeterminate. Pixels with such values can be flagged or masked in the computer as being invalid data and omitted from phase-unwrapping calculations. The masked pixels can then be replaced by a local average of valid pixels by a final smoothing operation performed on the final unwrapped phase data.

11.5.2 Phase-Unwrapping Techniques

The basic principle of phase-unwrapping is to 'integrate' the wrapped phase ϕ (in units of 2π) along a path through the data. At each pixel the phase gradient is calculated by differentiation:

$$\Delta\phi = \phi_n - \phi_{n-1} \qquad (11.63)$$

where n is the pixel number. If $|\Delta\phi|$ exceeds a certain threshold such as π, then a phase fringe edge (2π discontinuity) is assumed. This phase jump is corrected by the addition or subtraction of 2π according to the sign of $\Delta\phi$.

Although phase-unwrapping has been performed by analogue circuits as part of an electronic phase measurement process (Mertz 1983), most workers use digital processing techniques to apply this unwrapping process as a logical sequence of program steps. The most common principle used to correct for missed 2π phase jumps is based on the fact that the phase difference between any two points measured by integrating phase along a path between the two points is independent of the route chosen as long as the route does not pass through a phase discontinuity. Thus phase-unwrapping methods may be divided into path-dependent methods and path-independent methods.

11.5.3 Path-Dependent Methods

The simplest of the phase-unwrapping methods involves a sequential scan through the data, line by line, see Figure 11.15. At the end of each line, the phase difference between the last pixel and the pixel on the line below is determined and the line below is scanned in the reverse direction. In other words a two-dimensional data array is treated like a folded one-dimensional data set. This approach is successful when applied to high-quality data, but more complex variations are necessary in the presence of noise. These include multiple scan directions (Robinson and Williams 1986), spiral scanning (Vrooman and Maas 1989) and counting around defects (Huntley 1989).

One of the most popular approaches that is often used to avoid phase errors propagating through the data array is to unwrap the regions of 'good' pixel data first. The 'bad' pixels with high measurement uncertainty are then unwrapped but data propagation errors are then confined to small regions. This is referred to as pixel queuing. A common method is to place the pixel address on a processing queue so that the pixels having the smallest phase difference between neighbouring pixels are processed first (Schorner

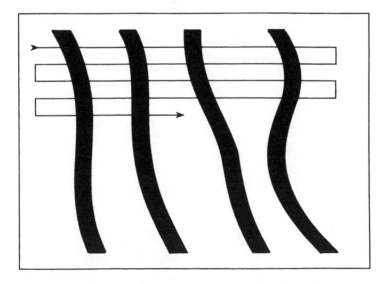

Figure 11.15   Line by line sequential scanning path

et al. 1991). This approach means that regions of small phase gradients and low noise data are unwrapped first.

Another method of defining good pixels is to examine the interference fringe contrast (visibility) at the pixel under consideration. It is important with these techniques that the order of pixel processing is such that the unwrapping process propagates along connected paths through the array. This can be achieved in different ways (Schorner *et al.* 1991; Towers *et al.* 1991).

Another variation of path-dependent methods is to segment the image into regions containing no phase ambiguities (Kwon *et al.* 1987; Gierloff 1987) or to divide the data array into square tiles or sub-arrays (Towers *et al.* 1991; Stephenson *et al.* 1994). Then the phase data at the edges of adjacent regions or tiles are compared and arranged based on the difference value that most edge pixels agree on.

11.5.4 Path-Independent Methods

A path-independent method due to Ghiglia *et al.* (1987) (Ghiglia and Romero 1994) is performed in the following way: In a 3×3 mask, the phase difference between the phase p of the central pixel and its four nearest neighbours in the horizontal and vertical directions are calculated. If one of the differences is greater than π in absolute value, $+2\pi$ or -2π is added to p dependent on the majority of the four differences being positive or negative. When there are two positive and two negative differences, an arbitrary decision is taken to add 2π. When none of the absolute differences exceed π, then p remains unchanged. Figure 11.16(a, b) shows an array of wrapped phase data before and after application of one iteration of this algorithm. After several repetitions (iterations) a checker board pattern is seen to spread across the entire array (Figure 11.16(c). One then comes to a point where further repetitions do not change the array further but where every second iteration results in an identical array. At this stage a global iteration is performed by

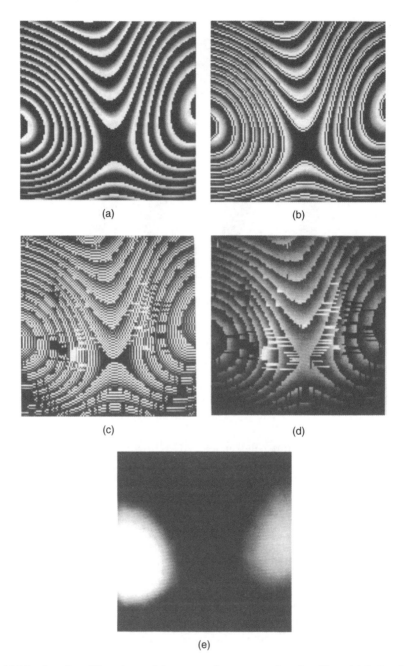

Figure 11.16 A series of iterations of the automation-unwrapping algorithm; (a) before iteration; (b) after the first local iteration; (c) after ten local iterations, (d) the first global iteration (averaged state); (e) complete phase unwrapping after 710 total local iterations and 13 total global iterations. (From Ghiglia, *et al.* 1987. Reproduced by permission of The Optical Society of America and by courtesy of D. C. Ghiglia)

replacing each pixel value by the value of the mean of each pair of pixels in the current and the preceding array. After application of this step, the new phase array resembles the starting array (of wrapped data) except that one less fringe is present (Figure 11.16(d)). The effect has been to unwrap one fringe. Repeated application of this process (local iteration followed by global iteration) removes one fringe at a time until the entire array is unwrapped (Figure 11.16(e)).

As can be easily realized, this algorithm is processing intensive. It is, however, very immune to noise and artifacts.

For more details about phase unwrapping, see Robinson (1993); Judge and Bryanston-Cross (1993).

11.5.5 Temporal Phase Unwrapping

As mentioned in Section 11.5.1, errors in the phase unwrapping will propagate from the defect or hole in the data through the rest of the data array. One way to circumvent this problem is to apply temporal phase unwrapping. The basic idea behind this method is that the phase at each pixel is measured as a function of time. Unwrapping is then carried out along the time axis for each pixel independent of the others. This method is applicable to a subclass of interferometry applications in which phase changes occur over time, for example deformation analysis in which the phase change is proportional to surface displacement.

Let the phase at a certain pixel at time t be denoted $\varphi(t)$. Assume that between the initial state and the final state, the phase has undergone s number of steps so that $t = 0, 1, 2, \ldots, s$. By finding the number N of phase jumps (change of 2π) between $\varphi(0)$ and $\varphi(s)$, the phase in the final state is then found by adding $N \cdot 2\pi$ to $\varphi(s)$, which in fact is nothing else than the old trusty method of counting fringes!

In practice the phase $\varphi(t)$ at time t can be found by phase-stepping interferometry as described in Section 11.3. Consider, for example, the four-frame technique with $\alpha_i = 0$, $\pi/2$, π, $3\pi/2$, which gives

$$\varphi(t) = \tan^{-1}\left(\frac{I_4(t) - I_2(t)}{I_1(t) - I_3(t)}\right) = \tan^{-1}\left(\frac{\Delta I_{42}(t)}{\Delta I_{13}(t)}\right) \tag{11.64}$$

where $\Delta I_{ij}(t) = I_i(t) - I_j(t)$.

Then we test for phase discontinuities by taking

$$\Delta\varphi(t) = \varphi(t) - \varphi(t-1) \tag{11.65}$$

Another solution is to calculate $\Delta\varphi(t)$ directly by the identity

$$\Delta\varphi(t) = \tan^{-1}\left[\frac{\Delta I_{42}(t)\Delta I_{13}(t-1) - \Delta I_{13}(t)\Delta I_{42}(t-1)}{\Delta I_{13}(t)\Delta I_{13}(t-1) + \Delta I_{42}(t)\Delta I_{42}(t-1)}\right] \equiv \varphi(t) - \varphi(t-1) \tag{11.66}$$

Since $\varphi(t)$ from Equation (11.64) lies in the range $-\pi$ to $+\pi$, $\Delta\varphi(t)$ from Equation (11.65) will be in the range -2π to 2π. The $\Delta\varphi(t)$ values calculated from Equation (11.66), however, lie in the range $-\pi$ to $+\pi$. This means that provided the deformation per time interval is sufficiently small (i.e., a temporal sampling rate of at least two samples per cycle), the phase $\varphi(s)$ can be calculated by summing the phase differences with no further unwrapping required:

$$\varphi(s) = \sum_{t=1}^{s} \Delta\varphi(t) \qquad (11.67)$$

This unwrapping procedure is therefore intrinsically simple, with no conditional jumps required in the program implementation.

PROBLEMS

11.1 Suppose we have measured the intensities f_1, f_2, f_4 and f_5 at the pixels $x_1 = -2$, $x_2 = -1$, $x_3 = 0$, $x_4 = 1$ and $x_5 = 2$ respectively. We want to find the fringe maximum by fitting these values to a quadratic curve. Use the formula developed in Appendix D to find an expression for the position x_p of the fringe maximum.

11.2 We want to compare the accuracy when fitting five points and when fitting three points to a quadratic curve. Assume that the intensity distribution along the pixel position x, ideally is given by

$$I(x) = 100 \left[1 + \cos\left(\frac{2\pi}{5} x \right) \right]$$

i.e. the fringe period is five pixels with a maximum at $x = 0$.

   Now assume that all the measured grey level values follows this distribution (to the nearest integer grey value) except the value measured at $x = 1$, which deviates by 20 percent from the ideal value.

   Find the measurement error in pixels when using three-point and five-point curve fitting.

11.3 The phase in a three-frame technique is given by

$$f = \tan\phi = \left(\frac{I_2 - I_3}{I_2 - I_1} \right)$$

Consider I_1, I_2 and I_3 as three mutually independent and normally distributed variables with the same standard deviation σ_I.

(a) Apply the usual law of error estimation to calculate the relative error σ_f/f.

(b) Find an expression for $g = \sin\phi$ and calculate in the same way σ_g/g. What quantity seems to be less sensitive to errors in I_1, I_2 and I_3?

11.4 Consider a four-frame technique with the following phase steps: $\alpha_i = -3\pi/4, -\pi/4, \pi/4, 3\pi/4$. Compute the matrices A, A^{-1} and B. Compute ϕ.

12

Computerized Optical Processes

12.1 INTRODUCTION

For almost 30 years, the silver halide emulsion has been first choice as the recording medium for holography, speckle interferometry, speckle photography, moiré and optical filtering. Materials such as photoresist, photopolymers and thermoplastic film have also been in use. There are two main reasons for this success. In processes where diffraction is involved (as in holographic reconstruction), a transparency is needed. The other advantage of film is its superior resolution. Film has, however, one big disadvantage; it must undergo some kind of processing. This is time consuming and quite cumbersome, especially in industrial applications.

Electronic cameras (vidicons) were first used as a recording medium in holography at the beginning of the 1970s. In this technique, called TV holography or ESPI (electronic speckle pattern interferometry), the interference fringe pattern is reconstructed electronically. At the beginning of the 1990s, computerized 'reconstruction' of the object wave was first demonstrated. This is, however, not a reconstruction in the ordinary sense, but it has proven possible to calculate and display the reconstructed field in any plane by means of a computer. It must be remembered that the electronic camera target can never act as a diffracting element. The success of the CCD-camera/computer combination has also prompted the development of speckle methods such as digital speckle photography (DSP).

The CCD camera has one additional disadvantage compared to silver halide films – its inferior resolution; the size of a pixel element of a 1317×1035 pixel CCD camera target is 6.8 μm. When used in DSP, the size σ_s of the speckles imaged onto the target must be greater than twice the pixel pitch p, i.e.

$$2p \leq \sigma_s = (1 + m)\lambda F \tag{12.1}$$

where m is the camera lens magnification and F the aperture number (see Equation (8.9)).

When applied to holography, the distance d between the interference fringes must according to the Nyquist theorem (see Section 5.8) be greater than $2p$:

$$2p \leq d = \frac{\lambda}{2\sin(\alpha/2)} \tag{12.2}$$

Assuming $\sin \alpha \approx \alpha$, this gives

$$\alpha \leq \frac{\lambda}{2p} \tag{12.3}$$

where α is the maximum angle between the object and reference waves and λ is the wavelength. For $p = 6.8$ μm this gives $\alpha = 2.7°$ ($\lambda = 0.6328$ μm).

In this chapter we describe the principles of digital holography and digital speckle photography. We also include the more mature method of TV holography.

12.2 TV HOLOGRAPHY (ESPI)

In this technique (also called *electronic speckle pattern interferometry*, ESPI) the holographic film is replaced by a TV camera as the recording medium (Jones and Wykes 1989). Obviously, the target of a TV camera can be used neither as a holographic storage medium nor for optical reconstruction of a hologram. Therefore the reconstruction process is performed electronically and the object is imaged onto the TV target. Because of the rather low resolution of a standard TV target, the angle between the object and reference waves has to be as small as possible. This means that the reference wave is made in-line with the object wave. A typical TV holography set-up therefore looks like that given in Figure 12.1. Here a reference wave modulating mirror (M_1) and a chopper are included, which are necessary only for special purposes in vibration analysis (see Section 6.9).

The basic principles of ESPI were developed almost simultaneously by Macovski *et al.* (1971) in the USA, Schwomma (1972) in Austria and Butters and Leendertz (1971) in England. Later the group of Løkberg (1980) in Norway contributed significantly to the field, especially in vibration analysis (Løkberg and Slettemoen 1987).

When the system in Figure 12.1 is applied to vibration analysis the video store is not needed. As in the analysis in Section 6.9, assume that the object and reference waves on the TV target are described by

$$u_o = U_o e^{i\phi_o} \tag{12.4a}$$

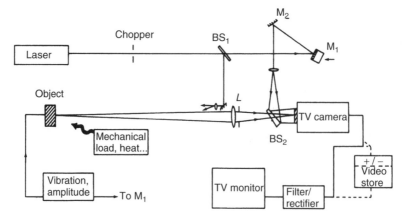

Figure 12.1 TV-holography set-up. From Lokberg 1980. (Reproduced by permission of Prof. O. J. Løkberg, Norwegian Institute of Technology, Trondheim)

and

$$u = U\mathrm{e}^{\mathrm{i}\phi} \tag{12.4b}$$

respectively. For a harmonically vibrating object we have (see equation (6.47) for $g = 2$)

$$\phi_0 = 2kD(x)\cos\omega t \tag{12.5}$$

where $D(x)$ is the vibration amplitude at the point of spatial coordinates x and ω is the vibration frequency. The intensity distribution over the TV-target becomes

$$I(x, t) = U^2 + U_0^2 + 2UU_0\cos[\phi - 2kD(x)\cos\omega t] \tag{12.6}$$

This spatial intensity distribution is converted into a corresponding time-varying video signal. When the vibration frequency is much higher than the frame frequency of the TV system ($\frac{1}{25}$ s, European standard), the intensity observed on the monitor is proportional to Equation (12.6) averaged over one vibration period, i.e. (cf. Equation (6.49))

$$I = \overline{U^2} + \overline{U_0^2} + \overline{2UU_0\cos\phi}\, J_0(2kD(x)) \tag{12.7}$$

where J_0 is the zeroth-order Bessel function and the bars denote time average. Before being displayed on the monitor, the video signal is high-pass filtered and rectified. In the filtering process, the first two terms of Equation (12.7) are removed. After full-wave rectifying we thus are left with

$$I = 2|\overline{UU_0\cos\phi}\, J_0(2kD(x))| \tag{12.8}$$

Actually, ϕ represents the phase difference between the reference wave and the wave scattered from the object in its stationary state. The term $UU_0\cos\phi$ therefore represents a speckle pattern and the J_0-function is said to modulate this speckle pattern. Equation (12.8) is quite analogous to Equation (6.51) except that we get a $|J_0|$-dependence instead of a J_0^2-dependence. The maxima and zeros of the intensity distributions have, however, the same locations in the two cases. A time-average recording of a vibrating turbine blade therefore looks like that shown in Figure 12.2(a) when applying ordinary holography, and that in Figure 12.2(b) when applying TV holography. We see that the main difference in the two fringe patterns is the speckled appearance of the TV holography picture.

When applied to static deformations, the video store in Figure 12.1 must be included. This could be a video tape or disc, or most commonly, a frame grabber (see Section 10.2) in which case the video signal is digitized by an analogue-to-digital converter. Assume that the wave scattered from the object in its initial state at a point on the TV target is described by

$$u_1 = U_0\mathrm{e}^{\mathrm{i}\phi_0} \tag{12.9}$$

After deformation, this wave is changed to

$$u_2 = U_0\mathrm{e}^{\mathrm{i}(\phi_0 + 2kd)} \tag{12.10}$$

Holography   ESPI

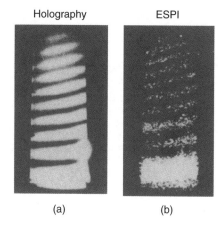

(a)   (b)

Figure 12.2 (a) Ordinary holographic and (b) TV-holographic recording of a vibrating turbine blade. (Reproduced by permission of Prof. O. J. Løkberg, Norwegian Institute of Technology, Trondheim)

where d is the out of plane displacement and where we have assumed equal field amplitudes in the two cases. Before deformation, the intensity distribution on the TV target is given by

$$I_1 = U^2 + U_o^2 + 2UU_o \cos(\phi - \phi_o) \tag{12.11}$$

where U and ϕ are the amplitude and phase of the reference wave. This distribution is converted into a corresponding video signal and stored in the memory. After the deformation, the intensity and corresponding video signal is given by

$$I_2 = U^2 + U_o^2 + 2UU_o \cos(\phi - \phi_0 - 2kd) \tag{12.12}$$

These two signals are then subtracted in real time and rectified, resulting in an intensity distribution on the monitor proportional to

$$I_1 - I_2 = 2UU_0|[\cos(\phi - \phi_0) - \cos(\phi - \phi_0 - 2kd)]|$$
$$= 4UU_0|\sin(\phi - \phi_0 - kd)\sin(kd)| \tag{12.13}$$

The difference signal is also high-pass filtered, removing any unwanted background signal due to slow spatial variations in the reference wave. Apart from the speckle pattern due to the random phase fluctuations $\phi - \phi_0$ between the object and reference fields, this gives the same fringe patters as when using ordinary holography to static deformations. The dark and bright fringes are, however, interchanged, for example the zero-order dark fringe corresponds to zero displacement.

This TV holography system has a lot of advantages. In the first place, the cumbersome, time-consuming development process of the hologram is omitted. The exposure time is quite short ($\frac{1}{25}$ s), relaxing the stability requirements, and one gets a new hologram each $\frac{1}{25}$ s. Among other things, this means that an unsuccessful recording does not have the same serious consequences and the set-up can be optimized very quickly. A lot of loading conditions can be examined during a relatively short time period. Time-average recordings of vibrating objects at different excitation levels and different frequencies are easily performed.

The interferograms can be photographed directly from the monitor screen or recorded on video tape for later analysis and documentation. TV holography is extremely useful for applications of the reference wave modulation and stroboscopic holography techniques mentioned in Section 6.9. In this way, vibration amplitudes down to a couple of angstroms have been measured. The method has been applied to a lot of different objects varying from the human ear drum (Løkberg *et al.* 1979) to car bodies (Malmo and Vikhagen 1988).

When analysing static deformations, the real-time feature of TV holography makes it possible to compensate for rigid-body movements by tilting mirrors in the illumination beam path until a minimum number of fringes appear on the monitor.

12.3 DIGITAL HOLOGRAPHY

In ESPI the object was imaged onto the target of the electronic camera and the interference fringes could be displayed on a monitor. We will now see how the image of the object can be reconstructed digitally when the unfocused interference (between the object and reference waves) field is exposed to the camera target. The experimental set-up is therefore quite similar to standard holography.

The geometry for the description of digital holography is shown in Figure 12.3. We assume the field amplitude $u_o(x, y)$ of the object to be existing in the xy-plane. Let the hologram (the camera target) be in the $\xi\eta$-plane a distance d from the object. Assume that a hologram given in the usual way as (cf. Equation (6.1))

$$I(\xi, \eta) = |r|^2 + |u_o|^2 + ru_o^* + r^*u_o \qquad (12.14)$$

is recorded and stored by the electronic camera. Here u_o and r are the object and reference waves respectively. In standard holography the hologram is reconstructed by illuminating the hologram with the reconstruction wave r. This can of course not be done here. However, we can simulate $r(\xi, \eta)$ in the $\xi\eta$-plane by means of the computer and therefore also construct the product $I(\xi, \eta)r(\xi, \eta)$. In Chapter 4 we learned that if the field amplitude distribution over a plane is given, then the field amplitude propagated to another point

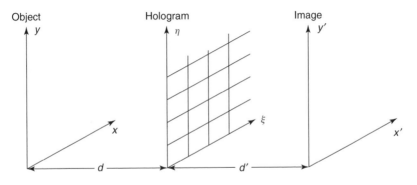

Figure 12.3

in space is found by summing the contributions from the Huygens wavelets over the aperture. To find the reconstructed field amplitude distribution $u_a(x', y')$ in the $x'y'$-plane we therefore apply the Rayleigh–Sommerfeld diffraction formula (Equation (4.7)):

$$u_a(x', y') = \frac{1}{i\lambda} \iint I(\xi, \eta) r(\xi, \eta) \frac{e^{ik\rho}}{\rho} \cos \Omega d\xi d\eta \tag{12.15}$$

with

$$\rho = \sqrt{d'^2 + (\xi - x')^2 + (\eta - y')^2} \tag{12.16}$$

We therefore should be able to calculate $u_a(x', y')$ in the $x'y'$-plane at any distance d' from the hologram plane. There are, however, two values of d' of most practical interest: (1) $d' = -d$ where the virtual image is located (see Section 6.4), and (2) $d' = d$, the location of the real image, provided the reference wave is a plane wave. As found in Section 6.4, this demands that the reference and reconstruction waves are identical. With today's powerful computers it is straightforward to calculate the integral in Equation (12.15). However, with some approximations and rearrangements of the integrand, the processing speed can be increased considerably. Below we discuss how to approach this problem.

The first method for solving Equation (12.15) is to apply the Fresnel approximation as described in Section 1.7. That is, to retain the first two terms of a binomial expansion of ρ and put $\cos \Omega = 1$. This gives

$$u_a(x', y') = \frac{\exp\{ikd'\}}{i\lambda d'} \iint I(\xi, \eta) r(\xi, \eta) \exp\left\{\frac{ik}{2d'}[(\xi - x')^2 + (\eta - y')^2]\right\} d\xi d\eta$$

$$= \frac{\exp\{ikd'\} \exp\{i\pi d'\lambda(u^2 + v^2)\}}{i\lambda d'} \iint I(\xi, \eta) r(\xi, \eta)$$

$$\times \exp\left\{\frac{i\pi}{d'\lambda}(\xi^2 + \eta^2)\right\} \exp\{-2i\pi(v\xi + \mu\eta)\} d\xi d\eta \tag{12.17}$$

where we have introduced the spatial frequencies

$$u = \frac{x'}{d\lambda} \text{ and } v = \frac{y'}{d\lambda} \tag{12.18}$$

Equation (12.17) can be written as

$$u_a = z\mathscr{F}\{I \cdot r \cdot w\} \tag{12.19}$$

The reconstructed field is therefore given as the Fourier transform of $I(\xi, \eta)$ multiplied by $r(\xi, \eta)$ and a quadratic phase function

$$w(\xi, \eta) = \exp\left\{\frac{i\pi}{d'\lambda}(\xi^2 + \eta^2)\right\} \tag{12.20}$$

The evaluated integral is multiplied by a phase function

$$z(u, v) = \exp\{ikd'\}\exp\{i\pi d'(u^2 + v^2)\} \tag{12.21}$$

In most applications $z(u, v)$ can be neglected, e.g. when only the intensity is of interest, or if only phase differences matter, as in holographic interferometry.

$\mathscr{F}\{f(\xi, \eta)w(\xi, \eta)\}$ is often referred to as a Fresnel transformation of $f(\xi, \eta)$. When $d' \to \infty$, $w(\xi, \eta) \to 1$ and the Fresnel transform reduces to a pure Fourier transform.

A spherical wave from a point $(0, 0, -d')$ is described by

$$r(\xi, \eta) = U_r \exp\left\{-\frac{i\pi}{d'\lambda}(\xi^2 + \eta^2)\right\} \tag{12.22}$$

By using this as the reconstruction wave, $r \cdot w = \text{constant}$, and again we get a pure Fourier transform. This case is called lensless Fourier transform holography. Although this method gives a more efficient computation, we lose the possibility for numerical focusing by varying the distance d', since it vanishes from the formula. In Figure 12.4 the Fresnel method is applied.

In the second method we first note that the diffraction integral, Equation (12.15), can be written as

$$u_a(x', y') = \iint I(\xi, \eta)r(\xi, \eta)g(x', y', \xi, \eta)\mathrm{d}\xi\,\mathrm{d}\eta \tag{12.23}$$

where

$$g(x', y', \xi, \eta) = \frac{1}{i\lambda}\frac{\exp\{ik\sqrt{d'^2 + (\xi - x')^2 + (\eta - y')^2}\}}{\sqrt{d'^2 + (\xi - x')^2 + (\eta - y')^2}} \tag{12.24}$$

which means that $g(x', y', \xi, \eta) = g(x' - \xi, y' - \eta)$ and therefore Equation (12.23) can be written as a convolution

$$u_a = (I \cdot r) \otimes g \tag{12.25}$$

From the convolution theorem (see Appendix B) we therefore have

$$\mathscr{F}\{u_a\} = \mathscr{F}\{I \cdot r\}\mathscr{F}\{g\} \tag{12.26}$$

By taking the inverse Fourier transform of this result, we get

$$u_a = \mathscr{F}^{-1}\{\mathscr{F}\{I \cdot r\}\mathscr{F}\{g\}\} \tag{12.27}$$

The Fourier transform of g can be derived analytically (Goodman 1996):

$$G(u, v) = \mathscr{F}\{g\} = \exp\left\{\frac{2\pi i d'}{\lambda}\sqrt{1 - (\lambda u)^2 - (\lambda v)^2}\right\} \tag{12.28}$$

and therefore

$$u_a(x', y') = \mathscr{F}^{-1}\{\mathscr{F}\{I \cdot r\} \cdot G\} \tag{12.29}$$

which saves us one Fourier transform.

Figure 12.4 Numerical reconstruction of the real image using the Fresnel method. The bright central spot is due to the spectrum of the plane reference wave. The object was a 10.5 cm high, 6.0 cm wide white plaster bust of the composer J. Brahms placed 138 cm from the camera target. Reproduced by courtesy of O. Skotheim (2001)

Holographic interferometry. An important application of digital holography is in the field of holographic interferometry. Standard methods (see Chapter 6) rely on the extraction of the phase from interference fringes. Digital holography has the advantage of providing direct access to phase data in the reconstructed wave field. Denoting the reconstructed real (or virtual) wave as

$$u_a = U e^{i\varphi} \tag{12.30}$$

we get

$$\varphi = \tan^{-1} \frac{\text{Im}\{u\}}{\text{Re}\{u\}} \tag{12.31}$$

This is a wrapped phase and we have to rely on unwrapping techniques as described in Section 11.5. By reconstructing the real wave of the object in states 1 and 2 (e.g. between a deformation) we can extract the two phase maps φ_1 and φ_2 and calculate the phase difference $\varphi = \varphi_1 - \varphi_2$.

12.4 DIGITAL SPECKLE PHOTOGRAPHY

In Section 8.4.2 we learned how to measure the displacement vector from a double-exposed specklegram by illuminating the specklegram by a direct laser beam and observing the resulting Young fringes on a screen (see Figure 8.10). In Section 8.5 we gave a more detailed explanation of this phenomenon. The intensities I_1 and I_2 in the first and second recording we wrote as $I_1(x, y) = I(x, y)$ and $I_2(x, y) = I(x + d, y)$. This could be done because we assumed the speckle displacement to be uniform within the laser beam illuminated area and for simplicity we assumed the displacement to be in the x-direction. The Fourier transforms of I_1, I_2 and I were denoted $J_1(u, v)$, $J_2(u, v)$ and $J(u, v)$ respectively. We found (Equation (8.38)) that

$$J_2(u, v) = J_1(u, v) \cdot e^{i2\pi ud} = J(u, v) \cdot e^{i2\pi ud} \tag{12.32}$$

Now we discuss another technique called Digital Speckle Photography (DSP). Here the specklegrams are recorded by an electronic camera. In practice, the image is divided into subimages with a size of, e.g., 8×8 pixels. Within each subimage, the speckle displacement is assumed to be constant. Assume that I_1 and I_2 are the intensities recorded in a particular subimage. The corresponding Fourier transforms are then easily calculated by a computer. Let us call this step 1 of our procedure. In step 2 we calculate a new spectrum given as

$$F(u, v) = \frac{J_1 \cdot J_2}{|J_1 \cdot J_2|} |J_1 \cdot J_2|^\alpha = \frac{J_1 \cdot J_2^*}{|J_1 \cdot J_2|^{1-\alpha}} \tag{12.33}$$

By using the result from Equation (12.32), we get

$$F(u, v) = |J(u, v)|^{2\alpha} e^{i2\pi ud} \tag{12.34}$$

To this we apply another Fourier transform operation (step 3):

$$\mathscr{F}\{F(u, v)\} = \int_{-\infty}^{\infty} \int |J(u, v)|^{2\alpha} e^{-i2\pi[u(\xi-d)+v\eta]} \, du \, dv = G_\alpha(\xi - d, \eta) \tag{12.35}$$

where

$$G_\alpha(\xi, \eta) = \int_{-\infty}^{\infty} \int |J(u, v)|^{2\alpha} e^{-i2\pi(u\xi+v\eta)} \, du \, dv \tag{12.36}$$

In practice, $G_\alpha(\xi - d, \eta)$ emerges as an expanded impulse or correlation peak located at $(d, 0)$ in the second spectral domain. By this method we have obtained the cross-correlation between I_1 and I_2. Therefore this procedure gives a more direct method for detecting the displacement d than does the Young fringe method. The parameter α controls the width of the correlation peak. Optimum values range from $\alpha = 0$ for images characterized by a high spatial frequency content and a high noise level, to $\alpha = 0.5$ for low noise images with less fine structure. For $\alpha > 0.5$ the high frequency noise is magnified, resulting in an unreliable algorithm.

The last two steps of this procedure cannot be done optically but are easily performed in a computer. The local displacement vector for each subimage is found by the above procedure and thereby the 2-D displacement for the whole field can be deduced. DSP is not restricted to laser speckles. On the contrary, white light speckles are superior to laser speckles when measuring object deformations, mainly because they are more robust to decorrelation.

A versatile method for creating white light speckles when measuring object contours or deformations is to project a random pattern onto the surface by means of an addressable video projector. DSP has also been used in combination with X-rays to measure internal deformations (Synnergren and Goldrein 1999). Here a plane of interest in the material is seeded with grains of an X-ray absorbing material and a speckled shadow image is cast on the X-ray film.

13

Fibre Optics in Metrology

13.1 INTRODUCTION

With a carrier frequency of some 10^{14} Hz, light has the potential of being modulated at much higher frequencies than radio waves. Since the mid-1960s the idea of communication through optical fibres has developed into a vital branch of electro-optics. Great progress has been made and this is now an established technique in many communication systems. From the viewpoint of optical metrology, optical fibres are an attractive alternative for the guiding of light. An even more important reason for studying optical fibres is their potential for making new types of sensors.

13.2 LIGHT PROPAGATION THROUGH OPTICAL FIBRES

More extensive treatments on optical fibres can be found in Senior (1985), Palais (1998), Keiser (1991) and Yu and Khoo (1990).

Figure 13.1 shows the basic construction of an optical fibre. It consists of a central cylindrical core with refractive index n_1, surrounded by a layer of material called the cladding with a lower refractive index n_2. In the figure a light ray is incident at the end of the fibre at an angle θ_0 to the fibre axis. This ray is refracted at an angle θ_1 and incident at the interface between the core and the cladding at an angle θ_2. From Snell's law of refraction we have

$$n_0 \sin \theta_0 = n_1 \sin \theta_1 \qquad (13.1)$$

where n_0 is the refractive index of the surrounding medium. From the figure, we see that

$$\theta_1 = \frac{\pi}{2} - \theta_2 \qquad (13.2)$$

If θ_2 is equal to the critical angle of incidence (cf. Section 9.5), we have

$$\sin \theta_2 = \frac{n_2}{n_1} \qquad (13.3)$$

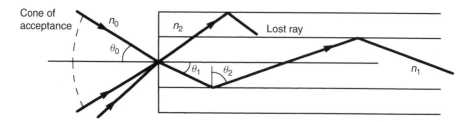

Figure 13.1   Basic construction of an optical fibre

which combined with Equations (13.1) and (13.2) gives

$$\theta_0 \equiv \theta_a = \sin^{-1}\left[\frac{\sqrt{n_1^2 - n_2^2}}{n_0}\right] \qquad (13.4)$$

For $\theta_0 < \theta_a$ the light will undergo total internal reflection at the interface between the core and the cladding and propagate along the fibre by multiple reflections at the interface, ideally with no loss. For $\theta_0 > \theta_a$ some of the light will transmit into the cladding and after a few reflections, most of the light will be lost.

This is the principle of light transmission through an optical fibre. The angle θ_a is an important parameter when coupling of the light into a fibre, usually given by its numerical aperture NA:

$$NA = n_0 \sin\theta_a = \sqrt{n_1^2 - n_2^2} \qquad (13.5)$$

In practice, coupling of the light into the fiber can be accomplished with the help of a lens, see Figure 13.2(a) or by putting the fibre in close proximity to the light source and

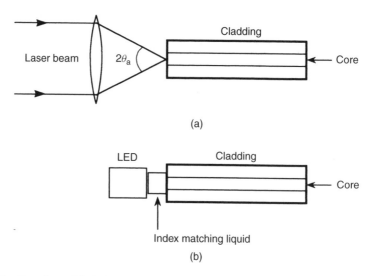

(a)

(b)

Figure 13.2   Coupling of light into a fibre by means of (a) a lens and (b) index-matching liquid

linking them with an index-matching liquid to reduce reflection losses, Figure 13.2(b). When using the method in Figure 13.2(a), it is important to have the angle of the incident cone less than θ_a to get maximum coupling efficiency.

The above description of light propagation through an optical fibre is not fully complete. To gain better understanding, the fibre must be treated as a waveguide and the electro-magnetic nature of the light must be taken into account. If a waveguide consisting of a transparent layer between two conducting walls is considered, the electric field across the waveguide will consist of interference patterns between the incident and reflected fields, or equivalently, between the incident field and its mirror image, see Figure 13.3. The path-length difference Δl between these fields is seen from the figure to be

$$\Delta l = d \sin \theta \tag{13.6}$$

where d is the waveguide diameter and θ is the angle of the incident beam. From the boundary conditions for such a waveguide we must have destructive interference at the walls, i.e. the path-length difference must be equal to an integral number of half the wavelength:

$$\Delta l = m \frac{\lambda}{2} \tag{13.7}$$

which gives

$$\sin \theta = \frac{m\lambda}{2d} \tag{13.8}$$

where m is an integer. Thus we see that only certain values of the angle of incidence are allowed. Each of the allowed beam directions are said to correspond to different modes of wave propagation in the waveguide. The field distribution across the waveguide for the lowest-order guided modes in a planar dielectric slab waveguide are shown in Figure 13.4. This guide is composed of a dielectric core (or slab) sandwiched between dielectric claddings of lower refractive index. As can be seen, the field is non-zero inside the

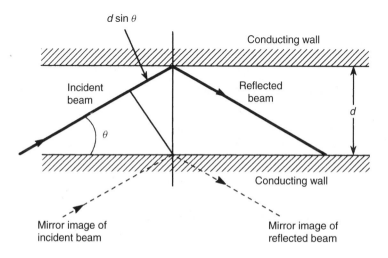

Figure 13.3   A conducting slab waveguide

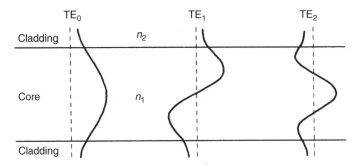

Figure 13.4   Electric field distribution of the lowest-order guided transversal modes in a dielectric slab waveguide

cladding. This is not in contradiction with the theory of total internal reflection (see Section 9.5) which predicts an evanescent wave decaying very rapidly in the cladding material.

The lowest number of modes propagating through the waveguide occurs when the angle of incidence is equal to θ_a. Then (assuming $n_0 = 1$ for air)

$$\sin \theta_a = \frac{m\lambda}{2d} = \sqrt{n_1^2 - n_2^2} \tag{13.9}$$

or

$$\frac{d}{\lambda} = \frac{m}{2\sqrt{n_1^2 - n_2^2}} \tag{13.10}$$

To have only the lowest-order mode ($m = 0$) propagating through the waveguide, we therefore must have

$$\frac{d}{\lambda} < \frac{1}{2\sqrt{n_1^2 - n_2^2}} \tag{13.11}$$

An exact waveguide theory applied to an optical fibre is quite complicated, but the results are quite similar. The condition for propagating only the lowest-order mode in an optical fibre then becomes

$$\frac{d}{\lambda} < \frac{2.405}{2\pi\sqrt{n_1^2 - n_2^2}} = \frac{2.405}{2\pi(NA)} = \frac{0.383}{NA} \tag{13.12}$$

A fibre allowing only the lowest-order mode to propagate is called a single-mode fibre, in contrast to a multimode fibre which allows several propagating modes.

13.3 ATTENUATION AND DISPERSION

That light will propagate through a fibre by multiple total internal reflections without loss is an idealization. In reality the light will be attenuated. The main contributions to attenuation is scattering (proportional to λ^{-4}) in the ultraviolet end of the spectrum and absorption in the infra-red end of the spectrum. Therefore it is only a limited part of

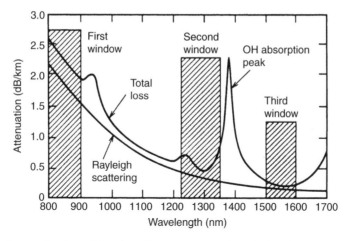

Figure 13.5 Attenuation in a silica glass fibre versus wavelength showing the three major wavelength regions at which fibre systems are most practical. (From Palais, J. C. (1998) Fiber Optic Communications (4th edn), Prentice Hall, Englewood Cliffs, N.J.) Reproduced by permission of Prentice Hall Inc.)

the electromagnetic spectrum where fibre systems are practical. Figure 13.5 shows the attenuation as a function of wavelength for silica glass fibres. Here are also shown the three major wavelength regions at which fibre systems are practical. These regions are dictated by the attenuation, but also by the light sources available.

Another source of loss in fibre communication systems is dispersion. Dispersion is due to the fact that the refractive index is not constant, but depends on the wavelength, i.e. $n = n(\lambda)$. In fibre systems one talks about material dispersion and waveguide dispersion. Here we will briefly mention material dispersion. That the refractive index varies with wavelength means that a light pulse from a source of finite spectral width will broaden as it propagates through the fibre due to the different velocities for the different wavelengths. This effect has significant influence on the information capacity of the fibre. The parameter describing this effect is the pulse spread per unit length denoted $\Delta\tau/L$ where $\Delta\tau$ is the difference in travel time for two extreme wavelengths of the source's spectral distribution through the length L. This gives

$$\frac{\Delta\tau}{L} = \Delta\left(\frac{1}{v_g}\right) \tag{13.13}$$

In dispersive media a light pulse propagates at the group velocity (Senior 1985) defined by

$$v_g = \frac{d\omega}{d\beta} \tag{13.14}$$

With the relations

$$\omega = kc = \frac{2\pi c}{\lambda} \tag{13.15a}$$

$$\beta = kn = \frac{2\pi n}{\lambda} \tag{13.15b}$$

we get

$$\frac{1}{v_g} = \frac{d\beta}{d\omega} = \frac{d\beta}{d\lambda}\frac{d\lambda}{d\omega} = \left(\frac{-\lambda^2}{2\pi c}\right)2\pi\left(\frac{1}{\lambda}\frac{dn}{d\lambda} - \frac{n}{\lambda^2}\right) = \frac{1}{c}\left(n - \lambda\frac{dn}{d\lambda}\right) \qquad (13.16)$$

This gives

$$\frac{\Delta\tau}{L} = \Delta\left(\frac{1}{v_g}\right) = \Delta\left(\frac{n - \lambda dn/d\lambda}{c}\right) \qquad (13.17)$$

The pulse spread per unit length per wavelength interval $\Delta\lambda$ becomes

$$\frac{\Delta\tau}{L\Delta\lambda} = \frac{d}{d\lambda}\left(\frac{1}{v_g}\right) = \frac{d}{d\lambda}\left(\frac{n}{c} - \frac{\lambda}{c}\frac{dn}{d\lambda}\right) = -\frac{\lambda}{c}\frac{d^2 n}{d\lambda^2} \qquad (13.18)$$

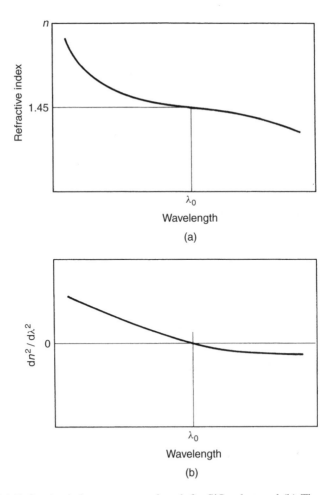

Figure 13.6 (a) Refractive index versus wavelength for SiO_2 glass and (b) The second derivative of the curve in (a)

The material dispersion is defined as $M = (\lambda/c)(d^2n/d\lambda^2)$. The pulse spread per unit length then can be written as

$$\frac{\Delta\tau}{L} = -M\,\Delta\lambda \qquad (13.19)$$

The refractive index for pure silicon dioxide (SiO_2) glass used in optic fibres has the wavelength dependence shown in Figure 13.6(a). At a particular wavelength λ_0, there is an inflection point on the curve. Because of this, $d^2n/d\lambda^2 = 0$ at λ_0 as seen from the curve of the second derivative in Figure 13.6(b). For pure silica, the refractive index is close to 1.45 and the inflection point is near $\lambda_0 = 1.3$ µm. Therefore this wavelength is very suitable for long distance optical fibre communication.

13.4 DIFFERENT TYPES OF FIBRES

Another construction than the step-index (SI) fibre sketched in Figure 13.1 is the so-called graded-index (GRIN) fibre. It has a core material whose refractive index varies with distance from the fibre axis. This structure is illustrated in Figure 13.7. As should be easily realized, the light rays will bend gradually and travel through a GRIN fibre in the oscillatory fashion sketched in Figure 13.7(d). As opposed to an SI fibre, the numerical aperture of a GRIN fibre decrease with radial distance from the axis. For this reason, the coupling efficiency is generally higher for SI fibres than for GRIN fibres, when each has the same core size and the same fractional refractive index change.

Conventionally, the size of a fibre is denoted by writing its core diameter and then its cladding diameter (both in micrometers) with a slash between them. Typical dimensions

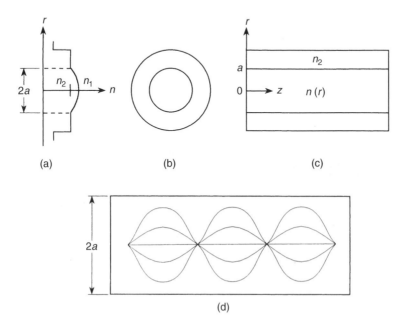

Figure 13.7 Graded index fibre: (a) refractive index profile; (b) end view; (c) cross-sectional view; and (d) ray paths along a GRIN fibre

of SI fibres are 50/125, 100/140 and 200/230 and typical dimensions of multimode GRIN fibres are 50/125, 62.5/125 and 85/125.

SI fibres have three common forms: (1) a glass core cladded with glass, (2) a silica glass core cladded with plastic (termed plastic-cladded silica (PCS) fibres), and (3) a plastic core cladded with another plastic. All-glass fibres have the lowest losses and the smallest pulse spreading, but also the smallest numerical aperture. PCS fibres have higher losses and larger pulse spreads and are suitable for shorter links, normally less than a few hundred metres. Their higher NA increase the coupling efficiency. All-plastic fibres are used for path lengths less than a few tens of meters. Their high NA gives high coupling efficiency. Single-mode fibres have the highest information capacity. GRIN fibres can transmit at higher information rates than SI fibres. Table 13.1 shows representative numerical values of important properties for the various fibres. Somewhat different characteristics may be found when searching the manufacturers' literature.

Table 13.1   (From Palais, J. C. (1998) *Fiber Optic Communication* (4th edn), Prentice Hall, Englewood Cliffs, New Jersey). Reproduced by permission

Description	Core Diameter (μm)	NA	Loss (dB/km)	$\Delta(\tau/L)$ (ns/km)	Source	Wavelength (nm)
Multimode						
Glass						
SI	50	0.24	5	15	LED	850
GRIN	50	0.24	5	1	LD	850
GRIN	50	0.20	1	0.5	LED, LD	1300
PCS						
SI	200	0.41	8	50	LED	800
Plastic						
SI	1000	0.48	200	–	LED	580
Single mode						
Glass	5	0.10	4	<0.5	LD	850
Glass	10	0.10	0.5	0.006	LD	1300
Glass	10	0.10	0.2	0.006	LD	1550

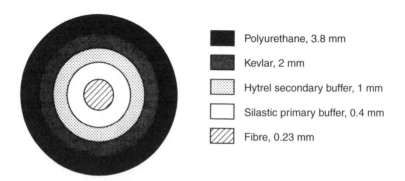

Polyurethane, 3.8 mm

Kevlar, 2 mm

Hytrel secondary buffer, 1 mm

Silastic primary buffer, 0.4 mm

Fibre, 0.23 mm

Figure 13.8   Light-duty, tight-buffer fibre cable (Siecor Corporation). The dimensions given are the diameters. (From Palais, J. C. (1998) *Fiber Optic Communications* (4th edn), Prentice Hall, Englewood Cliffs, N.J.) Reproduced by permission of Prentice Hall

The amount of protection against the environment of a fibre varies from one application to another. Various cable designs have been implemented. A representative light-duty cable is sketched in Figure 13.8. This cable weighs 12.5 kg/km and can withstand a tensile load of 400 N during installation and can be loaded up to 50 N in operation.

Fibre-optic communications developed very quickly after the first low-loss fibres were produced in 1970. Today, over 10 million km of fibre have been installed worldwide, numerous submarine fibre cables covering the Atlantic and Pacific oceans and many other smaller seas are operational. In addition, installation of fibre-optic local area networks (LANs) is increasing.

13.5 FIBRE-OPTIC SENSORS

Over the past few years, a significant number of sensors using optical fibres have been developed (Kyuma *et al.* 1982; Culshaw 1986; Udd (1991, 1993)). They have the potential for sensing a variety of physical variables, such as acoustic pressure, magnetic fields, temperature, acceleration and rate of rotation. Also sensors for measuring current and voltage based on polarization rotation induced by the magnetic field around conductors due to the Faraday effect in optical fibres have been developed. It should also be mentioned that a lot of standard optical equipment has been redesigned using optical fibres. The Laser Doppler velocimeter is an example where optical fibres have been incorporated to increase the versatility of the instrument.

Figure 13.9 shows some typical examples of fibre-optic sensors. In Figure 13.9(a) a thin semiconductor chip is sandwiched between two ends of fibres inside a steel pipe. The light is coming through the fibre from the left and is partly absorbed by the semi-conductor. This absorption is temperature-dependent and the amount of light detected at the end of the fibre to the right is therefore proportional to the temperature and

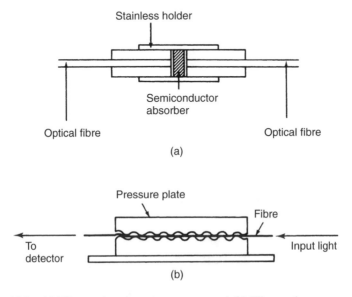

Figure 13.9   (a) Fibre-optic temperature sensor and (b) Fibre-optic pressure sensor

we have a fibre-optic temperature sensor. Figure 13.9(b) shows a simplified sketch of a pressure-sensing system. The optical fibre is placed between two corrugated plates. When pressure is applied to the plates, the light intensity transmitted by the fibre changes, owing to microbending loss. Such systems have also been applied as hydrophones and accelerometers.

Figure 13.10 shows the principle of a class of fibre-optic sensors based on interferometry. The fibres A and B can be regarded as either arm in a Mach–Zehnder interferometer. The detector will record an intensity which is dependent on the optical path-length difference through A and B. When, for example, fibre A is exposed to loads such as tension, pressure, temperature, acoustical waves, etc., the optical path length of A will change and one gets a signal from the detector varying as the external load.

Figure 13.11 shows a special application of optical fibres. In Figure 13.11(a) two fibre bundles, A and B, are mixed together in a bundle C in such a way that every second fibre in the cross-section of C comes from, say, bundle A. Figure 13.11(b) shows two neighbouring fibres, A and B. Fibre A emits a conical light beam. Fibre B will receive light inside a cone of the same magnitude. If a plane surface is placed a distance l in front

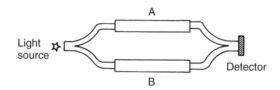

Figure 13.10   Interferometric fibre-optic sensor

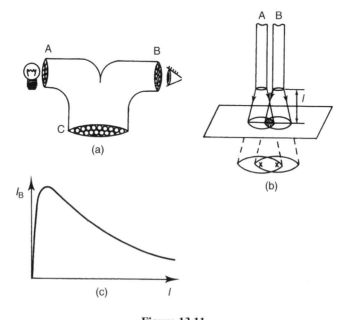

Figure 13.11

of the fibre ends, light will be scattered back and the amount of light received by fibre B will be proportional to the area of overlap between the two cones as illustrated in the figure. A curve describing the relation between the received light intensity I_B versus the distance l therefore will look like that given in Figure 13.11(c). For a distance l so long that the whole cross-section of fibre B is covered with light, I_B will have its maximum. A further increase in l will decrease the value of I_B.

The fibre bundles coupled together as in Figure 13.11(a) thus can be used as a non-contact distance sensor (Cook and Hamm 1979; Philips 1980). A light source emits light into bundle A and a detector is placed at the end of bundle B. By placing the end of bundle C close to a surface, the detector will give a signal which is proportional to the distance from the fibre end to the surface, as long as one is working within the linear portion of the left flank of the curve in Figure 13.11(c). The sensitivity limit of this type of sensor can be down to nanometres and it is especially well-suited to the measurement of small vibration amplitudes. By mounting a membrane in front of fibre bundle C, this equipment can be used for pressure measurements. In this way it has been applied for the measurement of blood pressure and electric fields inside big power cables.

One of the most important and potentially low-cost advantages of fibre-optic sensors is their multiplexing capability (Berthold 1993). For example, multiple fibre-optic accelerometers can be linked to a common transmission 'bus' fibre. These accelerometers can all be remotely interrogated in turn from a central processor without any intervening electronics. Either wavelength division (WDM) or time division (TDM) multiplexing methods could be employed. In contrast, commercially available piezo-electric accelerometers each requires its own charge amplifier/signal conditioner to achieve low-noise operation, and thus these devices are not compatible with a low-cost multiplexing architecture. A system based on wavelength division multiplexing is sketched in Figure 13.12. Here the different wavelengths from the light source (an LED) are directed to each accelerometer by the WDM units. At the receiving signal processing unit, the intensity of each separate wavelength is analysed.

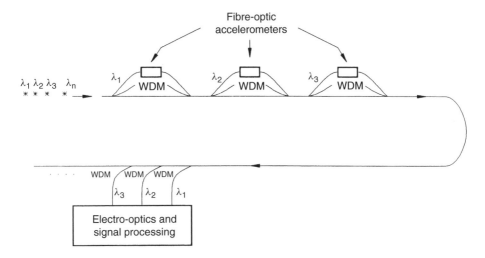

Figure 13.12   Multiplexing of fibre-optic sensors (accelerometers) using WDM

13.6 FIBRE-BRAGG SENSORS

Certainly the most important innovation of the 1990s in optical fibre sensing is the development of fibre grating sensors. Such a fibre-Bragg grating (FBG) is a periodic perturbation of the refractive index along the fibre length which is formed by exposure of the core of doped silica single-mode fibres by UV laser light using phase masks, or more efficiently, directly during the drawing process by an interference pattern of short laser pulses. The formation of permanent gratings in an optical fibre was first demonstrated by Hill *et al.* in 1978 (Hill *et al.* 1978, Kawasaki *et al.* 1978).

The working principle of FBGs is similar to a reflection volume hologram: see Section 6.6, Figure 6.4. When the object and reference waves are incident from opposite sides of the thick hologram emulsion, layers of metallic silver of the developed hologram are parallel to the hologram plane. By reconstructing the hologram in white light, the reconstructed wave would be single coloured with a wavelength equal to $\lambda = 2d$ where d is the layer spacing.

The refractive index variations in the core along the fibre length (the z-axis) can be described as

$$n(z) = n_c + \Delta n \cos qz \qquad (13.20)$$

where $q = 2\pi/d$ is the frequency and n_c is the refractive index of the unmodified core (typically $n_c = 1.46$). Δn is the amplitude of the refractive index variations and d is the grating period. Equation (13.20) could also describe the refractive index variations through the depth of a volume hologram. Let us start by analysing the situation shown in Figure 13.13. Here a plane wave is incident at an angle θ to the incremental planar layers orthogonal to the z-axis and reflected at each layer because of the refractive index change. We assume the reflectance to be small (i.e. the transmittance close to unity) so that the wave approximately maintains its amplitude as it penetrates the following layers of the medium.

If $\Delta r = (dr/dz)\Delta z$ is the incremental complex amplitude reflectance of the layer at position z, the total amplitude reflectance for the overall length L (see Figure 13.13) is

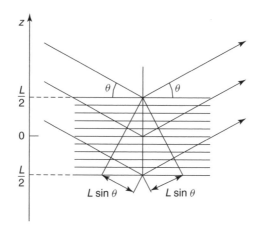

Figure 13.13   Reflections from planar layers

the sum of all incremental reflectances

$$r = \int_{-L/2}^{L/2} e^{i2kz \sin\theta} \frac{dr}{dz} dz \tag{13.21}$$

The phase factor $\exp\{i2kz \sin\theta\}$ is included since the reflected wave at a position z is advanced by a distance $2z \sin\theta$ relative to the reflected wave at $z = 0$. To find Δr in terms of Δn we must apply the Fresnel equations (see Section 9.4, Equation (9.22)). For light polarized normal to the plane of incidence (Equation (9.22b)) with $n_1 = n + \Delta n$, $n_2 = n$ and $\theta_1 = 90° - \theta$, we get

$$\Delta r = \frac{-1}{2n \sin^2\theta} \Delta n \tag{13.22}$$

where terms of second order in Δn are neglected. Similarly, for light polarized parallel to the plane of incidence (Equation (9.22a)) we get

$$\Delta r = \frac{-\cos 2\theta}{2n \sin^2\theta} \Delta n \tag{13.23}$$

For θ close to $0°$ or $90°$, both expressions will be approximately equal. Using Equations (13.20) and (13.22) we obtain

$$\frac{dr}{dz} = \frac{dr}{dn}\frac{dn}{dz} = \frac{-q}{2n \sin^2\theta} \Delta n \sin qz = r' \sin qz \tag{13.24}$$

where

$$r' = \frac{-q}{2n \sin^2\theta} \Delta n \tag{13.25}$$

Finally we substitute Equation (13.24) into (13.21) to find

$$r = r' \int_{-L/2}^{L/2} \sin(qz)e^{i2kz \sin\theta} dz = \frac{ir'}{2} \int_{-L/2}^{L/2} e^{i(2k \sin\theta - q)z} dz - \frac{ir'}{2} \int_{-L/2}^{L/2} e^{i(2k \sin\theta + q)z} dz \tag{13.26}$$

By evaluating the first integral we get

$$r = \frac{ir'}{2} L \, \text{sinc}\left[(q - 2k \sin\theta)\frac{L}{2\pi}\right] \tag{13.27}$$

The sinc-function has its maximum value of 1 when its argument is zero, i.e. when $q = 2k \sin\theta$ or

$$\frac{2\pi}{d} = 2\frac{2\pi}{\lambda} \sin\theta \tag{13.28}$$

which gives

$$\sin\theta = \frac{\lambda}{2d} \tag{13.29}$$

which is the same result as found in Section 6.6, Equation (6.29). This is the Bragg condition which also can be stated as a vector relation

$$\mathbf{k}_r = \mathbf{k} + \mathbf{q} \tag{13.30}$$

where $\mathbf{q} = (0, 0, q)$, $\mathbf{k} = (k \cos\theta, 0, -k \sin\theta)$ and $\mathbf{k}_r = (k \cos\theta, 0, k \sin\theta)$ are the wave vectors of the refractive index 'wave', the incident light wave, and the reflected light wave respectively.

The function in Equation (13.27) drops sharply and reaches its first zero when

$$(q - 2k \sin\theta)\frac{L}{2} = \pi \tag{13.31}$$

i.e. when

$$\sin\theta = \frac{\lambda}{2d} - \frac{\lambda}{2L} = \sin\theta_B - \frac{\lambda}{2L} \tag{13.32}$$

Since L is typically much greater than λ, this is a very small angular width. This is also the reason why we evaluated only the first integral of Equation (13.26). The second integral simply describes the symmetric situation when the plane wave is incident at an angle $-\theta$.

Our prime goal was to investigate what happens when a wave is incident from the left into an FBG. We get the answer by putting $\theta = 90°$ into Equation (13.29), giving $\lambda_B = 2d$, where λ_B is the Bragg wavelength inside the fibre. From Equation (13.32) we find the wavelength of the first minimum to be

$$\lambda_1 = \frac{2dL}{L-d} \tag{13.33}$$

and the spectral width

$$\Delta\lambda = 2(\lambda_1 - \lambda_B) = 2\left(\frac{2dL}{L-d} - 2d\right) \approx \frac{4d^2}{L} \tag{13.34}$$

Therefore when a wave $u_i(\lambda_i)$ is incident from the left into a fibre as in Figure 13.14, light with a narrow bandwidth is reflected from the refractive index variations, maximum reflectance occurring at the Bragg wavelength λ_B. The unreflected light is transmitted, and a typical intensity profile for the transmitted light is shown in Figure 13.15. (The intensity profile of the reflected light is found by turning the figure upside

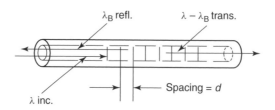

Figure 13.14   Reflection and transmission from an FBG

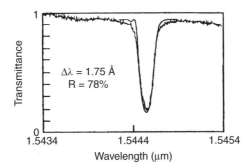

Figure 13.15 Transmission spectrum for a moderate reflectivity fibre grating. The grating length is about 10 mm as determined from measurements of the approximately Gaussian UV beam width. From Hill, K.O., and Meltz, G. (1997) Fiber Bragg grating technology fundamentals and overview, *J. Lightwave Technol.*, **15**, 1263–76. © IEEE. Reproduced by permission of The Institute of Electrical and Electronics Engineers, Inc

down!) Experiments have shown that the intensity profile is more like a Gaussian than a sinc-function. However, using $L = 10$ mm and $d = \lambda_B/2n = 1.5446/(2 \cdot 1.46) = 0.53$ μm, Equation (13.34) gives $\Delta\lambda = 1.12$ Å, which compares quite well with the experimental value of 1.75 Å.

FBGs have a lot of potential applications. Although research has concentrated on the development of Bragg grating-based fibre devices for use in fibre optic communication or fibre optic sensor systems, there are other potential applications in lidars, optical switching, optical signal processing and optical storage. A particularly interesting application is the Bragg grating dispersion compensator. As mentioned in Section 13.3, a light pulse sent through a fibre will broaden due to different velocities for the different wavelengths. By making an FBG with a variable (chirped) grating period, it should be possible to have the longer wavelength light reflected near the front of the grating while the shorter wavelength light is reflected near the back. Thus, the short wavelengths are delayed relative to the longer wavelengths. With a proper variation of the grating period, all wavelengths in the light pulse should exit the reflecting fibre at the same time and dispersion is equalized!

However, our main concern is the application of FBGs as sensors. The key feature of these sensors is that any change in fibre properties, such as strain, temperature or polarization which varies the modal index or grating pitch, will change the Bragg wavelength. A very important advantage of an FBG sensor is that it is wavelength-encoded. Shifts in the spectrum are independent of the light intensity and a unique property of each grating. With careful selection of the Bragg wavelengths, FBG sensors can be coupled in tandem without affecting the measurand of each other.

The sensitivity is governed by the fibre elastic, elasto-optic and thermo-optic properties and the nature of the load or strain applied to the structure that the fibre is attached to. The sensitivity to a particular measurand is the same as for other intrinsic sensors, such as a fibre interferometer. The shift $\Delta\lambda_B$ of the Bragg wavelength due to strain is given by

$$\Delta\lambda_B = \lambda_B \left\{ \varepsilon_l - \frac{n^2}{2}[p_{11}\varepsilon_t + p_{12}(\varepsilon_l + \varepsilon_t)] \right\} \tag{13.35}$$

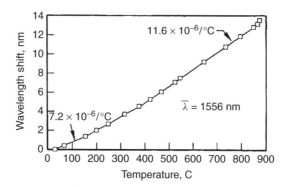

Figure 13.16 Bragg grating thermal sensitivity at elevated temperature. From Hill, K.O., and Meltz, G. (1997) Fiber Bragg grating technology fundamentals and overview, *J. Lightwave Technol.*, **15**, 1263–76. © IEEE. Reproduced by permission of The Institute of Electrical and Electronics Engineers, Inc

where the principal strains ε_l and ε_t are along and transverse to the fibre axis respectively. If the strain is homogeneous and isotropic, this simplifies to its more common form

$$\Delta\lambda_B = \lambda_B[1 - p_e]\varepsilon \approx 0.78\varepsilon \tag{13.36}$$

where the photoelectric contributions are collected into p_e defined by

$$p_e = \frac{n^2}{2}[p_{12} - \mu(p_{11} + p_{12})] \tag{13.37}$$

p_{11} and p_{12} are called the fibre Pockel's coefficients and μ is the Poisson ratio. Typical values for the sensitivity to axial strains are 1 nm/millistrain at 1300 and 0.64 nm/millistrain at 820 nm.

The temperature sensitivity is mainly due to the thermo-optic effect. Figure 13.16 shows $\Delta\lambda_B$ as a function of temperature at $\lambda_B = 1556$ nm. As can be seen, the response is almost linear. FBG sensors can also be used to measure acoustic signals. However, the sensitivity is quite low (of the order of 10^{-10} Pa^{-1}) because the glass fibre is very stiff. The sensitivity is also low for measuring electric and magnetic fields.

PROBLEMS

13.1 Consider an optical fibre designed to guide monochromatic light of $\lambda = 400$ nm. The diameter of the core is 10 μm. The refractive indices of the core and the cladding are $n_1 = 1.5$ and $n_2 = 1.35$. Calculate

(a) the allowable angles of the incident beams;

(b) the largest allowable angle of the incident beam;

(c) the actual acceptance angle of the fibre;

(d) the numerical aperture of the fibre.

13.2 The refractive index of a single-mode fibre core and its cladding are $n_1 = 1.5$ and $n_2 = 1.47$, and the wavelength of the light is $\lambda = 600$ nm. Calculate the core diameter required.

13.3 (a) Find the amount of pulse spreading in pure silica for an LED operating at 820 nm and having a 20 nm spectral width. The path is 10 km long and the material dispersion $M = 110$ ps/(nm km). (1 ps $= 10^{-12}$ s)

(b) Repeat the problem when $\lambda = 1.5$ μm and $\Delta\lambda = 50$ nm. $M = -15$ ps/(nm km).

(c) Repeat (a) and (b) when the source is a laser diode with a 1 nm spectral width.

13.4 Due to the pulse spread $\Delta\tau$, the wavelength components of a sinusoidally modulated beam will have different transit times. When the delay between the fastest and slowest wavelength is equal to half the period T of the modulated signal, the two components are in antiphase and will cancel each other. This sets a limit on the allowable pulse spread given by

$$\Delta\tau \leq \frac{T}{2}$$

which gives for the modulation frequency

$$f = \frac{1}{T} \leq \frac{1}{2\Delta\tau}$$

Calculate the frequency limit for a 10 km fibre link for the examples given in Problem 13.3.

13.5 The number of modes in a step-index fibre is given by

$$N = 2\left(\frac{\pi a}{\lambda}\right)^2 (n_1^2 - n_2^2)$$

where a is the core radius and λ is the free-space wavelength.

(a) Compute the number of modes for a fibre whose core diameter is 50 μm. Assume that $n_1 = 1.48$ and $n_2 = 1.46$. Let $\lambda = 820$ nm.

(b) Consider a SI fibre with $n_1 = 1.5$ and $n_2 = 1.485$ at 820 nm. If the core radius is 50 μm, how many modes can propagate.

(c) Repeat problem (b) if the wavelength is changed to 1.2 μm.

(d) What is the maximum core radius allowed for a glass fibre having $n_1 = 1.465$ and $n_2 = 1.46$ if the waveguide is to support only one mode at a wavelength of 1250 nm?

13.6 Consider the non-contact fibre-optic sensor illustrated in Figure 13.11. Assume that the intensity received by fibre B is proportional to $1/l^4$. Given that the area of overlap A between two circles of radius r and centre separation a is

$$A = 2a\sqrt{r^2 - \left(\frac{a}{2}\right)^2}$$

(a) Show that the received intensity I_B is indeed like that sketched in Figure 13.11(c). Let the distance at which the cones from fibre A and B start to overlap be l_0.

(b) Find the distance l_m where I_B is maximum in terms of l_0.

Appendix A

Complex Numbers

The extension of the notion of numbers from real to complex numbers consists of changing their representation from points on a line to points in a plane.

In Figure A.1, the complex number z is represented by a point with coordinates x and y along the real and imaginary coordinate axes. The unit along the imaginary axis is equal to $i = \sqrt{-1}$ such that z can be written as

$$z = x + iy \qquad (A.1)$$

where

$$x = \text{Re}\{z\} = \text{the real part of } z$$

$$y = \text{Im}\{z\} = \text{the imaginary part of } z$$

In polar coordinates (see Figure A.1), Equation (A.1) becomes

$$z = r(\cos\phi + i\sin\phi) \qquad (A.2)$$

where $r = |z| = $ the absolute value (modulus) or the length of z. We see that

$$r = \sqrt{x^2 + y^2} \qquad (A.3)$$

$$\tan\phi = y/x \qquad (A.4)$$

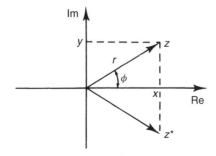

Figure A.1

By means of Euler's formula

$$e^{i\phi} = \cos\phi + i\sin\phi \tag{A.5}$$

z can be written as

$$z = re^{i\phi} \tag{A.6}$$

Those who are not familiar with the Euler formula can convince themselves by taking the differential

$$dz = r(-\sin\phi + i\cos\phi)\,d\phi$$
$$= ir(\cos\phi + i\sin\phi)\,d\phi = iz\,d\phi \tag{A.6a}$$

and then integrate

$$\int dz/z = i\int d\phi \tag{A.6b}$$

and the Euler formula is obtained.

The complex conjugate z^* of z is found by changing i into $-$i everywhere it occurs:

$$z^* = x - iy = r(\cos\phi - i\sin\phi) = re^{-i\phi} \tag{A.7}$$

The geometrical interpretation of complex conjugation is to find the mirror image of z about the real axis (see Figure A.1).

Addition (subtraction) of two complex numbers is done by simply adding (subtracting) the real and imaginary parts separately, viz.

$$z_1 \pm z_2 = (x_1 + iy_1) \pm (x_2 + iy_2) = (x_1 \pm x_2) + i(y_1 \pm y_2) \tag{A.8}$$

Multiplication and division are most easily performed using polar notation

$$z_1 z_2 = r_1 r_2 e^{i(\phi_1 + \phi_2)} \tag{A.9}$$

$$z_1 : z_2 = \frac{r_1}{r_2} e^{i(\phi_1 - \phi_2)} \tag{A.10}$$

From this, we find that

$$|z| = \sqrt{zz^*} = r \tag{A.11}$$

$$|z|^2 = zz^* = r^2 \tag{A.12}$$

For

$$z = z_1 + z_2 \tag{A1.13}$$

we get

$$|z|^2 = |z_1 + z_2|^2 = (z_1 + z_2)(z_1^* + z_2^*)$$
$$= |z_1|^2 + |z_2|^2 + z_1 z_2^* + z_1^* z_2$$
$$= r_1^2 + r_2^2 + 2r_1 r_2 \cos(\phi_1 - \phi_2) \tag{A.14}$$

Appendix B

Fourier Optics

B.1 THE FOURIER TRANSFORM

The two-dimensional Fourier transform of the function $g(x, y)$ is defined as

$$G(f_x, f_y) = \mathscr{F}\{g(x, y)\} = \int_{-\infty}^{\infty}\!\!\int g(x, y)e^{-i2\pi(f_x x + f_y y)}\, dx\, dy \qquad \text{(B.1)}$$

where from the inverse transform is given by

$$g(x, y) = \mathscr{F}^{-1}\{G(f_x, f_y)\} = \int_{-\infty}^{\infty}\!\!\int G(f_x, f_y)e^{i2\pi(f_x x + f_y y)}\, df_x\, df_y \qquad \text{(B.2)}$$

The following theorems can be proved by inserting into Equation (B.1).
Assume that

$$\mathscr{F}\{g(x, y)\} = G(f_x, f_y), \quad \text{and} \quad \mathscr{F}\{h(x, y)\} = H(f_x, f_y)$$

Then

1. *Linearity theorem*

$$\mathscr{F}\{\alpha g(x, y) + \beta h(x, y)\} = \alpha\mathscr{F}\{g(x, y)\} + \beta\mathscr{F}\{h(x, y)\} \qquad \text{(B.2a)}$$

2. *Similarity theorem*

$$\mathscr{F}\{g(ax, by)\} = \frac{1}{|ab|}G\left(\frac{f_x}{a}, \frac{f_y}{b}\right) \qquad \text{(B.2b)}$$

3. *Shift theorem*

$$\mathscr{F}\{g(x - a, y - b)\} = G(f_x, f_y)e^{-i2\pi(f_x a + f_y b)} \qquad \text{(B.2c)}$$

4. *Parseval's theorem*

$$\int_{-\infty}^{\infty}\!\!\int |g(x, y)|^2\, dx\, dy = \int_{-\infty}^{\infty}\!\!\int |G(f_x, f_y)|^2\, df_x\, df_y \qquad \text{(B.2d)}$$

5. *Convolution theorem*

$$\mathcal{F}\{g(x, y) \otimes h(x, y)\} = \mathcal{F}\left\{\int_{-\infty}^{\infty}\int g(\xi, \eta)h(x - \xi, y - \eta)\,d\xi\,d\eta\right\}$$
$$= G(f_x, f_y)H(f_x, f_y) \tag{B.2e}$$

6. *Autocorrelation theorem*

$$\mathcal{F}\{g(x, y) \odot g(x, y)\} = \mathcal{F}\left\{\int_{-\infty}^{\infty}\int g(\xi, \eta)g^*(\xi - x, \eta - y)\,d\xi\,d\eta\right\} = |G(f_x, f_y)|^2 \tag{B.2f}$$

7. *Fourier integral theorem*

$$\mathcal{F}\mathcal{F}^{-1}\{g(x, y)\} = \mathcal{F}^{-1}\mathcal{F}\{g(x, y)\} = g(x, y) \tag{B.2g}$$

A function and its transform can also be represented in polar coordinates such that $\mathcal{F}\{g(r, \theta)\} = G(\rho, \phi)$ where

$$x = r\cos\theta \quad f_x = \rho\cos\phi$$
$$y = r\sin\theta \quad f_y = \rho\sin\phi \tag{B.3}$$

An important class of functions are those which possess circular symmetry, that means they are independent of θ, such that

$$g(r, \theta) = g_R(r) \tag{B.4}$$

By substituting Equation (B.3) and (B.4) into (B.1) we get

$$G(\rho, \phi) = \int_0^{2\pi} d\theta \int_0^{\infty} r g_R(r)e^{-i2\pi r\rho(\cos\theta\cos\phi+\sin\theta\sin\phi)}\,dr \tag{B.5}$$
$$= \int_0^{\infty} r g_R(r)\,dr \int_0^{2\pi} e^{-2i\pi r\rho\cos(\theta-\phi)}\,d\theta$$

By applying the identity

$$J_0(a) = (1/2\pi) \int_0^{2\pi} e^{-ia\cos(\theta-\phi)}\,d\theta \tag{B.6}$$

where J_0 is the zeroth-order Bessel function, Equation (B.5) becomes

$$G(\rho) = 2\pi \int_0^{\infty} r g_R(r) J_0(2\pi r\rho)\,dr \tag{B.7}$$

Thus the Fourier transform of a circularly symmetric function is itself circularly symmetric. Equation (B.7) is referred to as the zeroth-order Hankel transform or alternatively as the Fourier-Bessel transform. Its inverse is given as

$$g_R(r) = 2\pi \int_0^\infty \rho G(\rho) J_0(2\pi r\rho) \, d\rho \tag{B.8}$$

Thus, for circularly symmetric functions there is no difference between the transform and inverse transform operations.

B.2 SOME FUNCTIONS AND THEIR TRANSFORMS

Below we give the definition of some frequently used functions. They are illustrated in Figure B.1 and some transform pairs are listed in Table B.1.

The delta function

$$\delta(x, y) = \begin{cases} \infty & \text{for } x = y = 0 \\ 0 & \text{otherwise} \end{cases} \tag{B.9}$$

$\delta(x, y)$ is a discontinuous function used for representing a point source or a focal point. Its integrated area is equal to 1 and it is usually represented by an arrow of height equal to 1. Mathematically, it can be defined as a limiting value. We might also take its transform properties as a definition:

$$\mathscr{F}\{\delta(x, y)\} = 1 \tag{B.10}$$

$$\mathscr{F}\{1\} = \delta(f_x, f_y) \tag{B.11}$$

$\delta(x, y)$ has the following properties:

$$\delta(ax, by) = \frac{1}{|ab|}\delta(x, y) \tag{B.12a}$$

Table B.1   Transform pairs

$g(x, y)$	$\mathscr{F}\{g(x, y)\}$
$\delta(x, y)$	1
$\cos(2\pi f_0 x)$	$\frac{1}{2}\delta(f_x - f_0) + \frac{1}{2}\delta(f_x + f_0)$
$\mathrm{rect}(x)\,\mathrm{rect}(y)$	$\mathrm{sinc}(f_x)\,\mathrm{sinc}(f_y)$
$\Lambda(x)\Lambda(y)$	$\mathrm{sinc}^2(f_x)\,\mathrm{sinc}^2(f_y)$
$\mathrm{comb}(x)\mathrm{comb}(y)$	$\mathrm{comb}(f_x)\mathrm{comb}(f_y)$
$\mathrm{circ}(r)$	$\dfrac{J_1(2\pi\rho)}{\rho}$
$\cos(2\pi f_0 r)$	$-\dfrac{1}{2\pi} f_0 (f_0^2 - \rho^2)^{3/2} \mathrm{circ}\left(\dfrac{\rho}{f_0}\right)$

$$\int_{-\infty}^{\infty} \int \delta(x, y)\, dx\, dy = 1 \tag{B.12b}$$

$$\int \int_{-\infty}^{\infty} g(\xi, \eta)\delta(x - \xi, y - \eta)\, d\xi\, d\eta = g(x, y) \tag{B.12c}$$

Equation (B.12c) is called the shifting property, which is very useful.

Rectangle function

$$\text{rect}(x) = \begin{cases} 1 & \text{for } |x| \le 1/2 \\ 0 & \text{otherwise} \end{cases} \tag{B.13}$$

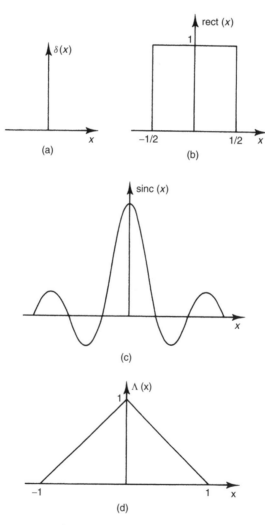

Figure B.1   (g) from J. W. Goodman, Introduction to Fourier Optics. © 1968. Reproduced by permission of McGraw-Hill Book Company.)

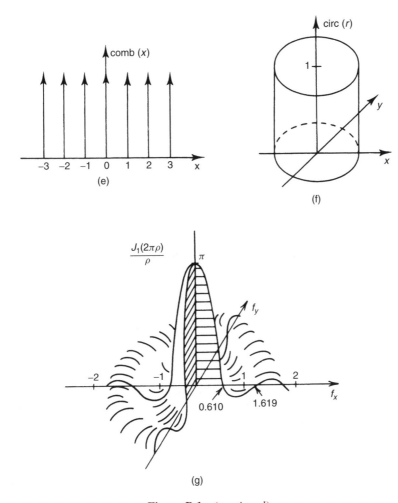

Figure B.1   (*continued*)

Sinc function

$$\text{sinc}(x) = \frac{\sin \pi x}{\pi x} \tag{B.14}$$

Triangle function

$$\Lambda(x) = \begin{cases} 1 - |x| & \text{for } |x| \le 1 \\ 0 & \text{otherwise} \end{cases} \tag{B.15}$$

Comb function

$$\text{comb}(ax) = \frac{1}{a} \sum_{n=-\infty}^{\infty} \delta\left(x - \frac{n}{a}\right) \tag{B.16}$$

Circle function

$$\text{circ}(r) = \begin{cases} 1 & \text{for } r \leq 1 \\ 0 & \text{otherwise} \end{cases} \tag{B.17}$$

The transform of $\text{circ}(r)$ (see Figure (B.1f)) is found by using Equation (B.7) and the identity

$$\int_0^x \xi J_0(\xi) d\xi = x J_1(x) \tag{B.18}$$

where J_1 is the first-order Bessel function. The other transform pairs are easily proved.

B.3 SOME IMPLICATIONS

In Sections 4.4, B.1 and B.2 we have established a powerful tool for the calculation of diffraction patterns and the analysis and synthesis of optical systems. Below, some important implications of the matter given in these sections are discussed. We omit the factor $K/i\lambda z$ in Equation (4.17) and regard the transfer from the diffraction plane to the observation plane as a pure Fourier transform. This we can do, since in calculating the intensity, $1/\lambda^2 z^2$ becomes an unimportant scale factor.

The linearity theorem

Assume that a diffracting screen is given by the transmittance function $g(x, y)$. Another screen, given by the transmittance function $h(x, y) = 1 - g(x, y)$ will be open at places where the first is non-transmitting and vice versa. Two such screens (objects) are called complementary. From the linearity theorem, we find that $\mathscr{F}\{h(x, y)\} = \delta(f_x, f_y) - \mathscr{F}\{g(x, y)\}$. This means that apart from a delta function (a light spot) at the origin, the diffraction patterns from two complementary objects will be equal. This fact is referred to as the Babinet principle and may be useful in cases where it is easier to calculate the diffraction pattern from the complementary object instead of the actual object.

The shift theorem

Consider an object with the transmittance function $g(x, y)$. The function $g(x - a, y - b)$ will represent the same object translated a distance a in the x-direction and a distance b in the y-direction. From the shift theorem we see that apart from a phase factor (which becomes equal to 1 by calculating the intensity), the diffraction pattern is unaffected by a translation in the xy-plane. This fact can be taken advantage of when measuring the dimensions of moving objects by making measurements on the diffraction pattern.

The convolution theorem

The convolution theorem tells that

$$\mathscr{F}\{g(x, y) \otimes h(x, y)\} = G(f_x f_y) H(f_x, f_y) \tag{B.19}$$

where we have introduced the shorthand notation for the convolution integral:

$$g(x, y) \otimes h(x, y) = \int_{-\infty}^{\infty} \int g(\xi, \eta) h(x - \xi, y - \eta) \, d\xi \, d\eta \qquad \text{(B.20)}$$

This implies that

$$\mathscr{F}\{g(x, y) h(x, y)\} = G(f_x, f_y) \otimes H(f_x, f_y) \qquad \text{(B.21)}$$

Geometrically, $g \otimes h$ is given as the area of overlap between g and h as a function of the position of g as g is translated from $-\infty$ to $+\infty$. Thus we have, for example, $\text{rect}(x) \otimes \text{rect}(x) = \Lambda(x)$. The convolution of a function with the δ-function becomes especially simple since $g(x, y) \otimes \delta(x, y) = g(x, y)$ (cf. Equation (B.12c)), and therefore

$$g(x, y) \otimes \delta(x - a, y - b) = g(x - a, y - b) \qquad \text{(B.22)}$$

The convolution theorem applied to Equation (B.22) gives

$$\begin{aligned}
\mathscr{F}\{g(x - a, y - b)\} &= \mathscr{F}\{g(x, y)\}\mathscr{F}\{\delta(x - a, y - b)\} \\
&= G(f_x, f_y) e^{-i2\pi(f_a a + f_y b)} \qquad \text{(B.23)}
\end{aligned}$$

which is another proof of the fact that the diffraction pattern from an object is unaffected (apart from a phase factor) by translating it in the object plane.

Appendix C

Fourier Series

A periodic function $g(x)$ with period L and fundamental frequency $f_0 = 1/L$ can be described as a sum of sines and cosines of frequencies kf_0, where $k = 1, 2, 3, \ldots$. There are three forms of such Fourier series.

(1) The trigonometric form:

$$g(x) = a_0 + \sum_{k=1}^{\infty} a_k \cos(k2\pi f_0 x) + b_k \sin(k2\pi f_0 x) \tag{C.1}$$

(2) The polar form:

$$g(x) = c_0 + \sum_{k=1}^{\infty} c_k \cos(k2\pi f_0 x + \theta_k) \tag{C.2}$$

(3) The exponential form:

$$g(x) = \sum_{k=-\infty}^{\infty} G_k e^{ik2\pi f_0 x} \tag{C.3}$$

The relations between the trigonometric and exponential forms are

$$G_0 = a_0, \quad G_k = \tfrac{1}{2}(a_k - ib_k), \quad G_{-k} = \tfrac{1}{2}(a_k + b_k) \tag{C.4}$$

By means of the orthogonal property of harmonic functions

$$\int_L \cos(m2\pi f_0 x) \sin(n2\pi f_0 x) \, dx = 0 \quad \text{(integer } m, n) \tag{C.5}$$

$$\int_L \exp(m2\pi f_0 x) \exp(-n2\pi f_0 x) \, dx = 0 \quad (m \neq n) \tag{C.6}$$

(where $\int_L$ means integration over one period), it is possible to show that the coefficients a_k, b_k and G_k are given by

$$a_k = \frac{2}{L} \int_L g(x) \cos(k2\pi f_0 x) \, dx \tag{C.7}$$

$$b_k = \frac{2}{L} \int_L g(x) \sin(k2\pi f_0 x)\, dx \tag{C.8}$$

$$G_k = \frac{1}{L} \int_L g(x) e^{-k2\pi f_0 x}\, dx \tag{C.9}$$

The following symmetry relations exist for the signal $g(x)$ and the coefficients:

(1) $g(x)$ is even: $b_k = 0$ and $G_k = 2a_k$ is purely real.

(2) $g(x)$ is odd: $a_k = 0$ and $G_k = -i2b_k$ is purely imaginary.

(3) $g(x) = -g(x \pm L/2)$ (halfwave symmetry): k is odd, i.e. $a_k = b_k = G_k = 0$ for k even.

The Fourier series provides a link between space and frequency through

$$g(x) = \sum_{k=-\infty}^{\infty} G_k \exp(ik2\pi f_0 x), \quad G_k = \frac{1}{L} \int_{-L/2}^{L/2} g(x) \exp(-ik2\pi f_0 x)\, dx \tag{C.10}$$

This relation may be displayed symbolically by the transform pair

$$g(x) \Leftrightarrow G_k$$

The following operational properties are easily proven:

(1) Spatial shift: $g(x \pm a) \Leftrightarrow G_k \exp(\pm ik2\pi f_0 a)$ \hfill (C.11)

(2) Scaling: $g(\alpha x) = \sum_{k=-\infty}^{\infty} G_k \exp(ik2\pi f_0 \alpha x)$ \hfill (C.12)

(3) Derivatives: $g'(x) \Leftrightarrow ik2\pi f_0 G_k$ \hfill (C.13)

(4) Integration: $\int_0^x g(x)\, dx \Leftrightarrow \dfrac{G_k}{ik2\pi f_0}$ (G_0 must be zero) \hfill (C.14)

Since we are familiar with the Fourier transform, consider a periodic function $g(x)$ of period L and fundamental frequency $f_0 = 1/L$. We form the function

$$g_0(x) = g(x) \operatorname{rect}\left(\frac{x}{L}\right) \tag{C.15}$$

i.e. the one-period version of $g(x)$ with the Fourier transform

$$G_0(f_x) = \mathscr{F}\{g_0(x)\} = \int_{-\infty}^{\infty} g_0(x) e^{-i2\pi f_x x}\, dx = \int_{-L/2}^{L/2} g(x) e^{-i2\pi f_x x}\, dx \tag{C.16}$$

From the definition of the coefficients of the Fourier series we have

$$G_k = \frac{1}{L} \int_{-L/2}^{L/2} g(x) e^{-i2k\pi f_0 x}\, dx = \frac{1}{L} \int_{-\infty}^{\infty} g(x) \operatorname{rect}\left(\frac{x}{L}\right) e^{-ik2\pi f_0 x}\, dx$$

$$= \frac{1}{L} \int_{-\infty}^{\infty} g_0(x) e^{-i2\pi (kf_0)x}\, dx = \frac{G_0(kf_0)}{L} \tag{C.17}$$

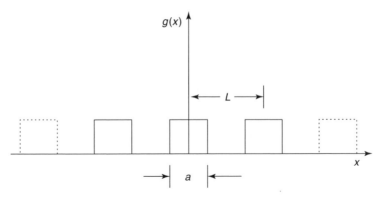

Figure C.1

which shows that the exponential coefficients G_k are equal to $G_0(kf_0)/L$ where G_0 is the Fourier transform of the one-period version of $g(x)$.

As an example, consider the Ronchi grating, Figure C.1, where $L = 2a$, $f_0 = 1/L$. We get

$$G_0(f_x) = \mathscr{F}\left\{\text{rect}\left(\frac{x}{a}\right)\right\} = a\,\text{sinc}(af_x) \tag{C.18}$$

$$G_k = \frac{G_0(kf_0)}{L} = \frac{a}{L}\,\text{sinc}(akf_0) = \frac{1}{2}\text{sinc}\left(\frac{k}{2}\right) \tag{C.19}$$

which gives

$$g(x) = \sum_{k=-\infty}^{\infty} G_k e^{ik2\pi f_0 x} = \frac{1}{2}\sum_{k=-\infty}^{\infty} \frac{\sin(k\pi/2)}{k\pi/2}e^{ik2\pi x/L} \tag{C.20}$$

or

$$g(x) = \frac{1}{2} + \sum_{k=1}^{\infty} \frac{\sin(k\pi/2)}{k\pi/2}\cos(k2\pi x/L) \quad \text{for } k = \text{odd} \tag{C.21}$$

This, we know, can also be written as

$$g(x) = \text{rect}\left(\frac{x}{a}\right) \otimes \text{comb}\left(\frac{x}{L}\right) = \text{rect}\left(\frac{2x}{L}\right) \otimes \text{comb}\left(\frac{x}{L}\right) \tag{C.22}$$

Appendix D

The Least-Squares Error Method

Consider the equation

$$f = a_0 g^{(0)} + a_1 g^{(1)} + a_2 g^{(2)} + \cdots + a_n g^{(n)} \tag{D.1}$$

where $g^{(i)}$ may be functions of, say, x and a_i are constant coefficients. ($g^{(i)}$ may be functions of $x_1, x_2, \ldots, x_{k-1}$ where $(x_1, \ldots, x_{k-1}, f)$ forms a k-tuple in k-dimensional space). Assume that a_i are unknown coefficients which can be determined by measuring coordinate pairs (x_j, f_j). To solve for the $n + 1$ unknown coefficients, we require at least $n + 1$ such observations. Such a system is said to be exactly determined. However, the solution of these exact systems is often ill-conditioned (numerically unstable) and it is usually good practice to overdetermine the system by specifying more control points than needed (and hence generate more simultaneous equations). However, an over-determined system does not have an exact solution and there are going to be some errors for some points. The idea, then, is to minimize these errors. A common approach is to minimize the sum of the square of each error, the so-called least-squares solution.

By making $m(m > n)$ observations, we get m equations which we may write in matrix form as

$$\begin{pmatrix} f_1 \\ f_2 \\ \vdots \\ f_m \end{pmatrix} = \begin{pmatrix} g_1^{(0)} & g_1^{(1)} & g_1^{(2)} & \cdots & g_1^{(n)} \\ g_2^{(0)} & g_2^{(1)} & g_2^{(2)} & \cdots & g_2^{(n)} \\ \vdots & \vdots & \vdots & & \vdots \\ g_m^{(0)} & g_m^{(1)} & g_m^{(2)} & \cdots & g_m^{(n)} \end{pmatrix} \begin{pmatrix} a_0 \\ a_1 \\ \vdots \\ a_n \end{pmatrix} + \begin{pmatrix} e_1 \\ e_2 \\ \vdots \\ e_m \end{pmatrix} \tag{D.2}$$

Where, since we not get an exact solution, we have included an error matrix e. Let us abbreviate this matrix equation to

$$f = Xa + e \tag{D.3}$$

To solve for a, we might think of multiplying Equation (D.3) by X^{-1}. Unfortunately, X is a non-square matrix and therefore cannot be inverted. We have

$$e = f - Xa \tag{D.4}$$

We form the sum of the square of each error by computing e^Te

$$e^Te = (f - Xa)^T(f - Xa) \tag{D.5}$$
$$= f^Tf - f^TXa - a^TX^Tf + a^TX^TXa$$

(For two matrices A and B we have $(AB)^T = B^TA^T$).

By taking the gradient of (e^Te) with respect to a we find how the errors change as the coefficients change

$$\partial(e^Te)/\partial(a) = 0 - 2X^Tf + 2X^TXa \tag{D.6}$$

where we have used the fact that

$$\frac{\partial}{\partial a}(f^TXa) = \frac{\partial}{\partial a}(a^TX^Tf) = X^Tf \tag{D.7a}$$

and

$$\frac{\partial}{\partial a}(a^TX^TXa) = 2X^TXa \tag{D.7b}$$

(Note that (e^Te) is a scalar and that the transpose of a scalar is equal to itself)

The sum of the square of each error is minimized when setting the gradient equal to zero, thus

$$2X^TXa - 2X^Tf = 0 \tag{D.8}$$
$$a = (X^TX)^{-1}X^Tf$$

$(X^TX)^{-1}X^T$ is commonly referred to as the pseudo-inverse of X.

As an example, consider the case where a quadratic curve

$$f(x) = a_0 + a_1x + a_2x^2 \tag{D.9}$$

is fitted to N observation points (x_j, f_j) where $j = 1, 2, \ldots, N$. We then have

$$X = \begin{bmatrix} 1 & x_1 & x_1^2 \\ 1 & x_2 & x_2^2 \\ \vdots & \vdots & \vdots \\ 1 & x_N & x_N^2 \end{bmatrix} \quad X^T = \begin{bmatrix} 1 & 1 & \cdots & 1 \\ x_1 & x_2 & \cdots & x_N \\ x_1^2 & x_2^2 & \cdots & x_N^2 \end{bmatrix} \tag{D.10}$$

$$X^TX = \begin{bmatrix} N & s_1 & s_2 \\ s_1 & s_2 & s_3 \\ s_2 & s_3 & s_4 \end{bmatrix} \tag{D.11}$$

where

$$s_1 = \sum x_j \quad s_2 = \sum x_j^2 \quad s_3 = \sum x_j^3 \quad s_4 = \sum x_j^4 \tag{D.12}$$

$$(X^T X)^{-1} = \frac{1}{\det (X^T X)} \begin{bmatrix} (s_2 s_4 - s_3^2) & -(s_1 s_4 - s_3 s_2) & (s_1 s_3 - s_2^2) \\ -(s_1 s_4 - s_2 s_3) & (N s_4 - s_2^2) & -(N s_3 - s_1 s_2) \\ (s_1 s_3 - s_2^2) & -(N s_3 - s_2 s_1) & (N s_2 - s_1^2) \end{bmatrix} \quad (D.13)$$

$$X^T f = \begin{bmatrix} \sum f_j \\ \sum x_j f_j \\ \sum x_j^2 f_j \end{bmatrix} \quad (D.14)$$

This is a general result.

Now assume that $f(x)$ is the intensity distribution along a row of pixels and that we want to determine the position x_p of its maximum. By differentiating Equation (D.9) we find this position to be

$$x_p = -\frac{a_1}{2a_2} \quad (D.15)$$

For simplicity, let's assume that we fit the distribution to three points (this is really not a good example since it gives an exactly determined system, but it is space-saving!) and that we have translated our coordinate system such that $x_1 = -1$, $x_2 = 0$ and $x_3 = 1$. Then we get $s_1 = 0$, $s_2 = 2$, $s_3 = 0$ and $s_4 = 2$, and

$$(X^T X)^{-1} = \frac{1}{\det (X^T X)} \begin{bmatrix} 4 & 0 & -4 \\ 0 & 2 & 0 \\ -4 & 0 & 6 \end{bmatrix} \quad (D.16)$$

$$X^T f = \begin{bmatrix} f_1 + f_2 + f_3 \\ -f_1 + f_3 \\ f_1 + f_3 \end{bmatrix} \quad (D.17)$$

This gives

$$\begin{bmatrix} a_0 \\ a_1 \\ a_2 \end{bmatrix} = \frac{1}{\det (X^T X)} \begin{bmatrix} 4 f_2 \\ 2(-f_1 + f_3) \\ 2 f_1 - 4 f_2 + 2 f_3 \end{bmatrix} \quad (D.18)$$

and

$$x_p = \frac{f_1 - f_3}{2 f_1 - 4 f_2 + 2 f_3} \quad (D.19)$$

The procedure will then be to first determine the intensity maximum to the nearest pixel x_2 and find the intensities of the first pixels to the left and right (x_1 and x_3) and then find the position of the maximum with subpixel accuracy according to Equation (D.19).

Appendix E

Semiconductor Devices

Solids are classified as insulators, semiconductors and metals. Every solid has its own characteristic energy band structure. Semiconductor materials at 0 K have basically the same structure as insulators – a filled valence band separated from an empty conduction band by a band gap containing no allowed energy states: see Figure E.1. The difference lies in the size of the band gap E_g, which is much smaller in semiconductors than in insulators. Metals have a partially filled conduction band at all temperatures and therefore have a large conductivity.

Semiconductor materials are found in column IV and neighbouring columns in the periodic table: see Table E.1. The column IV semiconductors are called elemental semiconductors, while the combinations seen in the table make up the compound semiconductors. The two-element (binary) compounds such as GaN, GaP and GaAs are common in light-emitting diodes (LEDs). Binary compounds have a fixed bandgap and wavelength: see Table E.2. Adding a third element changes the bandgap. Three-element (ternary) compounds such as GaAsP, and four-element (quaternary) compounds such as InGaAsP, can therefore be grown to provide added flexibility in choosing material properties. Such mixtures are often labelled in the manner $Al_xGa_{1-x}As$, where the x denotes the fraction of one element, in this case aluminium.

With increasing temperature, some electrons in the valence band of a semiconductor will be excited into the empty conduction band where they act as mobile charge carriers.

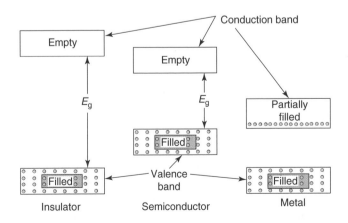

Figure E.1   Typical band structures at 0 K

Table E.1 Common semiconductor materials: (a) part of the Periodic Table where semiconductors occur; (b) elemental and compound semiconductors

(a)	II	III	IV	V	VI
		B	C	N	
		Al	Si	P	S
	Zn	Ga	Ge	As	Se
	Cd	In		Sb	Te
(b)	Elemental	IV compounds	Binary III–V compounds	Binary II–VI compounds	
	Si	SiC	AlP	ZnS	
	Ge	SiGe	AlAs	ZnSe	
			AlSb	ZnTe	
			GaN	CdS	
			GaP	CdSe	
			GaAs	CdTe	
			GaSb		
			InP		
			InAs		
			InSb		

Table E.2 Some elemental and III–V binary semiconductors and their bandgap energies E_g at $T = 300$ K and corresponding wavelengths $\lambda_g = hc/E_g$

Material	Bandgap energy E_g(eV)	Bandgap wavelength λ_g (μm)
Ge	0.66	1.88
Si	1.11	1.15
AlP	2.45	0.52
AlAs	2.16	0.57
AlSb	1.58	0.75
GaP	2.26	0.55
GaAs	1.42	0.87
GaSb	0.73	1.70
InP	1.35	0.92
InAs	0.36	3.5
InSb	0.17	7.3

In doing so, they leave behind an empty quantum state, allowing the remaining electrons to exchange places. This motion can be regarded as a motion in the opposite direction of a hole of opposite charge +e. The two charge carriers (electron and hole) are free to drift under the effect of an electric field, thereby generating an electric current.

The electrical and optical properties of semiconductors can be substantially altered by adding small controlled amounts of impurities, or dopants, which alter the concentration of mobile carriers by many orders of magnitude. Dopants with excess valence electrons (called donors) can be used to replace a small proportion of the normal atoms in the crystal and thereby create a predominance of mobile electrons; the material is then said to be an n-type semiconductor. Similarly, a p-type material can be made by using dopants with a deficiency of valence electrons, called acceptors. The result is a predominance of holes. Undoped semiconductors are referred to as intrinsic materials, whereas doped semiconductors are called extrinsic materials. The concentration of mobile electrons in an n-type semiconductor (called majority carriers) is far greater than the concentration of holes (called minority carriers). The opposite is true in p-type semiconductors, for which holes are majority carriers.

The thermal excitation of electrons from the valence into the conduction band results in the generation of electron-hole pairs (EHP). The reverse process, called electron-hole recombination, occurs when an electron decays from the conduction band to fill a hole in the valence band. The released energy may be given to an emitted photon, in which case the process is called radiative recombination.

Junctions between differently doped regions of the same semiconductor material are called homojunctions. (Junctions between different materials are called heterojunctions.) An important example is the p-n junction. When the two regions are brought into contact, the following sequence of events takes place (see Figure E.2):

(1) Electrons diffuse away from the n-region into the p-region, leaving behind positively charged ionized donor atoms. In the p-region the electrons recombine with abundant holes. Similarly, holes diffuse away from the p-region to the n-region where they recombine with mobile electrons.

(2) As a result, a narrow region on both sides of the junction becomes almost totally depleted of mobile charge carriers. This region is called the depletion layer. It contains only the fixed charges (positive ions on the n-side and negative ions on the p-side).

(3) The fixed charges create an electric field in the depletion layer which points from the n-side towards the p-side of the junction.

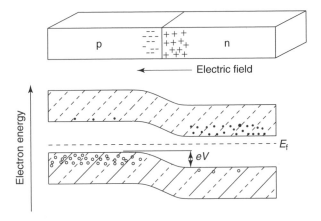

Figure E.2 A p-n junction in thermal equilibrium at $T > 0$ K. The depletion layer and energy-band diagram shown as functions of position

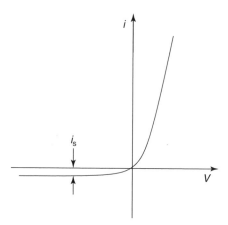

Figure E.3   Current–voltage characteristic of the ideal p-n junction diode

(4) An equilibrium condition is established that results in a built-in potential difference V_0, with the n-side having a higher potential than the p-side. No net current flows across the junction.

An externally applied potential will alter the potential difference between the p- and n-regions. If the junction is forward biased by applying a positive voltage V to the p-region, the height of the potential-energy hill is reduced by an amount eV. The excess majority carrier holes and electrons that enter the n- and p-regions respectively become minority carriers and recombine with the local majority carriers. This process is known as minority carrier injection.

If the junction is reverse biased, the height of the potential-energy hill is increased by eV which impedes the flow of majority carriers. A p-n junction therefore acts as a diode with a current–voltage characteristic as illustrated in Figure E.3.

References and Further Reading

1 BASICS

Born, M., and Wolf, E. (1999) *Principles of Optics*, 3rd edn, Pergamon Press, Oxford.

Fowles, G.R. (1989) *Introduction to Modern Optics*, 2nd edn, Dover, New York.

Hecht, E. (2001) *Optics*, 4th edn, Addison-Wesley, Reading, MA.

Hecht, E., and Zajac, A. (1974) *Optics*, Addison-Wesley, Reading, MA.

Klein, M.V., and Furtak, T.E. (1986) *Optics*, 2nd edn, Wiley, New York.

Longhurst, R.S. (1967) *Geometrical and Physical Optics*, 3rd edn, Wiley, New York.

Meyer-Arendt, J.R. (1972) *Introduction to Classical and Modern Optics*, Prentice-Hall, Englewood Cliffs, NJ.

Saleh, B.E.A., and Teich, M.C. (1991) *Fundamentals of Photonics*, Wiley, New York.

Young, H.D. (1968) *Fundamentals of Optics and Modern Physics*, McGraw-Hill, New York.

2 GAUSSIAN OPTICS

Gerrard, A., and Burch, J.M. (1994) *Introduction to Matrix Methods in Optics*, Dover, New York.

Hecht, E. (2001) *Optics*, 4th edn, Addison-Wesley, Reading, MA.

Klein, M.V., and Furtak, T.E. (1986) *Optics*, 2nd edn, Wiley, New York.

Laikin, M. (1991) *Lens Design*, Marcel Dekker, New York.

Smith, W.J. (2000) *Modern Optical Engineering*, 3rd edn, McGraw-Hill, New York.

3 INTERFERENCE

Baldwin, R.G., Gordon, G.B., and Rudè, A.F. (1971) *Hewlett-Packard Journal*, 14 December.

Born, M., and Wolf, E. (1999) *Principles of Optics*, 3rd edn, Pergamon Press, Oxford.

Cook, A.H. (1971) *Interference of Electromagnetic Waves*, Clarendon Press, Oxford.

Durst, F., Melling, A., and Whitelaw, J.H. (1981) *Principles and Practice of Laser-Doppler Anemometry*, Academic Press, London.

Gaal, P., Jani, P., and Czitrovszky, A. (1993). *Opt. Eng.*, **32**, 2574–7.

Hecht, E. (2001) *Optics*, 4th edn, Addison-Wesley, Reading, MA.

Klein, M.V., and Furtak, T.E. (1986) *Optics*, 2nd edn, Wiley, New York.

Puliafito, C.A., *et al.* (1996) *Optical Coherence Tomography of Ocular Diseases*, SLACK Inc., Thorofare.

Schmitt, J.M. (1999) OCT, a review, *IEEE J. Selected Topics in Quantum Electronics*, **5**.

Steel, W.H. (1984) *Interferometry*, 2nd edn, Cambridge University Press, Cambridge.

4 DIFFRACTION

Born, M., and Wolf, E. (1999) *Principles of Optics*, 3rd edn, Pergamon Press, Oxford.
Goodman, J.W. (1996) *Introduction to Fourier Optics*, 2nd edn, McGraw-Hill, New York.
Hecht, E. (2001) *Optics*, 4th edn, Addison-Wesley, Reading, MA.
Klein, M.V., and Furtak, T.E. (1986) *Optics*, 2nd edn, Wiley, New York.

5 LIGHT SOURCES AND DETECTORS

Barbe, D.F. (1975) *Proc. IEEE*, **63**, 38–67.
Barbe, D.F., and Campana, S.B. (1976) Aliasing and MTF effects in photosensor arrays, in Jespers, P.G., van de Wiele, F., and White, M.H. (eds) *Solid State Imaging*, Noordhoff, Leiden.
Barbe, D.F., and Campana, S.B. (1977) Imaging array using the charge-coupled concept, in Kazan, B. (ed.) *Advances in Image Pickup and Display*, Academic Press, New York.
Bjelkhagen, H.I. (1995) *Silver-Halide Recording Materials*, 2nd edn, Springer-Verlag, Berlin.
Boyle, W.S., and Smith, G.E. (1970) *Bell System Tech. J.*, **49**, 587–93.
Desmarais, L. (1998) *Applied Electro-Optics*, Prentice-Hall, New York.
Goodman, J.W. (1968) *Introduction to Fourier Optics*, McGraw-Hill, New York.
Grob, B. (1984) *Basic Television and Video Systems*, 5th edn, McGraw-Hill, New York.
Hecht, J. (1992) *The Laser Guidebook*, 2nd edn, McGraw-Hill, New York.
Javan, A., Bennett Jr., W.R., and Herriot, D.R. (1961) *Phys. Rev. Lett.*, **6**, 106–10.
Klein, M.V., and Furtak, T.E. (1986) *Optics*, 2nd edn, Wiley, New York.
Longhurst, R.S. (1967) *Geometrical and Physical Optics*, Wiley, New York.
Maiman, H. (1960) *Nature*, **187**, 493–4.
Oriel Corporation (1994) *Light Sources, Monochromators, Detection Systems*, Stratford, CT.
Saleh, B.E.A., and Teich, M.C. (1991) *Fundamentals of Photonics*, Wiley, New York.
Slater, P.N. (1980) *Remote Sensing, Optics and Optical Systems*, Addison-Wesley, Reading, MA.
Urbach, J.C., and Meier, R.W. (1966) *Appl. Opt.*, **5**, 666–7.
Wilson, J., and Hawkes, J.F.B. (1998) *Optoelectronics: An Introduction*, 3rd edn, Prentice-Hall, New York.

6 HOLOGRAPHY

Abramson, N. (1970) *Appl. Opt.*, **9**, 97–101.
Abramson, N. (1970) *Appl. Opt.*, **11**, 1143–7.
Abramson, N. (1981) *The Making and Evaluation of Holograms*, Academic Press, London.
Aleksoff, C.C. (1971) *Appl. Opt.*, **10**, 1329–41.
Aleksoff, C.C. (1974) in Erf, R. (ed.) *Holographic Nondestructive Testing*, Academic Press, New York.
Caulfield, H.J. (1979) *Handbook of Optical Holography*, Academic Press, New York.
Collier, R.J., Burckhardt, C.B., and Lin, L.H. (1971) *Optical Holography*, Academic Press, New York.
Dandliker, R., and Thalmann, R. (1985) *Opt. Eng.*, **24**, 824–31.
Erf, R. (ed.) (1974) *Holographic Nondestructive Testing*, Academic Press, New York.
Gabor, D. (1949) *Proc. Roy. Soc. Lond.*, **A197**, 454–87.
Goodman, J.W. (1996) *Introduction to Fourier Optics*, 2nd edn, McGraw-Hill, New York.
Hariharan, P. (1984) *Optical Holography*, Cambridge University Press, Cambridge.
Høgmoen, K., and Løkberg, O.J. (1976) *Appl. Opt.*, **16**, 1869–75.
Jones, R., and Wykes, C. (1989) *Holographic and Speckle Interferometry*, Cambridge University Press, Cambridge.

Kasper, J.E., and Feller, S.A. (1987) *The Complete Book of Holograms*, Wiley, New York.
Kreis, T. (1996) *Holographic Interferometry. Principles and Methods*, Wiley, New York.
Leith, E.N., and Upatnieks, J. (1962) *J. Opt. Soc. Am.*, **52**, 1123–30.
Løkberg, O.J. (1979) *Appl. Opt.*, **18**, 2377–84.
Metherell, A.F., Spinak, S., and Pisa, E.J. (1969) *J. Opt. Soc. Am.*, **59**, 1534.
Molin, N.-E., and Stetson, K.A. (1971) *Optik*, **33**, 399–422.
Vest, C.M. (1979) *Holographic Interferometry*, Wiley, New York.
Walles, S. (1969) *Ark. Fys.*, **40**, 299–403.
Wilson, A.D., Lee, C.H., Lominac, H.R., and Strope, D.H. (1971) *Exp. Mech.*, **11**, 1–6.

7 MOIRÉ METHODS. TRIANGULATION

Amidror, I. (2000) *The Theory of Moiré*, Kluwer Academic, New York.
Asundi, A.K., Chan, C.S., and Sajan, M.R. (1994) *Opt. Eng.*, **33**, 2760–9.
Cardenas-Garcia, J.F., Zheng, S., and Shen, F.Z. (1994) *Optics and Lasers in Engineering*, **21**, 77–98.
Chen, F., *et al.* (2000) *Opt. Eng.*, **39**, 10–22.
Dessus, B., and Leblanc, M. (1973) *Opto-Electronics*, **5**, 369–91.
Gåsvik, K.J. (1983) *Appl. Opt.*, **22**, 3543–8.
Gåsvik, K.J., and Fourney, M.E. (1986) *J. Appl. Mech.*, **53**, 652–6.
Harding, K.G., and Harris, J.S. (1983) *Appl. Opt.*, **22**, 856–61.
Hazell, C.R., and Niven, R.D. (1968) *Exp. Mech.*, **8**, 225–31.
Hung, Y.Y., *et al.* (2000) *Opt. Eng.*, **39**, 143–9.
Lianhua, J., *et al.* (2000) *Opt. Eng.*, **39**, 2119–23.
Liasi, E., and North, W.P.T. (1994) *Opt. Eng.*, **33**, 1200–5.
Lichtenberg, F.K. (1955) *Proc. SESA*, **12**, 83–98.
Post, D. (1971) *Appl. Opt.*, **10**, 901–7.
Post, D. (1982) *Opt. Eng.*, **21**, 458–67.
Sciammarella, C.A. (1972) *J. Phys. E: Scientific Instruments*, **5**, 833–45.
Sciammarella, C.A. (1982) *Exp. Mech.*, **22**, 418–33.
Takasaki, H. (1973) *Appl. Opt.*, **12**, 845–50.
Takasaki, H. (1982) *Optics and Lasers in Engineering*, **3**, 3–14.
Vest, C.M., and Sweeney, D.W. (1972) *Appl. Opt.*, **11**, 449–54.

8 SPECKLE METHODS

Burch, J.M., and Forno, C. (1975) *Opt. Eng.*, **14**, 178–85.
Burch, J.M., and Tokarski, J.M. (1968) *Optica Acta*, **15**, 101–11.
Butters, J.N., and Leendertz, J.A. (1971) *J. Phys. E: Scientific Instruments*, **4**, 277–9.
Dainty, J.C. (ed.) (1975) *Topics in Applied Physics 9: Laser Speckle and Related Phenomena*, Springer Verlag, Berlin.
Erf, R.K. (ed.) (1978) *Speckle Metrology*, Academic Press, New York.
Fourney, M.E. (1978) in Erf, R.K. (ed.) *Speckle Metrology*, Academic Press, New York.
Goodman, J.W. (1975) Statistical properties of laser speckle patterns, in Dainty, J.C. (ed.) *Topics in Applied Physics 9: Laser Speckle and Related Phenomena*, Springer Verlag, Berlin.
Goodman, J.W. (1985) *Statistical Optics*, John Wiley & Sons.
Hung, Y.Y. (1978) in Erf, R.K. (ed.) *Speckle Metrology*, Academic Press, New York.
Hung, Y.Y., and Taylor, C.E. (1973) *Soc. Photo-Opt. Instrum. Eng.*, **14**, 169–75.

Jones, R., and Wykes, C. (1989) *Holographic and Speckle Interferometry*, Cambridge University Press, Cambridge.

Leendertz, J.A., and Butters, J.N. (1973) *J. Phys. E: Scientific Instruments*, **6**, 1107–10.

Mohan, N.K., Saldner, H.O., and Molin, N.-E. (1993) *Opt. Lett.*, **18**, 1861–3.

Patten, R., *et al.* (2001) *Opt. Eng.*, **40**, 1438–40.

Saldner, H.O. (1994) *Electronic Holography and Shearography in Experimental Mechanics*, Licentiate thesis, Luleå University of Technology.

Shang, H.M., *et al.* (2000) *Opt. Eng.*, **39**, 23–31.

Sirohi, R.S. (1993) *Speckle Metrology*, Marcel Dekker, New York.

Yamaguchi, I. (1993) in Sirohi, R.S., *Speckle Metrology*, Marcel Dekker, New York.

9 PHOTOELASTICITY AND POLARIZED LIGHT

Brosseau, C. (2000) *Fundamentals of Polarized Light. A Statistical Optics Approach*, John Wiley & Sons.

Clarke, D., and Graininger, J.F. (1971) *Polarized Light and Optical Measurement*, Pergamon Press, Oxford.

Cloud, G.L. (1998) *Optical Methods for Engineering Analysis*, Cambridge University Press.

Fourney, M.E. (1968) *Exp. Mech.*, **8**, 33–8.

Gåsvik, K.J. (1976) *Exp. Mech.*, **16**, 146–50.

Hecht, E. (2001) *Optics*, 4th edn, Addison-Wesley, Reading, MA

Klein, M.V. (1970) *Optics*, Wiley, New York.

Kuske, A., and Robertson, G. (1974) *Photoelastic Stress Analysis*, Wiley, London.

Neal, W.E.J., and Fane, R.W. (1973) *J. Phys. E*, **6**, 409–16.

Passaglia, E. (ed.) (1964) *Ellipsometry in the Measurement of Surfaces and Their Films*, National Bureau of Standards, Washington, DC.

Ramesh, K. (2000) *Digital Photoelasticity. Advanced Techniques and Applications*, Springer-Verlag.

Simmons, J.W., and Guttmann, M.J. (1979) *States, Waves and Photons: A Modern Introduction to Light*, Addison-Wesley, Reading, MA.

Theocaris, P.S., and Gdoutos, E.E. (1979) *Matrix Theory of Photoelasticity*, Springer-Verlag, Berlin.

Tompkins, H.G. (1993) *A User's Guide to Ellipsometry*, Academic Press.

10 DIGITAL IMAGE PROCESSING

Ambardar, A. (1995) *Analog and Digital Signal Processing*, PWS Publishing Co.

Baxes, G.A. (1994) *Digital Image Processing. Principles and Applications*, John Wiley & Sons.

Canny, J. (1986) *IEEE Trans. Pattern Analysis and Machine Intelligence*, **8**, 679–98.

Faig, W. (1975) *Photogram. Eng. Remote Sensing*, **41**, 1479–86.

Fram, J.R., and Deutsch, E.S. (1975) *IEEE Trans. Computers*, **C-24**(6), 616–28.

Gonzales, R.C., and Woods, R.E. (2002) *Digital Image Processing*, 2nd edn, Addison-Wesley.

Iaroslavskii, L.P., and Eden, M. (1996) *Fundamentals of Digital Optics: Digital Signal Processing in Optics and Holography*, Springer-Verlag.

Lenz, R.K., and Tsai, R.Y. (1988) *IEEE Trans. Pattern Analysis and Machine Intelligence*, **10**, 713–20.

Marr, D., and Hildreth, E. (1980) *Proc. Roy. Soc. Lond.*, **B207**, 187–217.

Morrison, N. (1994) *Introduction to Fourier Analysis*, John Wiley & Sons.

Niblack, W. (1988) *An Introduction to Digital Image Processing*, Prentice-Hall, New York.

Pratt, W.K. (1991) *Digital Image Processing*, 2nd edn, Wiley, New York.

Prewitt, J.M.S. (1970) in Lipkin, B.S., and Rosenfeld, A. (eds) *Picture Processing and Psychopic-torics*, Academic Press, New York.

Roberts, L.G. (1965) in Tippett, J.T., *et al.* (eds) *Optical and Electro-optical Information Processing*, MIT Press, Cambridge, MA.

Rosenfeld, A., and Kak, A.C. (1982) *Digital Picture Processing*, 2nd edn, Academic Press, New York.

Seul, M. *et al.* (2000) *Practical Algorithms for Image Analysis*, Cambridge University Press, Cambridge.

Shih, S.-W.S., Hung, Y.-P., and Lin, W.-S. (1993) *Opt. Eng.*, **32**, 138–49.

Tsai, R.Y. (1987) *IEEE J. Robot. Automat.*, **RA-3**, 323–44.

Vernon, D. (1991) *Machine Vision*, Prentice-Hall, New York.

11 FRINGE ANALYSIS

Bareket, N. (1985) *Proc. SPIE*, **551**, 12–17.

Bone, D.J., Bachor, H.A., and Sandeman, R.J. (1986) *Appl. Opt.*, **25**, 1653–60.

Button, B.L., Cutts, J., Dobbins, B.N., Moxon, C.J., and Wykes, C. (1985) *Optics and Laser Technology*, **17**, 189–92.

Carré, P. (1966) *Metrologia*, **2**, 13–23.

Chang, M., Lin, P.P., and Tai, W.C. (1994) *Optics and Lasers in Engineering*, **20**, 163–76.

Choudry, A. (1981) *Appl. Opt.*, **20**, 1240–4.

Creath, K. (1986) *Proc. SPIE*, **680**, 19–28.

Creath, K. (1993) Chapter 4 in Robinson, D.W., and Reid, G.T. (eds) *Interferogram Analysis. Digital Fringe Pattern Measurement Techniques*, Institute of Physics Publishing, Bristol.

Der Hovanesian, J., and Hung, Y.Y. (1982) *Proc. SPIE*, **360**, 88–9.

Frankowski, G., Stobbe, I., Tischer, W., and Schillke, F. (1989) *Proc. SPIE*, **1121**, 89–100.

Gåsvik, K.J., and Robbersmyr, K.G. (1994) *Opt. Eng.*, **33**, 246–50.

Gåsvik, K.J., Robbersmyr, K.G., and Vadseth, T. (1989) *Proc. SPIE*, **1163**, 64–70.

Ghiglia, D.C., Mastin, G.A., and Romero, L.A. (1987) *J. Opt. Soc. Am.* (A), **4**, 267–80.

Ghiglia, D.C., and Priff, D. (1998) *Two-dimensional Phase Unwrapping*, John Wiley & Sons.

Ghiglia, D.C., and Romero, L.A. (1994) *J. Opt. Soc. Am.* (A), **11**, 107–17.

Gierloff, J.J. (1987) *Proc. SPIE*, **818**.

Greivenkamp, J.E. (1984) *Opt. Eng.*, **23**, 350–2.

Halliwell, N.A., and Pickering, C.J. (1993) Chapter 7 in Robinson, D.W., and Reid, G.T. (eds) *Interferogram Analysis. Digital Fringe Pattern Measurement Techniques*, Institute of Physics Publishing, Bristol.

Harding, K.G., Coletta, M.P., and Van Dommelen, C.H. (1988) *Proc. SPIE*, **1005**, 169–78.

Hariharan, P., Oreb, B.F., and Eiju, T. (1987) *Appl. Opt.*, **26**, 2504–5.

Huntley, J.M. (1989) *Appl. Opt. Lett.*, **28**, 3268–70.

Huntley, J.M., and Saldner, H.O. (1993) *Appl. Opt.*, **32**, 3047–52.

Ichioka, Y., and Inuiya, M. (1972) *Appl. Opt.*, **11**, 1507–14.

Jin, G., Bao, N., and Chung, P.S. (1994) *Opt. Eng.*, **33**, 2733–37.

Joenathan, C. (1994) *Appl. Opt.*, **33**, 4147–55.

Judge, T.R., and Bryanston-Cross, P.J. (1993) *Optics and Lasers in Engineering*, **21**, 199–239.

Kujawinska, M. (1993a) Chapter 5 in Robinson, D.W., and Reid, G.T. (eds) *Interferogram Analysis. Digital Fringe Pattern Measurement Techniques*, Institute of Physics Publishing, Bristol.

Kujawinska, M. (1993b) *Optics and Lasers in Engineering*, **19**, 261–8.

Kujawinska, M., and Robinson, D.W. (1988) *Appl. Opt.*, **27**, 312–20.

Kujawinska, M., Spik, A., and Robinson, D.W. (1989) in *Proc. FASIG 'Fringe Analysis 89'*, Loughborough, 4–5 April.

Kwon, O.Y., and Shough, D.M. (1985) *Proc. SPIE*, **599**, 273–9.

Kwon, O.Y., Shough, D.M., and Williams, R.A. (1987) *Opt. Lett.*, **12**, 855–7.
Larkin, K.G., and Oreb, B.F. (1992) *J. Opt. Soc. Am.* (A) **9**, 1740–8.
Lassahn, G.D., Lassahn, J.K., Taylor, P.L., and Deason, V.A. (1994) *Opt. Eng.*, **33**, 2039–44.
Malcolm, A.A., Burton, D.R., and Lalor, M.J. (1989) in *Proc. FASIG 'Fringe Analysis 89'*, Loughborough, 4–5 April.
Magee, E.P., and Welsh, B.M. (1994) *Opt. Eng.*, **33**, 3810–17.
Martinez-Celorio, R.A., *et al.* (2000) *Opt. Eng.*, **39**, 751–7.
Mertz, L. (1983) *Appl. Opt.*, **22**, 1535–9.
Morgan, C.J. (1982) *Opt. Lett.*, **7**, 368–70.
Morimoto, Y., and Fujisawa, M. (1994) *Opt. Eng.*, **33**, 3709–14.
Morimoto, Y., Gascoigne, H.E., and Post, D. (1994) *Opt. Eng.*, **33**, 2646–53.
Nugent, K.A. (1985) *Appl. Opt.*, **24**, 3101–5.
Perry Jr., K.E., and McKelvie, J. (1993) *Optics and Lasers in Engineering*, **19**, 269–84.
Quan, C., Bryanston-Cross, P.J., and Judge, T.R. (1993) *Optics and Lasers in Engineering*, **18**, 79–108.
Robinson, D.W. (1983) *Appl. Opt.*, **22**, 2169–76.
Robinson, D.W. (1993) Chapter 6 in Robinson, D.W., and Reid, G.T. (eds) *Interferogram Analysis. Digital Fringe Pattern Measurement Techniques*, Institute of Physics Publishing, Bristol.
Robinson, D.W., and Reid, G.T. (eds) (1993) *Interferogram Analysis. Digital Fringe Pattern Measurement Techniques*, Institute of Physics Publishing, Bristol.
Robinson, D.W., and Williams, D.C. (1986) *Optics Communications*, **57**, 26–30.
Rosvold, G.O. (1990) *Appl. Opt.*, **29**, 237–41.
Schorner, J., *et al.* (1991) *Optics and Lasers in Engineering*, **14**, 283–92.
Schwider, J., *et al.* (1983) *Appl. Opt.*, **22**, 3421–32.
Schwider, J., *et al.* (1993) *Opt. Eng.*, **32**, 1883–5.
Simova, E.S., and Stoev, K.N. (1993) *Opt. Eng.*, **32**, 2286–94.
Singh, H., and Sirkis, J.S. (1994) *Appl. Opt.*, **33**, 5016–20.
Smythe, R., and Moore, R. (1984) *Opt. Eng.*, **23**, 361–4.
Stephenson, P., Burton, D.R., and Lalor, M.J. (1994) *Opt. Eng.*, **33**, 3703–8.
Su, X., and Xue, L. (2001) *Opt. Eng.*, **40**, 637–43.
Surrel, Y. (1993) *Appl. Opt.*, **33**, 3598–600.
Takeda, M., Ina, H., and Kobayashi, S. (1982) *J. Opt. Soc. Am.*, **72**(1), 156–60.
Takeda, M., and Mutoh, K. (1983) *Appl. Opt.*, **22**, 3977–82.
Towers, D.P., Judge, T.R., and Bryanston-Cross, P.J. (1991) *Optics and Lasers in Engineering*, **14**, 239–82.
Vikhagen, E. (1990) *Appl. Opt.*, **29**, 137–44.
Vikram, C.S., Witherow, W.K., and Trolinger, J.D. (1993) *Appl. Opt.*, **32**, 6250–2.
Vrooman, H.A., and Maas, A.A. (1989) *Proc. FASIG 'Fringe Analysis 89'*, Loughborough, 4–5 April.
Womack, K.H. (1984) *Opt. Eng.*, **23**, 391–5.
Wyant, C.J. (1982) *Laser Focus*, May, 65–71.
Wyant, C.J., *et al.* (1984) *ASLE Trans.*, **27**, 101–13.
Yatagai, T. (1993) Chapter 3 in Robinson, D.W., and Reid, G.T. (eds) *Interferogram Analysis. Digital Fringe Pattern Measurement Techniques*, Institute of Physics Publishing, Bristol.
Yatagai, T., Idesawa, M., Yamaashi, Y., and Suzuki, M. (1982) *Opt. Eng.*, **21**, 901–6.
Yoshizawa, T., and Tomisawa, T. (1993) *Opt. Eng.*, **32**, 1668–74.
Yu, Q., Andresen, K., Osten, W., and Jüptner, W.P.O. (1994) *Opt. Eng.*, **33**, 1630–7.

12 COMPUTERIZED OPTICAL PROCESSES

Butters, J.N., and Leendertz, J.A. (1977) *Measurement and Control*, **4**, 349–54.
Chen, D.J., *et al.* (1993) *Appl. Opt.*, **32**, 1839–49.

Goodman, J.W. (1996) *Introduction to Fourier Optics*, 2nd edn, McGraw-Hill, New York.

Iaroslavskii, L.P., and Eden, M. (1996) *Fundamentals of Digital Optics: Digital Signal Processing in Optics and Holography*, Springer-Verlag.

Jones, R., and Wykes, C. (1989) *Holographic and Speckle Interferometry*, Cambridge University Press, Cambridge.

Kreis, T. (1996) *Holographic Interferometry. Principles and Methods*, John Wiley & Sons.

Kreis, T., and Jüptner, W. (1996) *Fringe '97 – Automatic Processing of Fringe Patterns*, Akademie Verlag.

Løkberg, O.J. (1980) *Phys. Technol.*, **11**, 16–22.

Løkberg, O.J., Høgmoen, K., and Holje, O.M. (1979) *Appl. Opt.*, **18**, 763–5.

Løkberg, O.J., and Slettemoen, G. (1987) in Wyant, J.C., and Shannon, R. (eds) *Applied Optics and Optical Engineering, Vol. X*, Academic Press, New York.

Macovski, A., Ramsey, S.D., and Schaefer, L.F. (1971) *Appl. Opt.*, **10**, 2722–7.

Malmo, J.T., and Vikhagen, E. (1988) *Exp. Tech.*, **12**, 28–30.

Rastogi, P.K. (ed.) (2001) *Digital Speckle Pattern Interferometry and Related Techniques*, John Wiley & Sons.

Schnars, U., and Jüptner, W. (1994) *Appl. Opt.*, **33**, 179–81.

Schwomma, O. (1972) Austrian patent no. 298830.

Sjødahl, M., and Synnergren, P. (1999) *Appl. Opt.*, **38**, 1990–7.

Skotheim, O. (2001) Diploma Thesis, Norwegian University of Science and Technology.

Synnergren, P., and Goldrein, H.T. (1999) *Appl. Opt.*, **38**, 5956–61.

13 FIBRE OPTICS IN METROLOGY

Berthold, III, J.W. (1993) Status and review of fiber optic sensors in industry, in Udd, E. (ed.) *Fiber Optic Sensors*, SPIE Critical Reviews CR 44, p. 271.

Cook, R.O., and Hamm, C.W. (1979) *Appl. Opt.*, **18**, 3230–82.

Culshaw, B. (1986) *J. Opt. Sensors*, **1**, 327.

Goff, D.R. (1999) *Fiber Optic Reference Guide: A Practical Guide to the Technology*, Focal Press.

Grattan, K.T.V., and Meggitt, B.T. (2000) *Optical Fiber Sensor Technology. Advanced Applications – Bragg Gratings and Distributed Sensors*, Kluwer Academic Publishers.

Hill, K.O., *et al.* (1978) *Appl. Phys. Lett.*, **32**, 647–9.

Hill, K.O., and Meltz, G. (1997) *J. Lightwave Technol.*, **15**, 1263–76.

Kashyap, R. (1999) *Fiber Bragg Gratings*, Academic Press.

Kawasaki, B.S., *et al.* (1978) *Opt. Lett.*, **3**, 66–8.

Keiser, G. (1991) *Optical Fiber Communications*, 2nd edn, McGraw-Hill, New York.

Kyuma, K., Tai, S., and Munoshita, M. (1982) *Optics and Lasers in Engineering*, **3**, 155–82.

Othonos, A., and Kalli, K. (1999) *Fiber Bragg Gratings: Fundamentals and Applications in Telecommunications*, Artech House.

Palais, J.C. (1998) *Fiber Optic Communications*, 4th edn, Prentice-Hall, Englewood Cliffs, NJ.

Philips, G.J. (1980) *Mech. Eng.*, July, 28.

Senior, J.M. (1985) *Optical Fiber Communications. Principles and Practice*, Prentice-Hall, Englewood Cliffs, NJ.

Udd, E. (ed.) (1991) *Fiber Optic Sensors: An Introduction for Engineers and Scientists*, Wiley, New York.

Udd, E. (1993) *Fiber Optic Sensors*, SPIE Critical Reviews CR 44.

Yu, F.T.S., and Khoo, I.C. (1990) *Principles of Optical Engineering*, Chapter 10, Wiley, New York.

Index